Argument-Driven Inquiry
in
CHEMISTRY

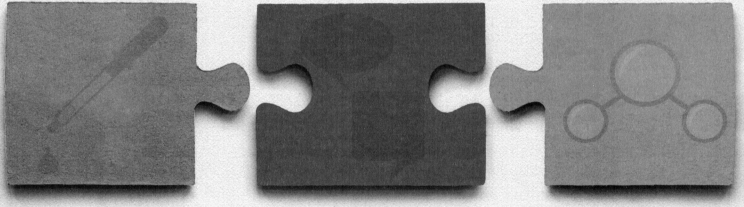

LAB INVESTIGATIONS
for GRADES 9–12

Argument-Driven Inquiry
in
CHEMISTRY

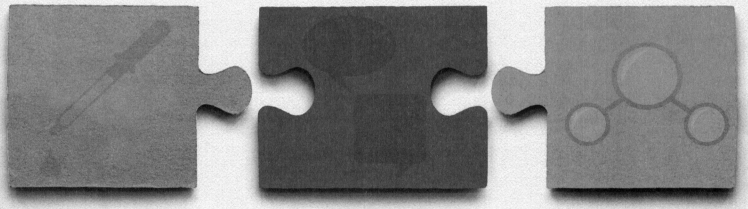

LAB INVESTIGATIONS
for GRADES 9–12

Victor Sampson, Peter Carafano, Patrick Enderle, Steve Fannin, Jonathon Grooms, Sherry A. Southerland, Carol Stallworth, and Kiesha Williams

NSTApress
National Science Teachers Association
Arlington, Virginia

National Science Teachers Association

Claire Reinburg, Director
Wendy Rubin, Managing Editor
Andrew Cooke, Senior Editor
Amanda O'Brien, Associate Editor
Amy America, Book Acquisitions Coordinator

ART AND DESIGN
Will Thomas Jr., Director

PRINTING AND PRODUCTION
Catherine Lorrain, Director

NATIONAL SCIENCE TEACHERS ASSOCIATION
David L. Evans, Executive Director
David Beacom, Publisher

1840 Wilson Blvd., Arlington, VA 22201
www.nsta.org/store
For customer service inquiries, please call 800-277-5300.

Copyright © 2015 by the National Science Teachers Association.
All rights reserved. Printed in the United States of America.
18 17 16 15   4 3 2 1

NSTA is committed to publishing material that promotes the best in inquiry-based science education. However, conditions of actual use may vary, and the safety procedures and practices described in this book are intended to serve only as a guide. Additional precautionary measures may be required. NSTA and the authors do not warrant or represent that the procedures and practices in this book meet any safety code or standard of federal, state, or local regulations. NSTA and the authors disclaim any liability for personal injury or damage to property arising out of or relating to the use of this book, including any of the recommendations, instructions, or materials contained therein.

Library of Congress Cataloging-in-Publication Data
Sampson, Victor, 1974- author.
  Argument-driven inquiry in chemistry : lab investigations for grades 9-12 / Victor Sampson [and 7 others].
    pages cm
Includes bibliographical references and index.
  ISBN 978-1-938946-22-6 -- ISBN 978-1-938946-54-7 (e-book) 1. Chemistry--Study and teaching (Secondary)--Activity programs.
2. Chemistry--Experiments. I. National Science Teachers Association. II. Title.
  QD43.S16 2014
  540.78--dc23
                          2014029558

  Cataloging-in-Publication Data for the e-book are also available from the Library of Congress.
  e-LCCN: 2014030242

CONTENTS

SECTION 1

Using Argument-Driven Inquiry

SECTION 2—Physical Sciences Core Idea 1.A
Structure and Properties of Matter

INTRODUCTION LABS

APPLICATION LABS

SECTION 3— Physical Sciences Core Idea 1.B
Chemical Reactions

INTRODUCTION LABS

SECTION 4—Appendixes

PREFACE

There is a push to change the way science is taught in the United States, called for by a different idea of what it means to know, understand, and be able to do science. As described in *A Framework for K–12 Science Education* (National Research Council [NRC] 2012) and the *Next Generation Science Standards* (NGSS Lead States 2013), science education should be structured to emphasize ideas *and* practices to

> ensure that by the end of 12th grade, *all* students have some appreciation of the beauty and wonder of science; possess sufficient knowledge of science and engineering to engage in public discussions on related issues; are careful consumers of scientific and technological information related to their everyday lives; are able to continue to learn about science outside school; and have the skills to enter careers of their choice, including (but not limited to) careers in science, engineering, and technology. (p. 1)

Instead of teaching with the goal of helping students learn facts and concepts, science teachers are now charged with helping their students become *proficient* in science by the time they graduate from high school. To allow for this proficiency, the NRC (2012) suggests that students need to understand four core ideas in the physical sciences, be aware of seven crosscutting concepts that span across the various disciplines of science, and learn how to participate in eight fundamental scientific practices in order to be considered proficient in science. These important practices, crosscutting concepts, and core ideas are summarized in Figure 1.

FIGURE 1

The three dimensions of the framework for the *NGSS*

Scientific Practices	**Crosscutting Concepts**
• Asking questions and defining problems	• Patterns
• Developing and using models	• Cause and effect: Mechanism and explanation
• Planning and carrying out investigations	• Scale, proportion, and quantity
• Analyzing and interpreting data	• Systems and system models
• Using mathematics and computational thinking	• Energy and matter: Flows, cycles, and conservation
• Constructing explanations and designing solutions	• Structure and function
• Engaging in argument from evidence	• Stability and change
• Obtaining, evaluating, and communicating information	

Physical Sciences Core Ideas

- PS1: Matter and its interactions
- PS2: Motion and stability: Forces and interactions
- PS3: Energy
- PS4: Waves and their applications in technologies for information transfer

Source: Adapted from NRC 2012, p. 3.

As described by the NRC (2012), new instructional approaches are needed to assist students in developing these proficiencies. This book provides 30 lab activities designed using an innovative approach to lab instruction called argument-driven inquiry (ADI). This approach and the labs based on it are aligned with the content, crosscutting concepts, and scientific practices outlined in Figure 1. Because the ADI model calls for students to give presentations to their peers, respond to questions, and then write, evaluate, and revise reports as part of each lab, the lab activities described in this book will also enable students to develop the disciplinary-based literacy skills outlined in the *Common Core State Standards* for English language Arts (NGAC and CCSSO 2010). Use of these labs, as a result, can help teachers align their instruction with current recommendations for making chemistry more meaningful for students and more effective for teachers.

References

National Governors Association Center for Best Practices and Council of Chief State School Officers (NGAC and CCSSO). 2010. *Common core state standards.* Washington, DC: NGAC and CCSSO.

National Research Council (NRC). 2012. *A framework for K–12 science education: Practices, crosscutting concepts, and core ideas.* Washington, DC: National Academies Press.

NGSS Lead States. 2013. *Next Generation Science Standards: For states, by states.* Washington, DC: National Academies Press. *www.nextgenscience.org/next-generation-science-standards.*

ACKNOWLEDGMENTS

The development of this book was supported by the Institute of Education Sciences, U.S. Department of Education, through grant R305A100909 to Florida State University. The opinions expressed are those of the authors and do not represent the views of the institute or the U.S. Department of Education.

ABOUT THE AUTHORS

Victor Sampson is an associate professor of STEM education and the director of the Center for Education Research in Mathematics, Engineering, and Science (CERMES) at The University of Texas at Austin (UT-Austin). He received a BA in zoology from the University of Washington, an MIT from Seattle University, and a PhD in curriculum and instruction with a specialization in science education from Arizona State University. Victor also taught high school biology and chemistry for nine years. He specializes in argumentation in science education, teacher learning, and assessment. To learn more about his work in science education, go to *www.vicsampson.com*.

Peter Carafano is an associate professor in the Science Department at Florida State University Schools, the K–12 research lab school for FSU, where he teaches high school and AP chemistry as well as health science. He began his career in education while working as a paramedic for the Delray Beach (FL) Fire Department. As a specialist in hazardous materials management and pre-hospital medical toxicology, Peter became a visiting instructor, lecturer, and author on hazardous materials and fire science chemistry. After retiring from the fire service in 2000, he obtained an MEd at FSU and began a new career as a teacher. Peter is also the director of the State of Florida Student Astronaut Challenge competition held at the Kennedy Space Center each year and the managing director of *The Journal of Emergency Medical Service Responders*, a student-driven and professionally peer-reviewed magazine.

Patrick Enderle is a research faculty member in CERMES at UT-Austin. He received his BS and MS in molecular biology from East Carolina University. Patrick spent some time as a high school biology teacher and several years as a visiting professor in the Department of Biology at East Carolina University. He then attended FSU, where he graduated with a PhD in science education. His research interests include argumentation in the science classroom, science teacher professional development, and enhancing undergraduate science education.

Steve Fannin is an assistant in research at the Florida Center for Research in Science, Technology, Engineering, and Mathematics (FCR-STEM) at FSU. Before taking his current position, he was a high school science teacher for 31 years. Steve received a BS in biological science from FSU and is a National Board Certified Teacher. He was honored with the Presidential Award for Excellence in Mathematics and Science Teaching in 2011.

Jonathon Grooms is a research scientist in CERMES at FSU. He received a BS in secondary science and mathematics teaching with a focus in chemistry and physics from FSU. Upon graduation, Jonathon joined FSU's Office of Science Teaching, where he directed the physical science outreach program Science on the Move. He also earned a PhD in science education from FSU.

Sherry A. Southerland is a professor at FSU and the co-director of FSU-Teach. FSU-Teach is a collaborative math and science teacher preparation program between the College of Arts and Sciences and the College of Education. She received a BS and an MS in biology from Auburn University and a PhD in curriculum and instruction from Louisiana State University, with a specialization in science education and evolutionary biology. Sherry has worked as a teacher educator, biology instructor, high school science teacher, field biologist, and forensic chemist. Her research interests include understanding the influence of culture and emotions on learning—specifically evolution education and teacher education—and understanding how to better support teachers in shaping the way they approach science teaching and learning.

Carol Stallworth is an honors and AP chemistry teacher at Lincoln High School in Tallahassee, Florida, and an adjunct chemistry professor at Tallahassee Community College. She received a BS and an MS in biochemistry from FSU. Carol has conducted biochemical research and wrote the article, "Cooperativity in Monomeric Enzymes With Single-Ligand Binding Sites," which appeared in the journal *Bioorganic Chemistry* in 2011.

Kiesha Williams is a chemistry teacher at Cypress Woods High School in Cypress, Texas. Before taking her current position, she taught chemistry at Florida State University Schools in Tallahassee, FL. Kiesha received a BS in chemistry and Spanish from the University of South Dakota and an MEd in science education from FSU. She has published two articles in *The Science Teacher*.

INTRODUCTION

The Importance of Helping Students Become Proficient in Science

The new aim of science education in the United States is for all students to become proficient in science by the time they finish high school. It is essential to recognize that science proficiency involves more than an understanding of important concepts, it also involves being able to *do* science. *Science proficiency*, as defined by Duschl, Schweingruber, and Shouse (2007), consists of four interrelated aspects. First, it requires an individual to know important scientific explanations about the natural world, to be able to use these explanations to solve problems, and to be able to understand new explanations when they are introduced to the individual. Second, it requires an individual to be able to generate and evaluate scientific explanations and scientific arguments. Third, it requires an individual to understand the nature of scientific knowledge and how scientific knowledge develops over time. Finally, and perhaps most important, an individual who is proficient in science should be able to participate in scientific practices (such as designing and carrying out investigations and arguing from evidence) and communicate in a manner that is consistent with the norms of the scientific community.

In the past decade, however, the importance of learning how to participate in scientific practices has not been acknowledged in the standards of many states. Many states have also attempted to make their science standards "more rigorous" by adding more content instead of designing them so they emphasize core ideas, scientific practices, and crosscutting concepts as described by the National Research Council (NRC) in *A Framework for K–12 Science Education* (NRC 2012). Unfortunately, the large number of benchmarks along with the pressure to cover them that results from the use of high-stakes tests has forced teachers "to alter their methods of instruction to conform to the assessment" (Owens 2009, p. 50). Teachers, as a result, tend to focus on content and neglect the practices of science inside the classroom. Teachers also tend to move through the science curriculum quickly to ensure that they "cover" all the standards before the students are required to take the high-stakes assessment. This trend takes us far afield from developing students' proficiency in the practices of science.

The current focus on covering all the standards, however, does not seem to be working. For example, *The Nation's Report Card: Science 2009* (National Center for Education Statistics 2011) indicates that only 21% of all 12th-grade students who took the National Assessment of Educational Progress in science scored at the proficient level. The performance of U.S. students on international assessments is even bleaker, as indicated by their scores on the science portion of the Programme for International Student Assessment (PISA). PISA is an international study that was launched by the Organisation for Economic Co-operation and Development (OECD) in 1997, with the goal of assessing education systems worldwide; more than 70 countries have participated in the study. The test is designed to assess reading, math, and science achievement and is given every three years. The mean score for students in the United States on the science portion of the PISA in 2012 is below the international mean (500), and there has been no significant change in the U.S. mean score since 2000; in fact, the U.S. mean score in 2012 is slightly less than it was in 2000 (OECD 2012; see Table 1). Students in countries such China, Korea, Japan, and Finland score significantly higher than students in the United States. These results suggest that U.S. students are not learning what they need to learn to become proficient in science, even though teachers are covering a great deal of material.

TABLE 1

PISA scientific literacy performance for U.S. students

Year	U.S. mean score*	U.S. rank/Number of countries assessed	Top three performers
2000	499	14/27	Korea (552) Japan (550) Finland (538)
2003	491	22/41	Finland (548) Japan (548) Hong Kong–China (539)
2006	489	29/57	Finland (563) Hong Kong–China (542) Canada (534)
2009	499	15/43	Japan (552) Korea (550) Hong Kong–China (541)
2012	497	36/65	Shanghai–China (580) Hong Kong–China (555) Singapore (551)

*The mean score of the PISA is 500 across all years.
Source: OECD 2012.

In addition to the poor performance of U.S. students on national and international assessments, empirical research in science education indicates that a curriculum that emphasizes breadth over depth and neglects the practices of science can actually hinder the development of science proficiency (Duschl, Schweingruber, and Shouse 2007; NRC 2005, 2008). As noted in the *Framework* (NRC 2012),

> K–12 science education in the United States fails to [promote the development of science proficiency], in part because it is not organized systematically across multiple years of school, emphasizes discrete facts with a focus on breadth over depth, and does not provide students with engaging opportunities to experience how science is actually done. (p. 1)

Based on their review of the available literature, the NRC recommends that science teachers spend more time focusing on key ideas to help students develop a more enduring understanding of science content. They also call for science teachers to start using instructional strategies that give students more opportunities to learn how to participate in the practices of science. Without this knowledge and these abilities, they argue, students will not be able to engage in public discussions about scientific issues related to their everyday lives, to be consumers of scientific information, and to have the skills needed to enter a science or science-related career. We think the school science laboratory is the perfect place to focus on key ideas and engage students in the practices of science and, as a result, help them develop the knowledge and abilities needed to be proficient in science.

How School Science Laboratories Can Help Foster the Development of Science Proficiency

Lab activities look rather similar in most high school classrooms (Hofstein and Lunetta 2004; NRC 2005). (We use the NRC's definition of a school science lab activity, which is "an opportunity for students to interact directly with the material world using the tools, data collection techniques, models, and theories of science" [NRC 2005, p. 3]). The teacher usually begins a lab activity by first introducing the students to a concept through a lecture or some other form of direct instruction. The teacher then gives the students a hands-on task to complete. To support students as they complete the task, teachers often provide students with a worksheet that includes a procedure explaining how to collect data, a data table to fill out, and a set of analysis questions to answer. The hope is that the experience gained through completion of the hands-on task and worksheet will illustrate, confirm, or otherwise verify the concept that was introduced to the students at the beginning of the activity. This type of approach, however, has been shown to be an ineffective way to help students understand the content under investigation, learn how to engage in important scientific practices (such as designing and carrying out an investigation, constructing explanations, or arguing from evidence), and develop scientific habits of mind (Duschl, Schweingruber, and Shouse 2007; NRC 2005). Thus, most lab activities, even if they are engaging or memorable, do little to promote science proficiency.

One way to address this problem is to widen the focus of lab instruction. A wider focus will require teachers to place more emphasis on "how we know" in chemistry (i.e., how new knowledge is generated and validated) in addition to "what we know" about matter and its interactions (i.e., the theories, laws, and unifying concepts). Science teachers will also need to focus more on the abilities and habits of mind that students need to have in order to construct and support scientific knowledge claims through argument and to evaluate the claims or arguments made by others (NRC 2012). The NRC calls for *argumentation* (defined as the process of proposing, supporting, evaluating, and refining claims) to play a more central role in the teaching and learning of science because argumentation is essential practice in science. The NRC (2012) provides a good description of the role argumentation plays in science:

> Scientists and engineers use evidence-based argumentation to make the case for their ideas, whether involving new theories or designs, novel ways of collecting data, or interpretations of evidence. They and their peers then attempt to identify weaknesses and limitations in the argument, with the ultimate goal of refining and improving the explanation or design. (p. 46)

In addition to changing the focus of instruction, teachers will also need to change the nature of lab instruction to promote and support the development of science proficiency. To change the nature of instruction, teachers need to make lab activities more authentic by giving students an opportunity to engage in scientific practices instead of giving them a worksheet with a procedure to follow and a data table to fill out. These activities, however, also need to be educative for students in order to help students develop the knowledge and abilities associated with science proficiency. Students need to receive feedback about how to improve, and teachers need to help students learn from their mistakes.

The argument-driven inquiry (ADI) instructional model (Sampson and Gleim 2009; Sampson, Grooms, and Walker 2009, 2011) was designed as a way to make lab activities more authentic and educative for students and thus help teachers promote and support the development of science proficiency. This instructional model reflects research about how people learn science (NRC 1999) and is also based on what is known about how to engage students in argumentation and other important scientific practices (Berland and Reiser 2009; Erduran and Jimenez-Aleixandre 2008; McNeill and Krajcik 2008; Osborne, Erduran, and Simon 2004; Sampson and Clark 2008).

Organization of This Book

The remainder of this book is divided into two parts. Part I begins with two text chapters describing the ADI instructional model and the development and components of the ADI lab investigations. Part II contains the lab investigations, including notes for the teacher, student handouts, and checkout questions. Four appendixes contain standards alignment matrixes, timeline and proposal options for the investigations, and a form for assessing the investigation reports.

Safety Practices in the Science Laboratory

It is important for science teachers to make hands-on and inquiry-based lab activities as safe as possible for students. Teachers therefore need to have proper safety equipment in the classroom (e.g., fume hoods, fire extinguishers, eyewash, and showers) and ensure that students use appropriate personal protective equipment (i.e., indirectly vented chemical-splash goggles meeting ANSI Z87.1 standard, chemical-resistant aprons and gloves) during all lab activities. Teachers also need to review and comply with all the local, state, and federal safety regulations as well as all safety policies and chemical storage and disposal protocols that have been established by the school district or school.

Throughout this book, safety precautions are provided for each investigation. Teachers should follow these safety precautions to provide a safer learning experience for students. The safety precautions associated with each activity are based, in part, on the use of the recommended materials and instructions, legal safety standards, and current professional safety practices. Selection of alternative materials or procedures for these activities may jeopardize the level of safety and therefore is at the user's own risk. We also recommend that students review *Safety in the Science Classroom, Laboratory, or Field Sites* (National Science Teacher Association [n.d.]) under the direction of the teacher before working in the laboratory for the first time. The students and their parent or guardians should then sign this document to acknowledge that they understand the safety procedures that must be followed during a lab activity.

References

Berland, L., and B. Reiser. 2009. Making sense of argumentation and explanation. *Science Education* 93 (1): 26–55.

Duschl, R. A., H. A. Schweingruber, and A. W. Shouse, eds. 2007. *Taking science to school: Learning and teaching science in grades K–8*. Washington, DC: National Academies Press.

Erduran, S., and M. Jimenez-Aleixandre, eds. 2008. *Argumentation in science education: Perspectives from classroom-based research*. Dordrecht, the Netherlands: Springer.

Hofstein, A., and V. Lunetta. 2004. The laboratory in science education: Foundations for the twenty-first century. *Science Education* 88: 28–54.

McNeill, K., and J. Krajcik. 2008. Assessing middle school students' content knowledge and reasoning through written scientific explanations. In *Assessing science learning: Perspectives from research and practice,* ed. J. Coffey, R. Douglas, and C. Stearns, 101–116. Arlington, VA: NSTA Press.

National Center for Education Statistics. 2011. *The nation's report card: Science 2009.* Washington, DC: U.S. Department of Education.

National Research Council (NRC). 1999. *How people learn: Brain, mind, experience, and school.* Washington, DC: National Academies Press.

National Research Council (NRC). 2005. *America's lab report: Investigations in high school science.* Washington, DC: National Academies Press.

National Research Council (NRC). 2008. *Ready, set, science: Putting research to work in K–8 science classrooms.* Washington, DC: National Academies Press.

National Research Council (NRC). 2012. *A framework for K–12 science education: Practices, crosscutting concepts, and core ideas.* Washington, DC: National Academies Press.

National Science Teachers Association (NSTA). n.d. *Safety in the science classroom, laboratory, or field sites. www.nsta.org/docs/SafetyInTheScienceClassroomLabAndField.pdf.*

Organisation for Economic Co-operation and Development (OECD). 2012. OECD Programme for International Student Assessment. *www.oecd.org/pisa.*

Osborne, J., S. Erduran, and S. Simon. 2004. Enhancing the quality of argumentation in science classrooms. *Journal of Research in Science Teaching* 41 (10): 994–1020.

Owens, T. 2009. Improving science achievement through changes in education policy. *Science Educator* 18 (2): 49–55.

Sampson, V., and D. Clark. 2008. Assessment of the ways students generate arguments in science education: Current perspectives and recommendations for future directions. *Science Education* 92 (3): 447–472.

Sampson, V., and L. Gleim. 2009. Argument-driven inquiry to promote the understanding of important concepts and practices in biology. *American Biology Teacher* 71 (8): 471–477.

Sampson, V., J. Grooms, and J. Walker. 2009. Argument-driven inquiry: A way to promote learning during laboratory activities. *The Science Teacher* 76 (7): 42–47.

Sampson, V., J. Grooms, and J. Walker. 2011. Argument-driven inquiry as a way to help students learn how to participate in scientific argumentation and craft written arguments: An exploratory study. *Science Education* 95 (2): 217–257.

SECTION 1
Using Argument-Driven Inquiry

CHAPTER 1
Argument-Driven Inquiry

Stages of Argument-Driven Inquiry

Each of the eight stages in the argument-driven inquiry (ADI) instructional model is designed to ensure that the experience is authentic (students have an opportunity to engage in the practices of science) *and* educative (students receive the feedback and explicit guidance that they need in order to improve on each aspect of science proficiency). Figure 2 (p. 4) summarizes the eight stages.

Stage 1: Identification of the Task and the Guiding Question; "Tool Talk"

In the ADI instructional model each lab activity begins with the teacher identifying a phenomenon to investigate and offering a guiding question for the students to answer. The goal of the teacher at this stage of the model is to capture the students' interest and provide them with a reason to complete the investigation. To aid in this, teachers should provide each student with a copy of the Lab Handout. The handout includes a brief introduction that describes a puzzling phenomenon or a problem to solve and provides a guiding question to answer. This handout also includes information about the medium they will use to present their argument (e.g., a whiteboard), some helpful tips on how to get started, and the criteria that will be used to judge argument quality (e.g., the sufficiency of the explanation and the quality of the evidence).

One engaging way to begin each lab is to select a different student to read each section of the handout aloud. After each section is read, the teacher should pause to clarify expectations, answer questions, and provide additional information as needed.

It is also important for the teacher to hold a "tool talk" during this stage, taking a few minutes to explain how to use specific lab equipment, specific indicators, computer simulations, or software to analyze data. Teachers need to hold a tool talk because students are often unfamiliar with lab equipment; even if they are familiar with the equipment, they will often use it incorrectly or in an unsafe manner. A tool talk can also be productive during this stage because students often find it difficult to design a method to collect the data needed to answer the guiding question (the task of stage 2) when they do not understand how to use the available materials. The teacher should also review specific safety protocols and precautions as part of the tool talk.

Once all the students understand the goal of the activity and how to use the available materials, the teacher should divide the students into small groups (we recommend three students per group to allow for optimal engagement) and move on to the second stage of the ADI model.

FIGURE 2

Stages of the argument-driven inquiry instructional model

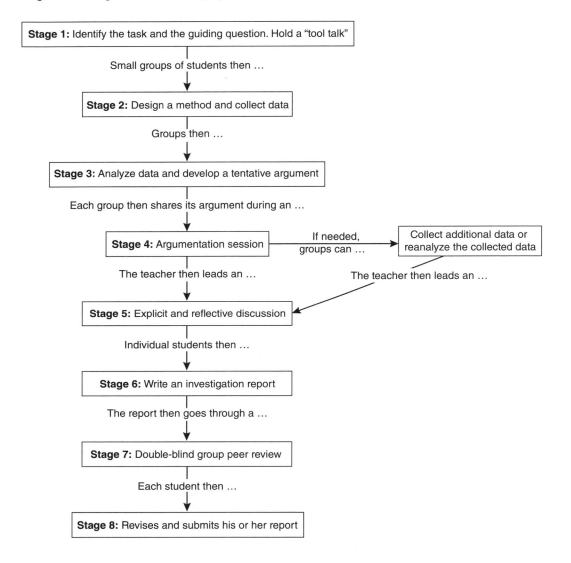

Stage 2: Designing a Method and Collecting Data

In this stage of the ADI model, small groups of students work together to (1) develop a method that they can use to gather the data needed to answer the guiding question and then (2) carry out the method. How students complete this stage depends on the nature of the investigation. Some investigations call for groups to answer the guiding question by designing a controlled experiment, whereas others require students to analyze an existing data set (e.g., a database, information sheets). To assist students with the process of designing their method, the teacher can have students complete an investigation proposal. These proposals guide students through the process of developing a method by encouraging them to think about what type of data they will need to collect, how to collect it, and how to analyze it. We have included three different investigation proposals in Appendix 3 (p. 509): Investigation Proposal A or Investigation Proposal B can be used when students need to design a method to test alternative explanations or claims; Investigation Proposal C can be used when students need to collect systematic observations and do not need to design a method to test alternative explanations or claims.

Stage 3: Data Analysis and Development of a Tentative Argument

The third stage of the instructional model calls for students to develop a tentative argument in response to the guiding question. The overall intent of this stage is to provide students with an opportunity to interact directly with the natural world (or in some cases with data drawn from the natural world) using appropriate tools and data collection techniques and to learn how to deal with the ambiguities of empirical work. This stage of the model also gives students a chance to learn why some methods work better than others and how the method used during a scientific investigation is based on the nature of the question and the phenomenon under investigation. At the end of this stage, students should have collected all the data they need to answer the guiding question.

Each group needs to be encouraged to first "make sense" of the measurements (temperature, mass, etc.) and/or observations (appearance, location, etc.) they collected during stage 2 of the model. Once the groups have analyzed and interpreted their data, they can then create their argument. The argument consists of a claim, the evidence they are using to support their claim, and a justification of their evidence (see Figure 3, p. 6). The *claim* is their answer to the guiding question. The *evidence* consists of the data (measurements or observations) they collected, an analysis of the data, and an interpretation of the analysis. The *justification of the evidence* is a statement that defends their choice of evidence by explaining why it is important and relevant, making the concepts or assumptions underlying the analysis and interpretation explicit.

The following example illustrates the three structural components of an argument that was made in response to the guiding question, "What type of metal are objects A, B, and C?"

FIGURE 3

Framework for the components of a scientific argument and criteria for evaluating the merits of the argument

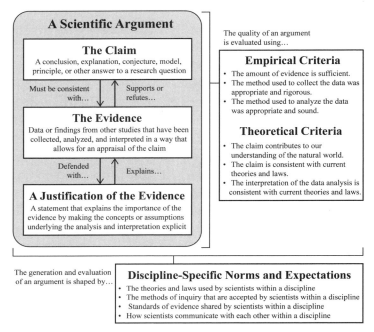

Claim: Objects A and B are tin. Object C is lead.

Evidence: The density of object A and object B is 7.44 g/cm³, which is almost the same as the density of tin (7.36 g/cm³). The density of object C is 11.12 g/cm³, which is almost the same as the density of lead (11.34 g/cm³).

Justification of evidence: Density is a physical property of matter and remains constant regardless of the amount of the object present. Therefore, density can be used to identify the substance that makes up an unknown object.

The claim in this argument provides an answer to the guiding question. The students used genuine evidence to support the claim by providing an analysis of the data collected (density of each substance) and an interpretation of the analysis (an inference based on known and unknown density values). Finally, the students provided a justification of the evidence in the argument by making explicit the underlying concept and assumptions (density as an inherent physical property) guiding the analysis of the data and the interpretation of the analysis.

It is important for students to understand that, in science, some arguments are better than others. An important aspect of science and scientific argumentation involves the evaluation of the various components of the arguments put forward by others. Therefore, the framework provided in Figure 3 also highlights two types of criteria that students can and should use to evaluate an argument in science: empirical criteria and theoretical criteria. *Empirical criteria* include

- how well the claim fits with all available evidence,
- the sufficiency of the evidence,
- the quality of the evidence,
- the appropriateness of the method used to collect the data, and
- the appropriateness of the method used to analyze the data.

Theoretical criteria refer to standards that are important in science but are not empirical in nature, including

- the sufficiency of the claim (i.e., Does it include everything needed?);
- the usefulness of the claim (e.g., Does it allow us to engage in new inquiries or understand a phenomenon?);
- the consistency of the claim and the reasoning in terms of other accepted theories, laws, or models; and
- the manner in which the data analysis was conducted.

What counts as quality within these different categories, however, varies from discipline to discipline (e.g., chemistry, physics, geology) and within the specific fields of each discipline (e.g., biochemistry, physical chemistry, organic chemistry). The variation is due to differences in the types of phenomena investigated, what counts as an accepted mode of inquiry (e.g., investigation vs. experimentation), and the theory-laden nature of scientific inquiry. It is important to keep in mind that "what counts" as a quality argument in science is discipline and field dependent.

To allow for the evaluation of the argument, each group of students should create their tentative argument in a medium that can easily be viewed by the other groups. We recommend using a 2' × 3' whiteboard or a large piece of butcher paper. Students should lay out each component of the argument on the board or paper. Figure 4 (p. 8) shows the general layout for a presentation of an argument, and Figure 5 (p. 8) provides an example of an argument crafted by students. Students can also create their tentative arguments using presentation software such as Microsoft's PowerPoint or Apple's Keynote and devote one slide to each component of an argument. The choice of medium is not important as long as students are able to easily modify the content of their argument as they work and it enables others to easily view their argument.

FIGURE 4 _____

Suggested layout of the components of an argument on a whiteboard

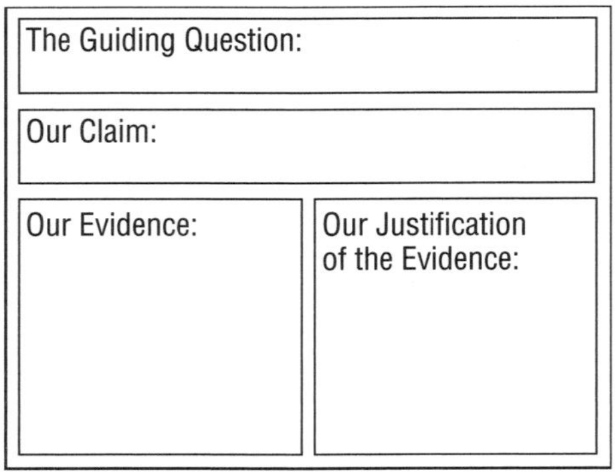

The Guiding Question:	
Our Claim:	
Our Evidence:	Our Justification of the Evidence:

FIGURE 5 _____

Example of an argument created by students

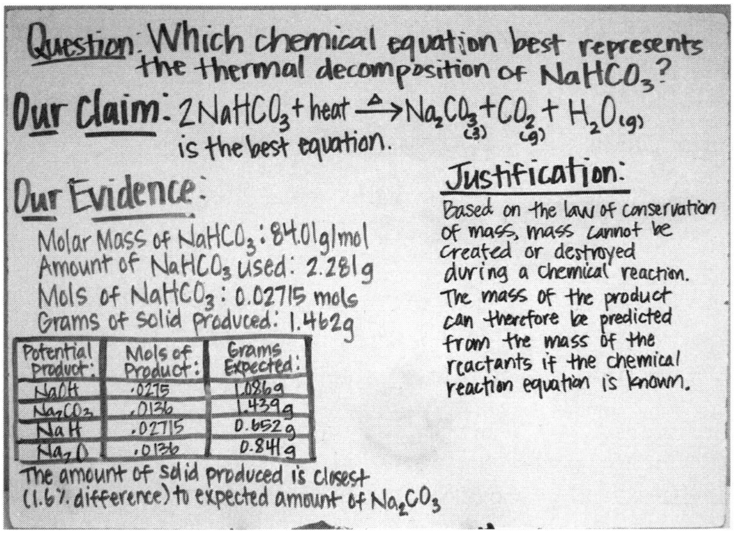

The intention of this stage of the model is to provide the student groups with an opportunity to make sense of what they are seeing or doing during the investigation. As students work together to create a tentative argument, they must talk with each other and determine how to analyze the data and how to best interpret the trends, differences, or relationships that they identify. They must also decide if the evidence (data that have been analyzed and interpreted) that they chose to include in their argument is relevant, sufficient, and convincing enough to support their claim. This, in turn, enables the groups of students to evaluate competing ideas and weed out any claim that is inaccurate, does not fit with all the available data, or contains contradictions.

This stage of the model is challenging for students because they are rarely asked to make sense of a phenomenon based on raw data, so it is important for the teacher to actively work to support their sense-making. The teacher should circulate from group to group to act as a resource person for the students, asking questions urging them to think about what they are doing and why. For example, the teacher should ask students probing questions to help them remember the goal of the activity (e.g., "What are you trying to figure out?"), to encourage them to think about whether or not the data are relevant (e.g., "Why is that measurement important?"), or to help them to remember to use rigorous criteria to evaluate the merits of a tentative idea (e.g., "Does that fit with all the data or what we know about exothermic reactions?"). It is also important to remember that students will struggle to develop arguments at the beginning of the year and will often rely on inappropriate criteria such as plausibility (e.g., "That sounds good to me.") or how data fit with personal experience (e.g., "That's what I saw on TV once.") as they attempt to make sense of their data. However, over time and with enough practice students will improve their skills. This is an important principle underlying the ADI instructional model.

Stage 4: Argumentation Session

In this stage, each group is given an opportunity to share, evaluate, and revise their tentative arguments with the other groups. This stage is included in the model for three reasons:

1. Scientific argumentation (i.e., arguing from evidence) is an important practice in science because critique and revision leads to better outcomes.

2. Research indicates that students learn more about the content and develop better critical thinking skills when they are exposed to the alternative ideas, respond to the questions and challenges of other students, and evaluate the merits of competing ideas (Duschl, Schweingruber, and Shouse 2007; NRC 2012).

3. Students learn how to distinguish between ideas using rigorous scientific criteria and are able to develop scientific habits of mind (e.g., treating ideas with initial skepticism, insisting that the reasoning and assumptions be made explicit, insisting that claims be supported by valid evidence) during the argumentation sessions.

Thus, this stage provides students with an opportunity to learn from and about scientific argumentation.

It is important to note, however, that supporting and promoting productive interactions between students inside the classroom can be difficult because this type of discussion is foreign to most students when they first begin participating in ADI. Students are therefore required to generate their arguments in a medium that can be seen by others to aid these interactions. By looking at whiteboards, paper, or slides, students tend to focus their attention on evaluating evidence rather than attacking the source of the ideas. This strategy often makes the discussion more productive and makes it easier for student to identify and weed out faulty ideas.

To allow all of the groups to share their arguments, we recommend that one member of each group stay at that group's lab station to share the group's argument while the other members of that group go to the other lab stations one at a time to listen to and critique the arguments developed by their classmates (see Figure 6, right, and Figure 7, p. 10). This type of format ensures that all ideas are heard and more students are actively involved in the

FIGURE 6

A high school student presents her group's argument to another group during the argumentation session.

FIGURE 7

Students participate in the argumentation session.

process, things that are often missed in a whole-class presentation format.

Just as is the case in earlier stages, it is important for the teacher to be involved (without leading) the discussions during the argumentation session. Once again, the teacher should move from group to group to keep students on task and model good scientific argumentation. The teacher can ask the presenter questions such as, "How did you analyze the available data?" or "Were there any data that did not fit with your claim?" to encourage students to use empirical criteria to evaluate the quality of the arguments. The teacher can also ask the presenters to explain how the claims they are presenting fit with the theories, laws, or models of science or to explain why the evidence they used is important. In addition, the teacher can also ask the students who are listening to the presentation questions such as "Do you think their analysis is accurate?" or "Do you think their interpretation is sound?" or even "Do you think their claim fits with what we know about X?"; the purpose of the teacher's questions is to remind them to use empirical and theoretical criteria to evaluate an argument during the discussions. Overall, the goal of the teacher at this stage of the lesson is to encourage students to think about how they know what they know and why some claims are more valid or acceptable in science. This stage of the model, however, is not the time to tell the students that they are right or wrong.

Stage 5: Explicit and Reflective Discussion

This stage of the ADI model provides a context for teachers to explain the nature of scientific knowledge and how this knowledge develops over time. Current research suggests that students only develop an appropriate understanding of the nature of science (NOS) and the nature of scientific inquiry (NOSI) when teachers discuss these concepts in an explicit fashion (Abd-El-Khalick and Lederman 2000; Akerson, Abd-El-Khalick, and Lederman 2000; Lederman and Lederman 2004; Lederman et al. 2014; Schwartz, Lederman, and Crawford 2004). In stage 5, the original student groups reconvene and discuss what they learned by interacting with individuals from the other groups during the argumentation session. Afterward, students can modify their tentative argument as needed or conduct an additional analysis of the data.

After their modifications are complete, the teacher should lead a whole-class discussion in which several students from different groups are encouraged to explain what they learned about the phenomenon under investigation. This discussion enables the teacher to ensure that the class reaches a scientifically acceptable conclusion. The teacher can also discuss any issues that were a common challenge for the groups during stages 2 and 3 of the activity.

Teachers should discuss one or two crosscutting concepts (e.g., the relationship between structure and function, or the importance of identifying and explaining patterns in nature) during this stage; the students' experiences during the lab investigation can be used as a concrete example of each concept. In addition, teachers should discuss one or two NOS (e.g., the difference between laws and theories, or the role that creativity and imagination play in science) or NOSI (e.g., what does and does not count as an experiment in science, or the different methods that scientists use to answer different types of questions) concepts that were prominent in the activity, again using the students' experiences during the investigation to illustrate these important concepts.

Stage 6: Writing the Investigation Report

Stage 6 is included in the ADI model because writing is an important part of doing science. Scientists must be able to read and understand the writing of others as well as evaluate its worth. They also must be able to share the results of their own research through writing. In addition, writing helps students learn how to articulate their thinking in a clear and concise manner, encourages metacognition, and improves student understanding of the content (Wallace, Hand, and Prain 2004). Finally, and perhaps most important, writing makes each student's thinking visible to the teacher (which facilitates assessment) and enables the teacher to provide students with the educative feedback they need in order to improve.

In stage 6, each student is required to write an individual investigation report using his or her group's argument. The report, which can be written during class or can be assigned as homework, should be centered on three fundamental questions:

1. What were you trying to do and why?

2. What did you do and why?

3. What is your argument?

An important component of this process is to encourage students to use tables or graphs to help organize the data they gathered and require them to reference the tables or graphs in the body of the report. This method allows them to learn how to construct an explanation, argue from evidence, and communicate information. It also enables students to master the disciplinary-based writing skills outlined in the *Common Core State Standards,* in English language arts (*CCSS ELA*; NGAC and CCSSO 2010). The report can be written during class or it can be assigned as homework.

The format of the report is designed to emphasize the persuasive nature of science writing and to help students learn how to communicate in multiple modes (words, figures, tables, and equations or formulas). The three-question format is well aligned with the components of a traditional lab reports (i.e., the introduction, procedure, results and discussion) but allows students to see the important role argument plays in science. We strongly recommend that teachers *limit the length of the investigation report* to two double-spaced pages or one single-spaced page. This limitation encourages students to write in a clear and concise manner, since there is little room for extraneous information. Requiring a short report is less intimidating to students than requiring a lengthier report, and it lessens the work required in the subsequent stages.

Stage 7: Double-Blind Group Peer Review

In this stage, each student is required to submit to the teacher two to four typed copies of the investigation report. Students do not place their names on the reports; instead, each student uses an identification number (assigned by the teacher) to maintain anonymity; this ensures that reviews are based on the ideas presented and not the person presenting the ideas. The students once again work in their original groups, and each group receives three or four sets of reports (i.e., the reports written by three or four different authors from other groups) and one peer-review guide for each author's report (see Appendix 4, p. 513). The peer-review guide lists specific criteria that are to be used by the group as they cooperatively evaluate the quality of each section of the investigation report as well as the mechanics of the writing. There is also space for the reviewers to provide the author with feedback about how to improve the report.

Reviewing each report as a group is an important component of the peer-review process because it provides students with a forum to discuss "what counts" as high quality or acceptable and in doing so forces them to reach a consensus during the process. This method also helps prevent students from checking off "yes" for each criterion on the peer-review guide without thorough consideration of the merits of the paper. It is also important for students to provide constructive and specific feedback to the author when areas of the paper are found to not meet the standards established by the peer-review guide. The peer-review process provides students with an opportunity to read good and bad examples of reports. This helps the students learn new ways to organize and present information, which in turn will help them write better on subsequent reports.

This stage of the model also gives students more opportunities to develop the reading skills that are needed to be successful in science. Students, for example, must be able to determine the central ideas or conclusions of a text and determine the meaning of symbols, key terms, and other domain-specific words. In addition, students must also be able to assess the reasoning and evidence that an author includes in a text to support his or her claim and compare or contrast findings presented in a text to those from other sources. Students can develop all these skills, as well as the other discipline-based reading standards found in the *CCSS ELA*, when they are required to read and critically review reports written by their classmates.

Stage 8: Revision and Submission of the Investigation Report

The final stage in the ADI instructional model is to revise the report based on the suggestions given during the peer review. If the report met all the criteria, the student may simply submit the paper to the teacher with the original peer-reviewed "rough draft" and peer-review guide attached, ensuring that the student's name replaces the identification number. Students whose reports are found by the peer-review group to be acceptable also have the option to revise the reports if they so desire after reviewing the work of other students. However, if a report was found unacceptable by the group during peer review, the author is required to rewrite it using the reviewers' comments and suggestions as a guideline. Once the report is revised, it is turned in to the teacher for evaluation, with the original "rough draft" and the peer-review sheet attached. The author is required to explain what he or she did to improve each section of the report in response to the reviewers' suggestions (or explain why the author decided to ignore the reviewers' suggestion) in the author response section of the peer-review guide. The teacher can then provide a score in the instructor score column of the peer-review guide and use these ratings to assign an overall grade.

This approach provides students with a chance to improve their writing mechanics and develop their reasoning and understanding of the instructional content. It also offers students the added benefit of reducing academic pressure by providing support in obtaining the highest possible grade for their final product.

The Role of the Teacher During Argument-Driven Inquiry

If the ADI instructional model is to be successful and student learning is to be optimized, the role of the teacher during a lab activity designed using this model must be different than the teacher's role during a more traditional lab. The teacher *must* act as a resource for the students—rather than as a director—as they work through each stage of the activity, encouraging students to think about *what they are doing* and *why they made that decision* throughout the process. This encouragement should take the form of probing questions that teachers ask as they walk around the classroom, such as "Why do you want to set up your equipment that way?" or "What type of data will you need to collect to be able to answer that question?" Teachers must refrain from telling or showing students how to "properly" conduct the investigation. Teachers must emphasize the need to maintain high standards for a scientific investigation by requiring students to use rigorous standards for "what counts" as a good method or a strong argument in the context of science. Finally, and perhaps most important, for an ADI activity to be successful, teachers must be willing to let students try and fail; students can learn from their mistakes with guidance from teachers. Teachers should not try to make the investigations included in this book "student-proof" by providing additional directions to ensure that students do everything right the first time. We have found that students often learn more from an ADI lab activity when they design a poor method to collect data or analyze their results in an inappropriate manner because their classmates quickly point out these mistakes during the argumentation session stage of the model and it leads to more teachable moments.

Because the teacher's role in an ADI lab is different from what typically happens in laboratories, we have created a chart describing teacher behaviors that are consistent and inconsistent with each stage of the instructional model (see Table 2). This table is organized by stage because what the students and the teacher need to accomplish during each stage is different. It might be helpful to keep this table handy as a guide when first attempting to implement the lab activities found in the book.

TABLE 2

Teacher behaviors during the stages of the ADI instructional model

Stage	Teacher behaviors	
	Consistent with ADI model	**Inconsistent with ADI model**
1: Identification of the task and the guiding question; "tool talk"	• Sparks students' curiosity • "Creates a need" for students to design and carry out an investigation • Organizes students into collaborative groups • Supplies students with the materials they will need • Holds a "tool talk" to show students how to use equipment and/or to illustrate proper techniques • Reviews relevant safety precautions and protocols • Provides students with hints	• Does not have students read the Lab Handout • Tells students that there is one correct answer • Tells students what they "should expect to see" or what results "they should get"
2: Designing a method and collecting data	• Encourages students to ask questions as they design their investigations • Asks groups questions about their method (e.g., "Why do you want to do it this way?") and the type of data they expect from that design • Reminds students of the importance of specificity when completing their investigation proposal	• Gives students a procedure to follow • Does not question students about their method or the type of data they expect to collect • Approves vague or incomplete investigation proposals

Table 2 (*continued*)

Stage	Teacher behaviors	
	Consistent with ADI model	**Inconsistent with ADI model**
3: Data analysis and development of a tentative argument	• Reminds students of the research question and the components of a scientific argument • Requires students to generate an argument that provides and supports a claim with genuine evidence (data + an analysis of the data + an interpretation of the analysis) • Asks students what opposing ideas or rebuttals they might anticipate • Encourages students to justify their evidence with scientific concepts	• Requires only one student to be prepared to discuss the argument • Moves to groups to check on progress without asking students questions about why they are doing what they are doing • Does not interact with students (uses the time to catch up on other responsibilities) • Tells students that their claim is right
4: Argumentation session	• Reminds students of appropriate behaviors during discussions • Reminds students to critique ideas, not people • Encourages students to ask peers questions • Keeps the discussion focused on the elements of the argument • Encourages students to use appropriate criteria for determining what does and does not count	• Allows students to criticize or tease each other • Asks questions about students' claims before other students can ask • Allows students to use inappropriate criteria for determining what does and does not count
5: Explicit and reflective discussion	• Discusses the content at the heart of the investigation and important theories, laws, or principles that students can use to justify their evidence when writing their investigation reports • Explains one or two crosscutting concepts using what happened during the lab investigation as an example • Highlights one or two aspects of the nature of science and/or scientific inquiry using what happened during the lab investigation as examples • Encourages students to identify strengths and limitations of their investigations • Discusses ways that students could improve future investigations	• Provides a lecture on the content • Skips over the discussion about the nature of science and the nature of scientific inquiry to save time • Tells students "what they should have learned" or "this is what you all should have figured out"

Table 2 (*continued*)

Stage	Teacher behaviors	
	Consistent with ADI model	**Inconsistent with ADI model**
6: Writing the investigation report	• Reminds students about the audience, topic, and purpose of the report • Provides the peer-review guide in advance • Provides examples of a high-quality report and an unacceptable report • Takes time to write the report in class in order to scaffold the process of writing each section of the report	• Has students write only a portion of the report • Allows students to write the report as a group • Does not require students to write the report, in order to save time
7: Double-blind group peer review	• Reminds students of appropriate behaviors for the peer-review process • Ensures that all groups are giving a quality and fair peer review to the best of their ability • Encourages students to remember that while grammar and punctuation are important, the main goal is an acceptable scientific claim with supporting evidence and a justification of the evidence • Ensures that students provide genuine feedback to the author when they identify a weakness or an omission. • Holds the reviewers accountable.	• Allows students to make critical comments about the author (e.g., "This person is stupid") rather than their work (e.g., "This claim needs to be supported by evidence") • Allows students to just check off "Yes" on each item without providing a critical evaluation of the report
8: Revision and submission of the investigation report	• Requires students to edit their reports based on the reviewers' comments • Requires students to respond to the reviewers' ratings and comments • Has students complete the checkout questions after they have turned in their report	• Allows students to turn in a report without a completed peer-review guide • Allows students to turn in a report without revising it first

References

Abd-El-Khalick, F., and N. G. Lederman. 2000. Improving science teachers' conceptions of nature of science: A critical review of the literature. *International Journal of Science Education* 22: 665–701.

Akerson, V., F. Abd-El-Khalick, and N. Lederman. 2000. Influence of a reflective explicit activity-based approach on elementary teachers' conception of nature of science. *Journal of Research in Science Teaching* 37 (4): 295–317.

Duschl, R. A., H. A. Schweingruber, and A. W. Shouse, eds. 2007. *Taking science to school: Learning and teaching science in grades K–8*. Washington, DC: National Academies Press.

Lederman, N. G., and J. S. Lederman. 2004. Revising instruction to teach the nature of science. *The Science Teacher* 71 (9): 36–39.

Lederman, J., N. Lederman, S. Bartos, S. Bartels, A. Meyer, and R. Schwartz. 2014. Meaningful assessment of learners' understanding about scientific inquiry: The Views About Scientific Inquiry (VASI) questionnaire. *Journal of Research in Science Teaching* 51 (1): 65–83.

National Governors Association Center for Best Practices and Council of Chief State School Officers (NGAC and CCSSO). 2010. *Common core state standards*. Washington, DC: NGAC and CCSSO.

National Research Council (NRC). 2012. *A framework for K–12 science education: Practices, crosscutting concepts, and core ideas*. Washington, DC: National Academies Press.

Schwartz, R. S., N. Lederman, and B. Crawford. 2004. Developing views of nature of science in an authentic context: An explicit approach to bridging the gap between nature of science and scientific inquiry. *Science Education* 88: 610–645.

Wallace, C., B. Hand, and V. Prain, eds. 2004. *Writing and learning in the science classroom*. Boston: Kluwer Academic Publishers.

CHAPTER 2
Lab Investigations

This book includes 30 chemistry lab investigations designed around the argument-driven inquiry (ADI) instructional model. The investigations are not meant to replace an existing curriculum but rather to transform the laboratory component of a chemistry course. A teacher can use these investigations as a way to introduce students to new content ("introduction labs") or as a way to give students an opportunity to apply a theory, law, or unifying concept introduced in class in a novel situation ("application labs"). To facilitate curriculum and lesson planning, the lab investigations have been aligned with *A Framework for K–12 Science Education* (NRC 2012); the *Common Core State Standards,* in English language arts (*CCSS ELA*) and mathematics (*CCSS Mathematics*); and various aspects of the nature of science (NOS) and the nature of scientific inquiry (NOSI) (Abd-El-Khalick and Lederman 2000; Lederman et al. 2002; Lederman et al. 2014). The matrixes in Appendix 1 (p. 494) illustrate these alignments.

Many of the ideas for the investigations in this book came from existing resources; however, we modified the activities from those resources to fit with the focus and nature of the ADI instructional model. This model provides teachers with a way to transform classic or traditional lab activities into authentic and educative investigations that enable students to become more proficient in science.

Once we created the ADI lab investigations, they were reviewed for content accuracy by several practicing chemists. Two chemistry teachers then piloted the investigations in several sections of a high school chemistry course (including general and honors sections). After the pilot year, each lab investigation and all related instructional materials (such as the investigation proposals and peer-review guide) were revised based on feedback from these two teachers as well as information from student assessments; these revisions were intended to increase the effectiveness of the lab investigations and related materials. The revised labs were then piloted and revised for a second time. The final iteration of each lab investigation is included in this book.

These lab investigations were developed as part of a three-year research project funded by the Institute of Education Sciences through grant R305A100909. The goal of this project, which took place at Florida State University and Florida State University Schools (a K–12 laboratory school), was to refine the ADI instructional model, develop a set of ADI lab activities for a variety of science disciplines, and examine what students learn when they complete eight or more ADI labs over the course of a school year. Our research indicates that students have much better inquiry and writing skills after participating in at least eight ADI lab investigations and make substantial gains in their understanding of important chemistry content and concepts related to NOS and NOSI (Grooms, Sampson, and

Carafano 2012; Sampson, Grooms, and Enderle 2012; Sampson et al. 2013). To learn more about the research associated with the ADI instructional model and how to use it in the classroom, visit *www.argumentdriveninquiry.com*.

Teacher Notes

Each chemistry teacher must decide when and how to use a laboratory to best support student learning. To help with this decision making, we have included Teacher Notes for each investigation. These notes include information about the purpose of the lab, the time needed to implement each stage of the model for that lab, the materials needed, and hints for implementation. There is also a Topic Connections section that shows how each ADI lab investigation is aligned with the NRC *Framework*, the *CCSS ELA* and/or the *CCSS Mathematics*, and NOS or NOSI concepts. In the following subsections, we will describe the information provided in each section of the Teacher Notes.

Purpose

This section of the Teacher Notes describes the main idea of the lab and indicates whether the activity is an introduction lab or an application lab. In either case, because of the ADI structure, very little emphasis needs to be placed on making sure the students "get the vocabulary" or "know their stuff" before the lab investigation begins. Instead, with the combination of the information provided in the Lab Handout, students' evolving understanding of the actual practice of science, and the various resources available to the students (i.e., the science textbook, the internet, and, of course, the teacher), students will develop a better understanding of the content as they work through the activity. This section also highlights the NOS or NOSI concepts that should be discussed during the explicit and reflective discussion stage.

The Content

This section of the Teacher Notes provides an overview of the concept that is being introduced to the students or that students will need to apply during the investigation. It also provides an answer to the guiding question of the investigation.

Timeline

Unlike most traditional laboratories, which can be completed in a single class period, ADI labs typically take three to five instructional days to complete. More time may be needed for the first few labs that your students conduct, but the time needed will be reduced as they become familiar with the practices employed in the ADI model (argumentation, designing investigations, writing reports). The "Timeline" section describes the instructional time (presented as a range) needed to complete each lab and refers to timeline options more fully described in Appendix 2 (p. 501). The figures in Appendix 2 show the day and stage(s) of the ADI model that ideally should be completed in class each day and outline the resulting products of each stage.

It is important to note that although the days are listed chronologically in the timeline options in Appendix 2), they do not necessarily have to fall on consecutive days. For example, when students need to write an investigation report, the teacher can allow them to have more than one night to complete the work, especially when they are getting used to what is expected of them. Also, some of the lab stages do not take an entire class period to complete, especially once students are acclimated to the ADI model.

Materials and Preparation

This section of the Teacher Notes describes the lab supplies and instructional materials (i.e., lab sheets, investigation proposals, and peer-review guide) needed to implement the lab activity. The lab supplies listed are designed for one group; however, multiple groups can share if resources are scarce. We have also included specific suggestions for some lab supplies that were found to work best during the pilot tests. However, if needed, substitutions can be made. Be sure to test all lab supplies before conducting the lab with the student, because new materials often have unexpected consequences.

We also explain in this section how to prepare the materials (including consumables) that the students will use during the investigation. Some labs require an hour or more of preparation time, so teachers should review this section at least 24 hours before conducting the investigation in the classroom.

Safety Precautions

This section of the Teacher Notes provides an overview of potential safety hazards as well as safety protocols that should be followed to make the laboratory safer for students. These are based on legal safety standards and current safety practices. Teachers should review and follow all local polices and protocols used within their school district and/or school (e.g., the district chemical hygiene plan, Board of Education safety policies).

Laboratory Waste Disposal

This section of the Teacher Notes provides recommendations for disposing of the laboratory waste that the students will generate during the lab investigation.

Topics for the Explicit and Reflective Discussion

This section is the "conceptual heart" of the laboratory. It provides an overview of important content to discuss, relevant crosscutting concepts, and facets of NOS/NOSI. Just as important, this section provides advice for teachers about how to encourage students to reflect on the strengths and limitations of their investigations and how to improve the design of their investigations in the future.

Hints for Implementing the Lab

These lab investigations have been tested by many teachers many times, and we have collected "hints" from these teachers for each stage of the ADI process. These hints can help you avoid some of the pitfalls teachers experienced during the testing and should make the investigation run smoothly. Tips for making the investigation safer are also included.

Topic Connections

This section, which is designed to inform curriculum and lesson planning, includes a table that highlights the scientific practices, crosscutting concepts, and core ideas from the NRC *Framework* that are aligned with the lab activity. The table also outlines supporting ideas, the *CCSS ELA* and the *CCSS Mathematics,* and the NOS/NOSI concepts addressed by the activity.

Instructional Materials

The instructional materials included in this book are reproducible copy masters that are designed to support students as they participate in an ADI lab investigation. The materials include Lab Handouts, Lab Reference Sheets, investigation proposals, the Peer-Review Guide and Instructor Scoring Rubric, and Checkout Questions. In the following subsections, we will provide an overview of these important materials.

Lab Handout

At the beginning of each lab investigation, each student should be given a copy of the Lab Handout. This handout provides information about the phenomenon that the students will investigate and a guiding question for the students to answer. The handout also provides hints for students to help them design their investigation in the Getting Started section, information about what to include in their tentative argument, safety precautions, and the requirements for the investigation report.

Lab Reference Sheet

Some lab investigations include an optional Lab Reference Sheet that provides additional information that the students can use as part of their investigation. Some of these sheets provide information about lab technique, and some include additional information about the theory, model, law, or concept at the heart of the investigation. If a teacher decides to use the Lab Reference Sheet, we recommend making a class set or providing one copy to each lab group.

Investigation Proposal

To help students design better investigations, we have developed and included three different types of investigation proposals in this book (see Appendix 3, p. 509). These investigation proposals are optional, but we have found that students design and carry out much better investigations when they are required to fill out a proposal and then get teacher feedback about their method before they begin. We provide recommendations in the Teacher Notes about which investigation proposal (A, B, or C) to use for a particular lab. If a teacher decides to use an investigation proposal as part of a lab, we recommend providing one copy for each group.

The Lab Handout for students has a heading asking if an investigation proposal is required, followed by boxes to check "yes" or "no." Teachers should make sure that students check the appropriate box when introducing the lab activity.

Peer-Review Guide and Instructor Scoring Rubric

The Peer-Review Guide and Instructor Scoring Rubric (see Appendix 4, p. 513) is designed to make explicit the criteria used to judge the quality of an investigation report. We recommend that teachers make one copy for each student and then provide the copies to the students before they begin writing their investigation reports. This will ensure that students understand how they will be evaluated. During stage 7 of the model (double-blind group peer review), each group should fill out the peer-review guide as they review the reports of their classmates (each group will need to review three or four different reports). The reviewers should rate the report on each criterion and provide advice to the author about ways to improve. Once the review is complete, the author needs to revise the report and respond to the reviewers' ratings and comments in the appropriate sections of the peer-review guide. The completed peer-review guide should be submitted to the teacher along with the final and first draft of the report for a final evaluation. To score the report, the teacher can simply fill out the instructor score column of the rubric and then total the scores.

Checkout Questions

To facilitate classroom assessment, we have included a set of Checkout Questions for each lab investigation. The questions target the key ideas, the crosscutting concepts, and the NOS/NOSI concepts addressed in the lab. Students should answer the questions on the same day they turn in their final reports. One copy of the Checkout Questions is needed for each student. The students should complete these questions on their own. Teacher can use the students' responses, along with the report, to determine if the students learned what they needed to during the lab, and then reteach as needed.

References

Abd-El-Khalick, F., and N. G. Lederman. 2000. Improving science teachers' conceptions of nature of science: A critical review of the literature. *International Journal of Science Education* 22: 665–701.

Grooms, J., V. Sampson, and P. Carafano. 2012. The impact of a new instructional model on high school science writing. Paper presented at the annual international conference of the American Educational Research Association, Vancouver.

Lederman, N. G., F. Abd-El-Khalick, R. L. Bell, and R. S. Schwartz. 2002. Views of nature of science questionnaire: Toward a valid and meaningful assessment of learners' conceptions of nature of science. *Journal of Research in Science Teaching* 39 (6): 497–521.

Lederman, J., N. Lederman, S. Bartos, S. Bartels, A. Meyer, and R. Schwartz. 2014. Meaningful assessment of learners' understanding about scientific inquiry: The Views About Scientific Inquiry (VASI) questionnaire. *Journal of Research in Science Teaching* 51 (1): 65–83.

National Governors Association Center for Best Practices and Council of Chief State School Officers (NGAC and CCSSO). 2010. *Common core state standards.* Washington, DC: NGAC and CCSSO.

National Research Council (NRC). 2012. *A framework for K–12 science education: Practices, crosscutting concepts, and core ideas.* Washington, DC: National Academies Press.

Sampson, V., P. Enderle, J. Grooms, and S. Witte. 2013. Writing to learn and learning to write during the school science laboratory: Helping middle and high school students develop argumentative writing skills as they learn core ideas. *Science Education* 97 (5): 643–670.

Sampson, V., J. Grooms, and P. Enderle. 2012. Using laboratory activities that emphasize argumentation and argument to help high school students learn how to engage in scientific inquiry and understand the nature of scientific inquiry. Paper presented at the annual international conference of the National Association for Research in Science Teaching, Indianapolis.

SECTION 2
Physical Sciences
Core Idea 1.A

Structure and Properties of Matter

Introduction Labs

LAB 1

Lab 1. Bond Character and Molecular Polarity: How Does Atom Electronegativity Affect Bond Character and Molecular Polarity?

Purpose

The purpose of this lab is to *introduce* students to the concepts of bonds and molecular polarity. This lab gives students an opportunity to use a computer simulation to explore the effect of atom electronegativity on bond character and molecular polarity. Students will also learn about the differences between laws and theories and between data and evidence in science.

The Content

Chemical compounds can generally be classified into two broad categories: molecular compounds and ionic compounds. Molecular compounds consist of atoms that are held together by covalent bonds. Ionic compounds, in contrast, are composed of positive and negative ions that are joined by ionic bonds. A covalent bond is formed when two atoms share one or more pairs of electrons. An ionic bond is formed when one or more of the electrons from one atom are transferred to another atom, which results in a positive ion and a negative ion. These ions then attract each other because they have opposite electrical charges. Figure 1.1 illustrates the difference between a covalent bond and an ionic bond.

A bond between two atoms, however, is never completely ionic or covalent. The character of a bond depends on how strongly each of the bonded atoms attracts electrons. *Electronegativity* refers to the relative ability of an atom to attract electrons. As shown in Table 1.1, the character and type of bond formed between two atoms can be predicted based on magnitude of the difference in electronegativity. Identical atoms have an electronegativity difference of zero, which means the electrons are equally shared between the two atoms. This type of bond is considered a nonpolar covalent bond. Atoms with a small difference in electronegativities (0.1–0.39) will share electrons unequally, resulting in a mostly covalent bond. Unequal sharing of electrons (0.4–1.7) results in a polar covalent bond. When there is a large difference in the electronegativity between two atoms (> 1.7), an electron is transferred and the result is a mostly ionic bond.

FIGURE 1.1

Difference between an ionic bond and a covalent bond

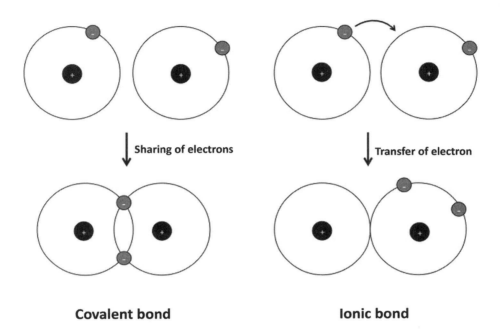

Sharing of electrons   Transfer of electron

Covalent bond   **Ionic bond**

TABLE 1.1

Electronegativity difference and bond character

Electronegativity difference	Bond character
>1.7	Mostly ionic
0.4–1.7	Polar covalent
0.1–0.39	Mostly covalent
0.0	Nonpolar covalent

   When a polar bond forms, the shared electrons are pulled toward one of the atoms. The shared electrons therefore spend more time around the atom with the greater electronegativity. The unequal sharing of electrons results in a partial charge at one side of the bond (see Figure 1.2, p. 30). The Greek letter delta (δ) is used to represent a partial charge. In a polar covalent bond, δ^- represents a partial negative charge and δ^+ represents a partial positive charge. As shown in Figure 1.2, δ^- and δ^+ can be added to an image of a molecule to represent the polarity of the covalent bond. Molecules that contain covalent bonds can be either polar or nonpolar; the type depends on the location and nature of the covalent bonds in the molecule.

FIGURE 1.2

Electronegativity difference between chlorine (Cl) and hydrogen (H)

The electronegativity of Cl is higher than that of H. Therefore, in a molecule containing Cl and H the shared electron is with the Cl atom more often than it is with the H atom. The delta symbols are used to indicate the partial charge at each side of the molecule from the unequal sharing of electrons.

Electronegativity of Cl = 3.16
Electronegativity of H = 2.20

Difference = 0.96

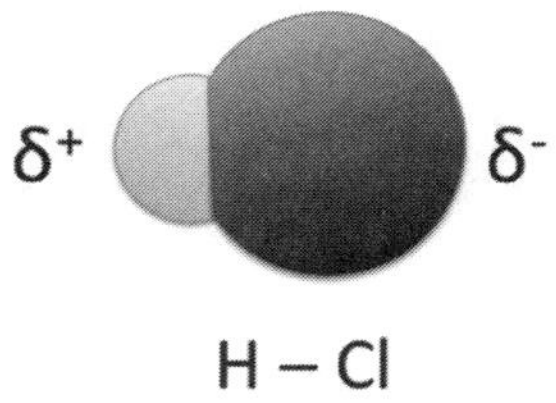

δ^+ δ^-

H – Cl

Timeline

The instructional time needed to complete this lab investigation is 130–200 minutes. Appendix 2 (p. 501) provides options for implementing this investigation over several class periods. Option C (200 minutes) should be used if students are unfamiliar with scientific writing because this option provides extra instructional time for scaffolding the writing process. You can scaffold the writing process by modeling, providing examples, and providing hints as students write each section of the report. Option D (130 minutes) should be used if students are familiar with scientific writing and have the skills needed to write an investigation report on their own. In option D, students complete stage 6 (writing the investigation report) and stage 8 (revising the investigation report) as homework.

Materials and Preparation

The materials needed to implement this investigation are listed in Table 1.2. The *Molecule Polarity* simulation was developed by PhET Interactive Simulations, University of Colorado (*http://phet.colorado.edu*), and is available at *http://phet.colorado.edu/en/simulation/molecule-polarity*. It is free to use and can be run online using an internet browser. You should access the website and learn how the simulation works before beginning the lab investigation. In addition, it is important to check if students can access and use the simulation from a school computer because some schools have set up firewalls and other restrictions on web browsing.

TABLE 1.2

Materials list

Item	Quantity
Computer with internet access	1 per group
Whiteboard, 2' × 3' *	1 per group
Lab handout	1 per student
Peer-review guide and instructor scoring rubric	1 per student

* As an alternative, students can use computer and presentation software such as Microsoft PowerPoint or Apple Keynote to create their arguments.

Safety Precautions

Remind students to follow all normal lab safety rules.

Laboratory Waste Disposal

No waste disposal is needed in this lab investigation.

Topics for the Explicit and Reflective Discussion

Concepts That Can Be Used to Justify the Evidence

To provide an adequate justification of their evidence, students must explain why they included the evidence in their arguments and make the assumptions underlying their analysis and interpretation of the data explicit. In this investigation, students can use the following concepts to help justify their evidence:

- Atomic structure
- Chemical bonds and bond types

We recommend that you discuss these fundamental concepts during the explicit and reflective discussion to help students make this connection.

How to Design Better Investigations

It is important for students to reflect on the strengths and weaknesses of the investigation they designed during the explicit and reflective discussion. Students should therefore be encouraged to discuss ways to eliminate potential flaws, measurement errors, or sources of bias in their investigations. To help students be more reflective about the design of their investigation, you can ask the following questions:

- What were some of the strengths of your investigation? What made it scientific?

- What were some of the weaknesses of your investigation? What made it less scientific?

- If you were to do this investigation again, what would you do to address the weaknesses in your investigation? What could you do to make it more scientific?

Crosscutting Concepts

This investigation is well aligned with three crosscutting concepts found in *A Framework for K–12 Science Education*, and you should review these concepts during the explicit and reflective discussion.

- *Patterns:* A major objective in chemistry is to identify patterns, such as the nature of the molecules that are formed when different types of atoms interact.

- *Systems and system models:* Scientists often need to use models to understand complex phenomena. In this investigation, students use a computer simulation to explore the effect of atom electronegativity on bond character and molecular polarity.

- *Structure and function:* The way an object is shaped or structured determines many of its properties and functions. The structure of a molecule, for example, determines if it is polar.

The Nature of Science and the Nature of Scientific Inquiry

This investigation is well aligned with two important concepts related to the *nature of science* (NOS) and the *nature of scientific inquiry* (NOSI), and you should review these concepts during the explicit and reflective discussion.

- *The difference between laws and theories in science:* A scientific law describes the behavior of a natural phenomenon or a generalized relationship under certain conditions; a scientific theory is a well-substantiated explanation of some aspect of the natural world. Theories do not become laws even with additional evidence; they explain laws. However, not all scientific laws have an accompanying explanatory theory. It is also important for students to understand that scientists do not discover laws or theories; the scientific community develops them over time.

- *The difference between data and evidence in science:* Data are measurements, observations, and findings from other studies that are collected as part of an investigation. Evidence, in contrast, is analyzed data and an interpretation of the analysis.

Hints for Implementing the Lab

- Learn how to use the online simulation before the lab begins. It is important for you to know how to use the simulation so you can help students when they get stuck or confused.

- A group of three students per computer tends to work well.

- Allow the students to play with the simulation as part of the tool talk before they begin to design their investigation. This gives students a chance to see what they can and cannot do with the simulation.

- Be sure that students record actual values (e.g., electronegativity differences and bond characters) and are not just attempting to hand draw what they see on the computer screen.

- Encourage students to take screenshots of the molecules they observe using the simulation and to use the images in their reports.

Topic Connections

Table 1.3, p. 34, provides an overview of the scientific practices, crosscutting concepts, disciplinary core ideas, and supporting ideas at the heart of this lab investigation. In addition, it lists NOS and NOSI concepts for the explicit and reflective discussion. Finally, it lists literacy and mathematics skills (*CCSS ELA* and *CCSS Mathematics*) that are addressed during the investigation.

LAB 1

TABLE 1.3

Lab 1 alignment with standards

Scientific practices	• Asking questions and defining problems • Developing and using models • Planning and carrying out investigations • Analyzing and interpreting data • Constructing explanations and designing solutions • Engaging in argument from evidence • Obtaining, evaluating, and communicating information
Crosscutting concepts	• Patterns • Systems and system models • Structure and function
Core idea	• PS1.A: Structure and properties of matter
Supporting ideas	• Atomic structure • Bond character • Molecular polarity
NOS and NOSI concepts	• Scientific laws and theories • Difference between data and evidence
Literacy connections (*CCSS ELA*)	• *Reading:* Key ideas and details, craft and structure, integration of knowledge and ideas • *Writing:* Text types and purposes, production and distribution of writing, research to build and present knowledge, range of writing • *Speaking and listening:* Comprehension and collaboration, presentation of knowledge and ideas
Mathematics connections (*CCSS Mathematics*)	• Reason abstractly and quantitatively • Look for and make use of structure

Lab Handout

Lab 1. Bond Character and Molecular Polarity: How Does Atom Electronegativity Affect Bond Character and Molecular Polarity?

Introduction

Chemists often classify chemical compounds into one of two broad categories. The first category is molecular compounds, and the second category is ionic compounds. Molecular compounds consist of atoms that are held together by covalent bonds. Ionic compounds, in contrast, are composed of positive and negative ions that are joined by ionic bonds. Covalent bonds are formed when atoms share one or more pairs of electrons. An ionic bond is formed when one or more electrons from one atom are transferred to another atom. The transfer of one or more electrons from one atom to another results in the formation of a positive ion and a negative ion. The ions then attract each other because they have opposite electrical charges.

The term *electronegativity* refers to a measure of an atom's tendency to attract electrons from other atoms. Atom electronegativity affects the nature or the character of the bond that will form between two atoms. The electronegativity of atoms also affects the electrical charge of a molecular compound. In some molecules, the electronegativity of the atoms that make up the molecule results in one side of the molecule having a partial negative electrical charge and the other side having a partial positive charge. When this happens, the molecule is described as being polar. Water is an example of a polar molecule because the oxygen side of the molecule has a partial negative charge and the hydrogen side of the molecule has a partial positive charge. Nonpolar molecules, in contrast, do not have electrical poles. Carbon dioxide is an example of a nonpolar molecule because both sides of the molecule have the same charge.

In this investigation, you will explore the relationship between the electronegativity of the atoms found within a chemical compound and the character of the bond that holds that compound together. You will also explore how atom electronegativity and molecular polarity are related.

Your Task

Use a computer simulation to explore the effect of atom electronegativity on bond character and molecular polarity.

LAB 1

The guiding question of this investigation is, **How does atom electronegativity affect bond character and molecular polarity?**

Materials

You will use an online simulation called *Molecule Polarity* to conduct your investigation. You can access the simulation by going to the following website: *http://phet.colorado.edu/en/ simulation/molecule-polarity*.

Safety Precautions

Follow all normal lab safety rules.

Investigation Proposal Required? ☐ Yes ☐ No

Getting Started

The *Molecule Polarity* simulation (see Figure L1.1) enables you to create molecules with different numbers of atoms in them and to adjust the electronegativity of each atom in the molecule. You can also view the partial charge of each side of the molecule, the electrostatic potential across the molecule, and the bond character. This information will allow you to explore how atom electronegativity affects bond character and molecular polarity.

To configure the simulation for this investigation, click on "Bond Character" in the View box and on "Electrostatic Potential" in the Surface box. This will allow you to explore how changing the electronegativity of atoms affects the nature of the bond that forms between them. It will also allow you to examine how atom electronegativity affects the electrical charge of a chemical compound. The other options, such as "Bond Dipole" and "Partial Charges" in the View box and "None" and "Electron Density" in the Surface box, should not be checked. Once the simulation is ready to use, you must determine what type of data you will need to collect, how you will collect the data, and how you will analyze the data to answer the guiding question.

To determine *what type of data you need to collect*, think about the following questions:

- What type of observations will you need to record during your investigation?
- When will you need to make these observations?

To determine *how you will collect the data*, think about the following questions:

- What types of molecules will you need to include in the simulation (i.e., molecules made up of two atoms, molecules made up of three atoms, or both)?
- What range of electronegativity values will you need to investigate?
- What types of comparisons will you need to make?

FIGURE L1.1 _____

A screenshot of the *Molecule Polarity* simulation

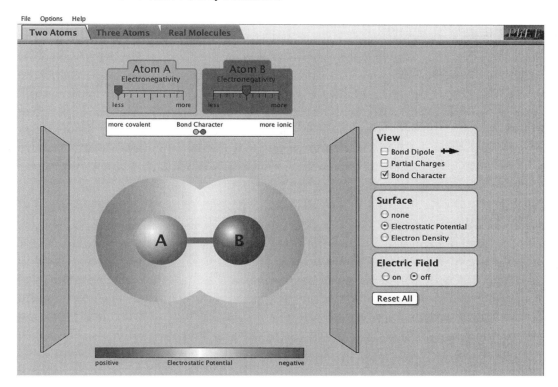

- How will you keep track of the data you collect and how will you organize it?

To determine *how you will analyze the data*, think about the following questions:

- What type of calculations will you need to make?
- What type of graph could you create to help make sense of your data?

Connections to Crosscutting Concepts, the Nature of Science, and the Nature of Scientific Inquiry

As you work through your investigation, be sure to think about

- the importance of looking for and identifying patterns,
- how models are used to study natural phenomena,
- how the structure of an object is related to its function,
- the difference between laws and theories in science, and
- the difference between data and evidence in science.

LAB 1

FIGURE L1.2

Argument presentation on a whiteboard

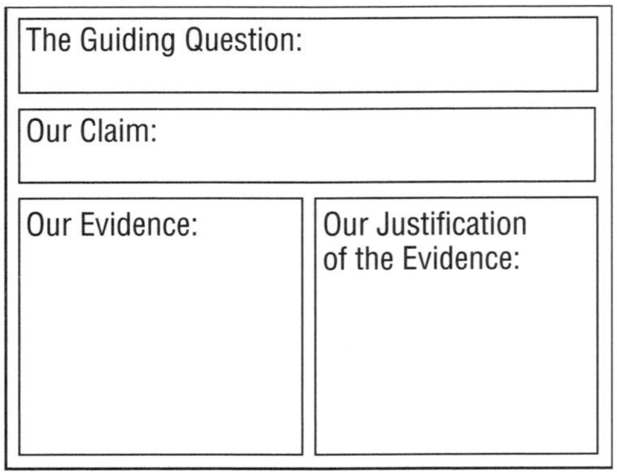

The Guiding Question:	
Our Claim:	
Our Evidence:	Our Justification of the Evidence:

Initial Argument

Once your group has finished collecting and analyzing your data, you will need to develop an initial argument. Your argument must include a *claim*, which is your answer to the guiding question. Your argument must also include *evidence* in support of your claim. The evidence is your analysis of the data and your interpretation of what the analysis means. Finally, you must include a *justification* of the evidence in your argument. You will therefore need to use a scientific concept or principle to explain why the evidence that you decided to use is relevant and important. You will create your initial argument on a whiteboard. Your whiteboard must include all the information shown in Figure L1.2.

Argumentation Session

The argumentation session allows all of the groups to share their arguments. One member of each group stays at the lab station to share that group's argument, while the other members of the group go to the other lab stations one at a time to listen to and critique the arguments developed by their classmates. The goal of the argumentation session is not to convince others that your argument is the best one; rather, the goal is to identify errors or instances of faulty reasoning in the initial arguments so these mistakes can be fixed. You will therefore need to evaluate the content of the claim, the quality of the evidence used to support the claim, and the strength of the justification of the evidence included in each argument that you see. To critique an argument, you might need more information than what is included on the whiteboard. You might therefore need to ask the presenter one or more follow-up questions, such as:

- What did your group do to analyze the data, and why did you decide to do it that way?
- Is that the only way to interpret the results of your group's analysis? How do you know that your interpretation of the analysis is appropriate?
- Why did your group decide to present your evidence in that manner?
- What other claims did your group discuss before deciding on that one? Why did you abandon those alternative ideas?
- How confident are you that your group's claim is valid? What could you do to increase your confidence?

Once the argumentation session is complete, you will have a chance to meet with your group and revise your original argument. Your group might need to gather more data or design a way to test one or more alternative claims as part of this process. Remember, your goal at this stage of the investigation is to develop the most valid or acceptable answer to the research question!

Report

Once you have completed your research, you will need to prepare an *investigation report* that consists of three sections that provide answers to the following questions:

1. What question were you trying to answer and why?

2. What did you do during your investigation and why did you conduct your investigation in this way?

3. What is your argument?

Your report should answer these questions in two pages or less. The report must be typed and any diagrams, figures, or tables should be embedded into the document. Be sure to write in a persuasive style; you are trying to convince others that your claim is acceptable or valid!

LAB 1

Lab 1. Bond Character and Molecular Polarity: How Does Atom Electronegativity Affect Bond Character and Molecular Polarity?

1. How does the electronegativity of atoms affect the polarity of a molecule?

2. Petroleum products are used for a variety of applications, including fueling automobiles, cooking, and heating homes. Petroleum is composed of a variety of molecules containing different numbers of carbon and hydrogen atoms bonded together. On several occasions, large amounts of petroleum have spilled into the ocean from shipwrecks. When the petroleum spills into the ocean it floats because it is less dense than water, but it also does not mix with the water due to the nature of its chemical bonds.

   Use what you know about bond types and electronegativity to explain why the petroleum molecules do not mix with water molecules.

3. The terms *data* and *evidence* have the same meaning in science.

   a. I agree with this statement

   b. I disagree with this statement

   Explain your answer, using an example from your investigation about bond character and molecular polarity.

4. Theories can become laws over time.

   a. I agree with this statement
   b. I disagree with this statement

   Explain your answer, using an example from your investigation about bond character and molecular polarity.

5. Scientists often use models to help them understand natural phenomena. Explain what a model is and why models are important, using an example from your investigation about bond character and molecular polarity.

6. Scientists often look for and attempt to explain patterns in nature. Explain why patterns are important, using an example from your investigation about bond character and molecular polarity.

LAB 2

Lab 2. Molecular Shapes: How Does the Number of Substituents Around a Central Atom Affect the Shape of a Molecule?

Purpose

The purpose of this lab is to *introduce* students to the valence shell electron pair repulsion (VSEPR) model. This lab gives students an opportunity to use a computer simulation to explore the factors that affect the three-dimensional shape of simple covalent compounds. Students will also learn about the difference between laws and theories and the wide range of methods that scientists can use during an investigation.

The Content

The three-dimensional arrangement of atoms in molecules is referred to as *molecular geometry*. The shape of a molecule is based on the specific orientation of bonding atoms. Chemists are interested in molecular geometry because it determines the behavior of molecules and is important for understanding the chemistry of vision, smell and odors, taste, drug reactions, and enzyme-controlled reactions. The VSEPR model is often used to predict the geometry of molecules. The valence shell is the outermost electron-occupied shell of an atom. This shell holds the electrons that are involved in bonding.

The fundamental principle of VSEPR is rather simple: Electron pairs around a central atom arrange themselves so they are as far apart as possible from each other. The repulsion between negatively charged electron pairs in bonds or as lone pairs causes them to spread apart as much as possible. Pairs of electrons in bonds are those electrons shared by the central atom and any atom to which it is bonded. Lone pairs of electrons are those pairs of electrons on an individual atom that are not shared with another atom. In a polyatomic molecule, several atoms are bonded to a central atom and the central atom may also include one or more pairs of lone electrons. Bonded atoms and lone pairs of electrons are often described as *substituents*.

According to the VSEPR model, the molecular geometry of a molecule can be determined by (1) creating a Lewis dot diagram to determine which atoms are bonded to each other and the total electron pairs involved in a molecule and (2) determining the electron pair geometry from the number of substituents. Molecules can then be divided into two groups. The first group consists of molecules that have *no* lone electron pairs, and the second group consists of molecules with one or more pairs of lone electrons. Molecules in the first group have a molecular geometry that is identical to the electron pair geometry. Molecules in the second group have a molecular geometry that is different from the electron pair geometry because only the positions of bonded atoms are used to describe the molecular geometry. Table 2.1 provides an overview of electron pair geometries and molecular geometries for some common combinations of substituents around the central atom.

TABLE 2.1

Electron pair geometries and molecular geometries for some common combinations of substituents around the central atom

	Substituents		Electron pair geometry			
Steric number*	Bonded atoms	Electron pairs	Name	Image	Molecular geometry	Examples
2	2	0	Linear		Linear	CO_2 BeH_2 HCN
3	3	0	Trigonal planar		Trigonal planar	BF_3 $COCl_2$
3	2	1	Trigonal planar		Bent	SO_2 O_3 NO_2
4	4	0	Tetrahedral		Tetrahedral	CH_4 CCl_4
4	3	1	Tetrahedral		Trigonal pyramidal	NH_3 NF_3 PCl_3
4	2	2	Tetrahedral		Bent	H_2O ClO_2

* Number of bonded atoms + number of electron pairs

LAB 2

Timeline

The instructional time needed to complete this lab investigation is 130–200 minutes. Appendix 2 (p. 501) provides options for implementing this lab investigation over several class periods. Option C (200 minutes) should be used if students are unfamiliar with scientific writing, because this option provides extra instructional time for scaffolding the writing process. You can scaffold the writing process by modeling, providing examples, and providing hints as students write each section of the report. Option D (130 minutes) should be used if students are familiar with scientific writing and have the skills needed to write an investigation report on their own. In option D, students complete stage 6 (writing the investigation report) and stage 8 (revising the investigation report) as homework.

Materials and Preparation

The materials needed to implement this investigation are listed in Table 2.2. The *Molecule Shapes* simulation was developed by PhET Interactive Simulations, University of Colorado (*http://phet.colorado.edu*), and is available at *http://phet.colorado.edu/en/simulation/molecule-shapes*. It is free to use and can be run online using an internet browser. You should access the website and learn how the simulation works before beginning the lab investigation. In addition, it is important to check if students can access and use the simulation from a school computer because some schools have set up firewalls and other restrictions on web browsing.

TABLE 2.2

Materials list

Item	Quantity
Computer with internet access	1 per group
Whiteboard, 2' × 3' *	1 per group
Lab handout	1 per student
Peer-review guide and instructor scoring rubric	1 per student

* As an alternative, students can use computer and presentation software such as Microsoft PowerPoint or Apple Keynote to create their arguments.

Safety Precautions

Remind students to follow all normal lab safety rules.

Laboratory Waste Disposal

No waste disposal is needed in this lab investigation.

Topics for the Explicit and Reflective Discussion

Concepts That Can Be Used to Justify the Evidence

To provide an adequate justification of their evidence, students must explain why they included the evidence in their arguments and make the assumptions underlying their analysis and interpretation of the data explicit. In this investigation, students can use the following concepts to help justify their evidence:

- Atomic structure
- Electron dot structures
- Chemical bonds

We recommend that you discuss these fundamental concepts during the explicit and reflective discussion to help students make this connection.

How to Design Better Investigations

It is important for students to reflect on the strengths and weaknesses of the investigation they designed during the explicit and reflective discussion. Students should therefore be encouraged to discuss ways to eliminate potential flaws, measurement errors, or sources of bias in their investigations. To help students be more reflective about the design of their investigation, you can ask the following questions:

- What were some of the strengths of your investigation? What made it scientific?
- What were some of the weaknesses of your investigation? What made it less scientific?
- If you were to do this investigation again, what would you do to address the weaknesses in your investigation? What could you do to make it more scientific?

Crosscutting Concepts

This investigation is well aligned with three crosscutting concepts found in *A Framework for K–12 Science Education,* and you should review these concepts during the explicit and reflective discussion.

- *Patterns:* A major objective in chemistry is to identify patterns, such as shapes of molecules associated with different numbers of substituents.
- *Systems and system models:* Scientists often need to use models to understand complex phenomena. In this investigation, for example, students use a computer simulation in order to explore the relationship between molecular geometry and the number of substituents found around a central atom.

- *Structure and function:* The way an object is shaped or structured determines many of its properties and functions. The shape of a molecule, for example, determines its function.

The Nature of Science and the Nature of Scientific Inquiry

This investigation is well aligned with two important concepts related to the *nature of science* (NOS) and the *nature of scientific inquiry* (NOSI), and you should review these concepts during the explicit and reflective discussion.

- *The difference between laws and theories in science:* A scientific law describes the behavior of a natural phenomenon or a generalized relationship under certain conditions; a scientific theory is a well-substantiated explanation of some aspect of the natural world. Theories do not become laws even with additional evidence; they explain laws. However, not all scientific laws have an accompanying explanatory theory. It is also important for students to understand that scientists do not discover laws or theories; the scientific community develops them over time.

- *Methods used in scientific investigations*: Examples of methods include experiments, systematic observations of a phenomenon, literature reviews, and analysis of existing data sets; the choice of method depends on the objectives of the research. There is no universal step-by step scientific method that all scientists follow; rather, different scientific disciplines (e.g., chemistry vs. biology) and fields within a discipline (e.g., organic vs. physical chemistry) use different types of methods, use different core theories, and rely on different standards to develop scientific knowledge.

Hints for Implementing the Lab

- Learn how to use the online simulation before the lab begins. It is important for you to know how to use the simulation so you can help students when they get stuck or confused.
- A group of three students per computer tends to work well.
- Allow the students to play with the simulation as part of the tool talk before they begin to design their investigation. This gives students a chance to see what they can and cannot do with simulation.
- Encourage students to take screenshots of the molecules they observe using the simulation and to use the images in their reports.

Topic Connections

Table 2.3 provides an overview of the scientific practices, crosscutting concepts, disciplinary core ideas, and supporting ideas at the heart of this lab investigation. In addition, it lists

NOS and NOSI concepts for the explicit and reflective discussion. Finally, it lists literacy skills (*CCSS ELA*) that are addressed during the investigation.

TABLE 2.3

Lab 2 alignment with standards

Scientific practices	<ul><li>Asking questions and defining problems</li><li>Developing and using models</li><li>Planning and carrying out investigations</li><li>Analyzing and interpreting data</li><li>Constructing explanations and designing solutions</li><li>Engaging in argument from evidence</li><li>Obtaining, evaluating, and communicating information</li></ul>
Crosscutting concepts	<ul><li>Patterns</li><li>Systems and system models</li><li>Structure and function</li></ul>
Core idea	<ul><li>PS1.A: Structure and properties of matter</li></ul>
Supporting ideas	<ul><li>Atomic structure</li><li>VSEPR</li><li>Molecular shape</li></ul>
NOS and NOSI concepts	<ul><li>Scientific laws and theories</li><li>Methods used in scientific investigations</li></ul>
Literacy connections (*CCSS ELA*)	<ul><li>*Reading:* Key ideas and details, craft and structure, integration of knowledge and ideas</li><li>*Writing:* Text types and purposes, production and distribution of writing, research to build and present knowledge, range of writing</li><li>*Speaking and listening:* Comprehension and collaboration, presentation of knowledge and ideas</li></ul>

LAB 2

Lab Handout

Lab 2. Molecular Shapes: How Does the Number of Substituents Around a Central Atom Affect the Shape of a Molecule?

Introduction

Molecules are three-dimensional entities and therefore should be depicted in three dimensions. We can translate the two-dimensional electron dot structure representing a molecule into a more useful three-dimensional rendering by using a tool known as the valance shell electron pair repulsion (VSEPR) model. According to this model, any given pair of valance shell electrons strives to get as far away as possible from all other electron pairs in the shell. This includes both nonbonding pairs of electrons and any bonding pair not taking part in a double or triple bond. Pairs of electrons in a multiple bond stay together because of their mutual attractions for the same two nuclei. It is this striving for maximum separation distance between electron pairs that determines the geometry of any molecule.

The VSEPR model allows us to use the electron dot structures of atoms to predict the three-dimensional geometry of simple molecules. This geometry is determined by considering the number of *substituents* surrounding a central atom. A substituent is any bonded atom or nonbonding pair of electrons. For example, the carbon on the methane molecule (CH_4) shown in Figure L2.1 has four substituents (four hydrogen atoms). The oxygen atom of water also has four substituents (two hydrogen atoms and two nonbonding pairs of electrons).

FIGURE L2.1 _____

Substituents around a central atom

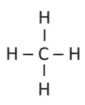

Methane

1 central atom (carbon)
4 substituents (4 hydrogen atoms)

Dots are used to indicate the space occupied by non-bonded electron pairs

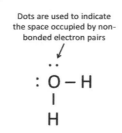

Water

1 central atom (oxygen)
4 substituents (2 hydrogen atoms and 2 pairs of non-bonding electrons)

Your Task

Use a computer simulation to develop a rule that you can use to predict the shape of a molecule based on the number of atoms and lone electron pairs (i.e., substituents) around a central atom.

The guiding question of this investigation is, **How does the number of substituents around a central atom affect the shape of a molecule?**

Materials

You will use an online simulation called *Molecule Shapes* to conduct your investigation. You can access the simulation by going to the following website: *http://phet.colorado.edu/en/simulation/molecule-shapes*.

Safety Precautions

Follow all normal lab safety rules.

Investigation Proposal Required?   ☐ Yes   ☐ No

Getting Started

The *Molecule Shapes* simulation (see Figure L2.2) models how the number of atoms and lone electron pairs around a central atom affects the shape of a molecule. With this simulation, you can decide how many atoms are found in the molecule, the nature of the bonds found between the atoms (single, double, or triple) and the number of lone electron pairs that are found around the central atom. Your goal is to develop a rule that you can use to predict the shape of a molecule based on the number of atoms and lone electron pairs (i.e., substituents) found around the central atom. Before you start using the simulation, however, you must determine what type of data you will need to collect, how you will collect the data, and how you will analyze the data to answer the guiding question.

FIGURE L2.2

A screenshot of the *Molecule Shapes* simulation

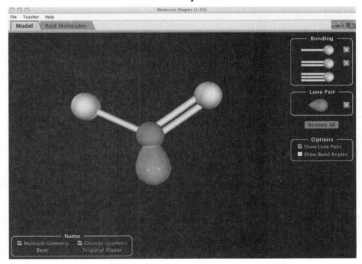

To determine *what type of data you need to collect*, think about the following questions:

- What type of observations will you need to record during your investigation?
- When will you need to make these observations?

To determine *how you will collect the data*, think about the following questions:

- What types of molecules will you need to examine using the simulation?
- What type of comparisons will you need to make?
- How will you keep track of the data you collect and how will you organize it?

To determine *how you will analyze the data*, think about the following questions:

- What type of calculations will you need to make?
- What type of graph could you create to help make sense of your data?

Connections to Crosscutting Concepts, the Nature of Science, and the Nature of Scientific Inquiry

As you work through your investigation, be sure to think about

- the importance of looking for and identifying patterns,
- how models are used to study natural phenomena,
- how the structure of an object is related to its function,
- the difference between laws and theories in science, and
- the wide range of methods that can be used by scientists during an investigation.

FIGURE L2.3

Argument presentation on a whiteboard

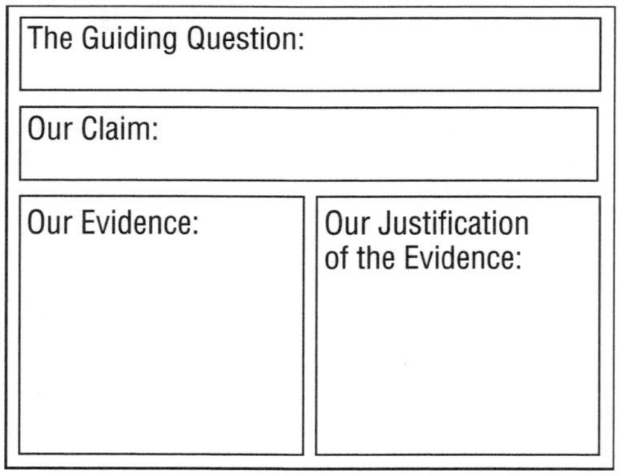

The Guiding Question:	
Our Claim:	
Our Evidence:	Our Justification of the Evidence:

Initial Argument

Once your group has finished collecting and analyzing your data, you will need to develop an initial argument. Your argument must include a *claim*, which is your answer to the guiding question. Your argument must also include *evidence* in support of your claim. The evidence is your analysis of the data and your interpretation of what the analysis means. Finally, you must include a *justification* of the evidence in your argument. You will therefore need to use a scientific concept or principle to explain why the evidence that you decided to use is relevant and important. You will create your initial argument on a whiteboard. Your whiteboard must include all the information shown in Figure L2.3.

Argumentation Session

The argumentation session allows all of the groups to share their arguments. One member of each group stays at the lab station to share that group's argument, while the other members of the group go to the other lab stations one at a time to listen to and critique the arguments developed by their classmates. The goal of the argumentation session is not to convince others that your argument is the best one; rather, the goal is to identify errors or instances of faulty reasoning in the initial arguments so these mistakes can be fixed. You will therefore need to evaluate the content of the claim, the quality of the evidence used to support the claim, and the strength of the justification of the evidence included in each argument that you see. To critique an argument, you might need more information than what is included on the whiteboard. You might therefore need to ask the presenter one or more follow-up questions, such as:

- What did your group do to analyze the data, and why did you decide to do it that way?
- Is that the only way to interpret the results of your group's analysis? How do you know that your interpretation of the analysis is appropriate?
- Why did your group decide to present your evidence in that manner?
- What other claims did your group discuss before deciding on that one? Why did you abandon those alternative ideas?
- How confident are you that your group's claim is valid? What could you do to increase your confidence?

Once the argumentation session is complete, you will have a chance to meet with your group and revise your original argument. Your group might need to gather more data or design a way to test one or more alternative claims as part of this process. Remember, your goal at this stage of the investigation is to develop the most valid or acceptable answer to the research question!

Report

Once you have completed your research, you will need to prepare an *investigation report* that consists of three sections that provide answers to the following questions:

1. What question were you trying to answer and why?
2. What did you do during your investigation and why did you conduct your investigation in this way?
3. What is your argument?

Your report should answer these questions in two pages or less. The report must be typed and any diagrams, figures, or tables should be embedded into the document. Be sure to write in a persuasive style; you are trying to convince others that your claim is acceptable or valid!

LAB 2

Lab 2. Molecular Shapes: How Does the Number of Substituents Around a Central Atom Affect the Shape of a Molecule?

1. Describe the basic principle of the valance shell electron pair repulsion (VSEPR) model.

Use what you know about the VSEPR model to answer questions 2 and 3.

2. Draw the Lewis dot structure for CO_2 in the space below.

What shape is a molecule of CO_2?

   a. Linear

   b. Trigonal planar

   c. Bent

   d. Tetrahedral

Explain your answer.

3. Draw the Lewis dot structure for CCl_4 in the space below.

What shape is a molecule of CCl_4?

   a. Linear

   b. Trigonal planar

   c. Bent

   d. Tetrahedral

Explain your answer.

4. All scientists follow the scientific method during an investigation.

   a. I agree with this statement.
   b. I disagree with this statement.

   Explain your answer, using an example from your investigation about molecular shapes.

5. Theories and laws are different kinds of scientific knowledge.

   a. I agree with this statement.
   b. I disagree with this statement.

   Explain your answer, using an example from your investigation about molecular shapes.

6. Scientists often use models to help them understand natural phenomena. Explain what a model is and why models are important, using an example from your investigation about molecular shapes.

7. Scientists often look for and attempt to explain patterns in nature. Explain why patterns are important, using an example from your investigation about molecular shapes.

8. In nature, the structure of an object is often related to the function or properties of that object. Explain why this is true, using an example from your investigation about molecular shapes.

LAB 3

Teacher Notes

Lab 3. Rate of Dissolution: Why Do the Surface Area of the Solute, the Temperature of the Solvent, and the Amount of Agitation That Occurs When the Solute and the Solvent Are Mixed Affect the Rate of Dissolution?

Purpose

The purpose of this lab is to *introduce* students to the concepts of solutes, solvents, solubility, and rate of dissolution. This lab gives students an opportunity to develop and use a model that can help them explain why the surface area of a solute, the temperature of the solvent, and the amount of agitation that occurs when the solute and the solvent are mixed affect the rate of dissolution. Students will also learn about what does and does not count as an experiment in science and why scientists need to be creative and have a good imagination to excel in science.

The Content

A *solution* is a uniform mixture of two or more pure substances. The substance that is dissolved to make a solution is called the *solute*, and the substance that does the dissolving is called the *solvent*. *Solubility*, which is defined as the amount of solute that will dissolve in a given amount of solvent at a particular temperature, depends on the physical properties of the solute, the physical properties of the solvent, and how the two substances interact when they are combined. The rate of dissolution, which is a measure of how fast a solid will dissolve in a liquid, is affected by three factors: (1) the surface area of the solute, (2) the temperature of the solvent, and (3) the amount of agitation that occurs when the solute and solvent are mixed.

The process of a solid dissolving in water is a surface phenomenon. To illustrate this process, consider what happens when salt is mixed with water. Free-moving water molecules randomly collide with the ions (Na^+ and Cl^-) on the surface of the crystal by means of ion-dipole attractive forces. These collisions gradually strip the surface ions away from the bulk crystal. As the surface ions are removed, the next layer of ions becomes the new surface layer. This interaction at the surface of the crystal continues until the crystal is completely dissolved or the solvent is saturated. Figure 3.1 provides an illustration of this process from a submicroscopic view.

FIGURE 3.1

A submicroscopic view of a salt crystal dissolving in water

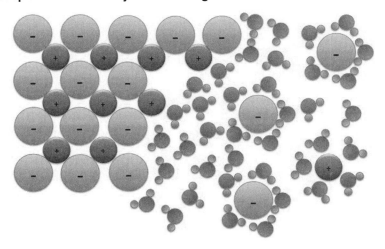

This surface collision model explains why the surface area of the solute, the temperature of the solvent, and the amount of agitation that occurs when the solute and the solvent are mixed affect the rate of dissolution. First, the rate of dissolution will increase as the particle size of the solute decreases. Smaller particles dissolve faster than larger ones because smaller particles have a greater surface area to volume ratio than larger particles. Second, the rate of dissolution will increase as the temperature of the solvent increases. A higher temperature corresponds to greater average kinetic energy of the molecules or ions, which means that the surface collisions happen at a faster rate and with more energy. Finally, greater agitation increases the rate of dissolution because stirring and shaking the solvent help move the ions or molecules that have been stripped from the surface of the solute away and bring in more molecules of solvent.

Timeline

The instructional time needed to complete this lab investigation is 180–250 minutes. Appendix 2 (p. 501) provides options for implementing this lab investigation over several class periods. Option E (250 minutes) should be used if students are unfamiliar with scientific writing because this option provides extra instructional time for scaffolding the writing process. You can scaffold the writing process by modeling, providing examples, and providing hints as students write each section of the report. Option F (180 minutes) should be used if students are familiar with scientific writing and have the skills needed to write an investigation report on their own. In option F, students complete stage 6 (writing the investigation report) and stage 8 (revising the investigation report) as homework.

LAB 3

Materials and Preparation

The materials needed to implement this investigation are listed in Table 3.1. The consumables and equipment can be purchased from a science supply company such as Carolina, Flinn Scientific, or Ward's Science. We recommend that you use a set routine for distributing and collecting the materials during the lab investigation. For example, the consumables and equipment for each group can be set up at each group's lab station before class begins, or one member from each group can collect them from a table or a cart when needed during class.

Safety Precautions

Remind students to follow all normal lab safety rules. Copper(II) sulfate is moderately toxic by ingestion and inhalation, and it is also a skin and respiratory tract irritant. You will therefore need to explain the potential hazards of working with copper(II) sulfate and how to work with hazardous chemicals. In addition, tell students to take the following safety precautions:

- Wear indirectly vented chemical-splash goggles and chemical-resistant gloves and aprons when they are collecting their data.
- Never put consumables in their mouth (including the rock candy used during this investigation).
- Use caution when working with hot plates, and keep them away from water and other liquids.
- Handle all glassware with care.
- Never return the consumables to stock bottles.
- Wash their hands with soap and water when they are done collecting the data.

Laboratory Waste Disposal

Solid copper(II) sulfate and sugar can be disposed of in a landfill. We recommend following Flinn laboratory waste disposal method 26a to dispose of these solids. Aqueous solutions of copper(II) sulfate and sugar can be disposed of down a drain if the drain is connected to a sanitation sewer system. Information about laboratory waste disposal methods is included in the Flinn Catalog and Reference Manual; you can request a free copy at *www.flinnsci.com*.

Topics for the Explicit and Reflective Discussion

Concepts That Can Be Used to Justify the Evidence

To provide an adequate justification of their evidence, students must explain why they included the evidence in their arguments and make the assumptions underlying their analysis and interpretation of the data explicit. In this investigation, students can use the following concepts to help justify their evidence:

Rate of Dissolution

Why Do the Surface Area of the Solute, the Temperature of the Solvent, and the Amount of Agitation That Occurs When the Solute and the Solvent Are Mixed Affect the Rate of Dissolution?

TABLE 3.1

Materials list

Item	Quantity
Consumables	
Copper(II) sulfate, $CuSO_4$—powder	1–2 g per group
Copper(II) sulfate, $CuSO_4$—fine crystal	1–2 g per group
Copper(II) sulfate, $CuSO_4$—medium crystal	1–2 g per group
Rock candy	5–10 pieces per group
Ice	As needed
Equipment and other materials	
Stopwatch	1 per group
Hot plate	1 per group
Electronic or triple beam balance	1 per group
Stirring rod or magnetic stirrer	1 per group
Graduated cylinder, 50 ml	1 per group
Beakers, 250 ml	2 per group
Beakers or Erlenmeyer flasks, 50 or 100 ml	4 per group
Thermometer (nonmercury) or temperature probe	1 per group
Spatula or chemical scoop	1 per group
Weighing paper or dishes	As needed
Mortar and pestle	1 per group
Investigation Proposal A (optional)	3 per group
Whiteboard, 2' × 3' *	1 per group
Lab handout	1 per student
Peer-review guide and instructor scoring rubric	1 per student

* As an alternative, students can use computer and presentation software such as Microsoft PowerPoint or Apple Keynote to create their arguments.

- Molecular-kinetic theory of matter
- The nature of ionic and covalent compounds
- Characteristics of solutions

We recommend that you discuss these fundamental concepts during the explicit and reflective discussion to help students make this connection.

How to Design Better Investigations

It is important for students to reflect on the strengths and weaknesses of the investigation they designed during the explicit and reflective discussion. Students should therefore be encouraged to discuss ways to eliminate potential flaws, measurement errors, or sources of bias in their investigations. To help students be more reflective about the design of their investigation, you can ask the following questions:

- What were some of the strengths of your investigation? What made it scientific?
- What were some of the weaknesses of your investigation? What made it less scientific?
- If you were to do this investigation again, what would you do to address the weaknesses in your investigation? What could you do to make it more scientific?

Crosscutting Concepts

This investigation is well aligned with two crosscutting concepts found in *A Framework for K–12 Science Education,* and you should review these concepts during the explicit and reflective discussion.

- *Cause and effect: Mechanism and explanation:* One of the main objectives of science is to identify and establish relationships between a cause and an effect. In this investigation, for example, students need to be able to determine how several different factors affect the rate of dissolution in order to understand how solutes dissolve in a solvent.
- *Systems and system models:* Scientists often need to use models to understand complex phenomena. In this investigation, for example, students develop a model to help explain what is happening when a solute dissolves in a solvent at the submicroscopic level.

The Nature of Science and the Nature of Scientific Inquiry

This investigation is well aligned with two important concepts related to the *nature of science* (NOS) and the *nature of scientific inquiry* (NOSI), and you should review these concepts during the explicit and reflective discussion.

Rate of Dissolution

Why Do the Surface Area of the Solute, the Temperature of the Solvent, and the Amount of Agitation That Occurs When the Solute and the Solvent Are Mixed Affect the Rate of Dissolution?

- *The importance of imagination and creativity in science:* Students should learn that developing explanations for or models of natural phenomena and then figuring out how they can be put to the test of reality is as creative as writing poetry, composing music, or designing skyscrapers. Scientists must also use their imagination and creativity to figure out new ways to test ideas and collect or analyze data.

- *The nature and role of experiments:* Scientists use experiments to test the validity of a hypothesis (i.e., a tentative explanation) for an observed phenomenon. Experiments include a test and the formulation of predictions (expected results) if the test is conducted and the hypothesis is valid. The experiment is then carried out and the predictions are compared with the observed results of the experiment. If the predictions match the observed results, then the hypothesis is supported. If the observed results do not match the prediction, then the hypothesis is not supported. A signature feature of an experiment is the control of variables to help eliminate alternative explanations for observed results.

Hints for Implementing the Lab

Allowing students to design their own procedures for collecting data gives students an opportunity to try, to fail, and to learn from their mistakes. However, you can scaffold students as they develop their procedure by having them fill out an investigation proposal. These proposals provide a way for you to offer students hints and suggestions without telling them how to do it. You can also check the proposals quickly during a class period.

- Students can use a mortar and pestle to grind the crystal of copper(II) sulfate and rock candy to fit their needs.

- Students should decide what controls to include in their investigation. To help save time, however, tell students that a "no mixing" control may not be practical for this investigation because it will take a long time for the large crystals to dissolve.

- Students should decide how to measure the rate of dissolution. The rate of dissolution can be measured in several ways, including the amount of time required to completely dissolve a solid and how much of the solid is dissolved after a given amount of time. Students should be able to explain why they decided to use one technique rather than another.

Topic Connections

Table 3.2 (p. 62) provides an overview of the scientific practices, crosscutting concepts, disciplinary core ideas, and supporting ideas at the heart of this lab investigation. In addition, it lists NOS and NOSI concepts for the explicit and reflective discussion. Finally, it lists literacy and mathematics skills (*CCSS ELA* and *CCSS Mathematics*) that are addressed during the investigation.

LAB 3

TABLE 3.2 _____

Lab 3 alignment with standards

Scientific practices	• Asking questions and defining problems • Developing and using models • Planning and carrying out investigations • Analyzing and interpreting data • Constructing explanations and designing solutions • Engaging in argument from evidence • Obtaining, evaluating, and communicating information
Crosscutting concepts	• Cause and effect: Mechanism and explanation • Systems and system models
Core idea	• PS1.A: Structure and properties of matter
Supporting ideas	• Solutes • Solvents • Solutions • Rate of dissolution • Solubility • Ions • Ionic compound • Covalent compound • Molecular-kinetic theory of matter
NOS and NOSI concepts	• Imagination and creativity in science • Nature and role of experiments
Literacy connections (*CCSS ELA*)	• *Reading:* Key ideas and details, craft and structure, integration of knowledge and ideas
	• *Writing:* Text types and purposes, production and distribution of writing, research to build and present knowledge, range of writing
	• *Speaking and listening:* Comprehension and collaboration, presentation of knowledge and ideas
Mathematics connections (*CCSS Mathematics*)	• Reason abstractly and quantitatively • Model with mathematics • Attend to precision • Look for and express regularity in repeated reasoning

Rate of Dissolution

Why Do the Surface Area of the Solute, the Temperature of the Solvent, and the Amount of Agitation That Occurs When the Solute and the Solvent Are Mixed Affect the Rate of Dissolution?

Lab Handout

Lab 3. Rate of Dissolution: Why Do the Surface Area of the Solute, the Temperature of the Solvent, and the Amount of Agitation That Occurs When the Solute and the Solvent Are Mixed Affect the Rate of Dissolution?

Introduction

A solution is a uniform mixture of two or more pure substances. The substance that is dissolved is called the solute, and the substance that does the dissolving is called the solvent. When a solid dissolves in a solvent, it is assumed that the solid dissociates into the elementary particles that make up that solid. The type of elementary particle depends on the nature of the solid. A covalent compound will dissociate into individual molecules when it is added to water, whereas an ionic compound will dissociate into positive and negative ions.

Copper(II) sulfate is an example of a substance that dissolves in water. Copper(II) sulfate is an ionic compound with the chemical formula $CuSO_4$. When it is added to water it dissociates into Cu^{2+} and SO_4^{2-} ions. Copper(II) sulfate is an important agricultural chemical. Solution of copper(II) sulfate is often sprayed on plants, including wheat, potatoes, tomatoes, grapes, and citrus fruits, to help prevent fungal diseases.

Solubility, which is defined as the amount of solute that will dissolve in a given amount of solvent at a particular temperature, depends on the nature of the solute and how it interacts with the solvent. For example, 47.93 g of copper(II) sulfate will dissolve in 100 grams of water at 70°C, but only 37.46 g of sodium chloride (NaCl) will dissolve in the same amount and temperature of water. The rate of dissolution, in contrast, is a measure of how fast a solute dissolves in a solvent. There are three factors that affect the rate of dissolution: (1) the surface area of the solute, (2) the temperature of the solvent, and (3) the amount of agitation that occurs when the solute and the solvent are mixed.

To create large amounts of solutions in short periods of time, it is important to understand not only how much of a solute will dissolve in a solvent at a given temperature but also why different factors affect how fast it will dissolve. You will therefore explore three factors that affect the rate of dissolution of copper(II) sulfate and then develop a conceptual model that you can use to explain your observations and predict the dissolution rates of other solutes under different conditions.

LAB 3

Your Task

Determine how the surface area of the solute, the temperature of the solvent, and the amount of agitation that occurs when the solute and the solvent are mixed affect the rate that copper(II) sulfate dissolves in water. Then develop a conceptual model that can be used to explain *why* these factors influence the rate of dissolution. Once you have developed your conceptual model, you will need to test it to determine if it allows you to predict the dissolution rate of another solute under various conditions.

The guiding question of this investigation is, **Why do the surface area of the solute, the temperature of the solvent, and the amount of agitation that occurs when the solute and the solvent are mixed affect the rate of dissolution?**

Materials

You may use any of the following materials during your investigation:

Consumables	Equipment
• $CuSO_4$—powder	• Stopwatch
• $CuSO_4$—fine crystal	• Hot plate
• $CuSO_4$—medium crystal	• Electronic or triple beam balance
• Rock candy	• Stirring rod or magnetic stirrer
	• 1 Graduated cylinder (50 ml)
	• 2 Beakers (each 250 ml)
	• 4 Beakers or Erlenmeyer flasks (each 50 or 100 ml)
	• Thermometer or temperature probe
	• Spatula or chemical scoop
	• Weighing paper or dishes
	• Mortar and pestle

Safety Precautions

Follow all normal lab safety rules. Copper(II) sulfate is a skin and respiratory irritant and is moderately toxic by ingestion and inhalation. Your teacher will provide important information about working with the chemicals associated with this investigation. In addition, take the following safety precautions:

- Wear indirectly vented chemical-splash goggles and chemical-resistant gloves and apron while in the laboratory.
- Never taste any of the chemicals (including the rock candy).
- Handle all glassware with care.
- Use caution when working with hot plates because they can burn skin. Hot plates also need to be kept away from water and other liquids.
- Wash your hands with soap and water before leaving the laboratory.

Investigation Proposal Required?   ☐ Yes   ☐ No

Getting Started

The first step in developing your model is to design and carry out a series of experiments to determine how the surface area of the solute, the temperature of the solvent, and the amount of agitation that occurs when the solute and solvent are mixed affect the rate of dissolution of copper(II) sulfate. To conduct these experiments, you must determine what type of data you will need to collect, how you will collect the data, and how you will analyze the data to answer the guiding question.

To determine *what type of data you need to collect*, think about the following questions:

- What type of measurements or observations will you need to record during each experiment?
- When will you need to make these measurements or observations?

To determine *how you will collect the data*, think about the following questions:

- What will serve as your independent variable?
- How will you vary the independent variable while holding other variables constant?
- What types of comparisons will you need to make?
- What will you do to reduce measurement error?
- How will you keep track of the data you collect and how will you organize it?

To determine *how you will analyze the data*, think about the following questions:

- What type of calculations will you need to make?
- What type of graph could you create to help make sense of your data?

Once you have carried out your series of experiments, your group will need to develop a conceptual model to explain why these three factors influence the rate of dissolution in the way that they do. The model also needs to be able to explain the nature of the interactions that are taking place between the solute and the solvent on the submicroscopic level.

The last step in this investigation is to test your model. To accomplish this goal, you can use rock candy to determine if your model leads to accurate predictions about the rates of dissolution for a covalent compound under different conditions. If you can use your model to make accurate predictions about the rate of dissolution of other types of solutes under different conditions, then you will be able to generate the evidence you need to convince others that the conceptual model you developed is valid.

LAB 3

Connections to Crosscutting Concepts, the Nature of Science, and the Nature of Scientific Inquiry

As you work through your investigation, be sure to think about

- the importance of developing causal explanations for observations,
- how models are used to help understand natural phenomena,
- the role of imagination and creativity in science, and
- the role of experiments in science.

FIGURE L3.1 _____

Argument presentation on a whiteboard

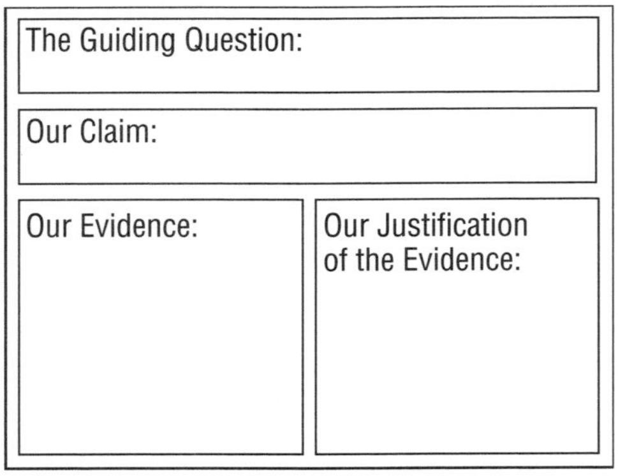

The Guiding Question:	
Our Claim:	
Our Evidence:	Our Justification of the Evidence:

Initial Argument

Once your group has finished collecting and analyzing your data, you will need to develop an initial argument. Your argument must include a *claim*, which is your answer to the guiding question. Your argument must also include *evidence* in support of your claim. The evidence is your analysis of the data and your interpretation of what the analysis means. Finally, you must include a *justification* of the evidence in your argument. You will therefore need to use a scientific concept or principle to explain why the evidence that you decided to use is relevant and important. You will create your initial argument on a whiteboard. Your whiteboard must include all the information shown in Figure L3.1.

Argumentation Session

The argumentation session allows all of the groups to share their arguments. One member of each group stays at the lab station to share that group's argument, while the other members of the group go to the other lab stations one at a time to listen to and critique the arguments developed by their classmates. The goal of the argumentation session is not to convince others that your argument is the best one; rather, the goal is to identify errors or instances of faulty reasoning in the initial arguments so these mistakes can be fixed. You will therefore need to evaluate the content of the claim, the quality of the evidence used to support the claim, and the strength of the justification of the evidence included in each argument that you see. To critique an argument, you might need more information than what is included on the whiteboard. You might therefore need to ask the presenter one or more follow-up questions, such as:

- How did your group collect the data? Why did you use that method?
- What did your group do to make sure the data you collected are reliable? What did you do to decrease measurement error?

Rate of Dissolution

Why Do the Surface Area of the Solute, the Temperature of the Solvent, and the Amount of Agitation That Occurs When the Solute and the Solvent Are Mixed Affect the Rate of Dissolution?

- What did your group do to analyze the data, and why did you decide to do it that way? Did you check your calculations?

- Is that the only way to interpret the results of your group's analysis? How do you know that your interpretation of the analysis is appropriate?

- Why did your group decide to present your evidence in that manner?

- What other claims did your group discuss before deciding on that one? Why did you abandon those alternative ideas?

- How confident are you that your group's claim is valid? What could you do to increase your confidence?

Once the argumentation session is complete, you will have a chance to meet with your group and revise your initial argument. Your group might need to gather more data or design a way to test one or more alternative claims as part of this process. Remember, your goal at this stage of the investigation is to develop the most valid or acceptable answer to the research question!

Report

Once you have completed your research, you will need to prepare an *investigation report* that consists of three sections that provide answers to the following questions:

1. What question were you trying to answer and why?

2. What did you do during your investigation and why did you conduct your investigation in this way?

3. What is your argument?

Your report should answer these questions in two pages or less. The report must be typed and any diagrams, figures, or tables should be embedded into the document. Be sure to write in a persuasive style; you are trying to convince others that your claim is acceptable or valid!

LAB 3

Lab 3. Rate of Dissolution: Why Do the Surface Area of the Solute, the Temperature of the Solvent, and the Amount of Agitation That Occurs When the Solute and the Solvent Are Mixed Affect the Rate of Dissolution?

Hikers often need more drinking water than they can carry when go on overnight trips. They can get water from rivers and streams as they hike but must purify it before they can drink it because most rivers and streams in the United States contain microorganisms that can cause serious illness. One of the safest and least expensive ways to purify water is to add an iodine tablet to it. Iodine purifies river and stream water because it kills the microorganisms in the water. The iodine tablet, however, must completely dissolve in the water before it is safe to drink.

1. What happens to a solute, such as a tablet of iodine, when it dissolves in water?

2. Use what you know about the factors that affect the rate of dissolution of a solute to suggest three things a thirsty backpacker could do to get an iodine tablet to dissolve faster in water. Be sure to explain why these three things will make the iodine tablet dissolve faster.

Rate of Dissolution

Why Do the Surface Area of the Solute, the Temperature of the Solvent, and the Amount of Agitation That Occurs When the Solute and the Solvent Are Mixed Affect the Rate of Dissolution?

3. Measuring the temperature of water is an example of an experiment.

   a. I agree with this statement.
   b. I disagree with this statement.

   Explain your answer, using an example from your investigation about rate of dissolution.

4. Scientists do not need to be creative or have a good imagination to excel in science.

   a. I agree with this statement.
   b. I disagree with this statement.

   Explain your answer, using an example from your investigation about rate of dissolution.

5. An important goal in science is to develop causal explanations for observations. Explain what a causal explanation is and why it is important, using an example from your investigation about rate of dissolution.

6. Scientists often use models to help them understand natural phenomena. Explain what a model is and why models are important, using an example from your investigation about rate of dissolution.

LAB 4

Teacher Notes

Lab 4. Molarity: What Is the Mathematical Relationship Between the Moles of a Solute, the Volume of the Solvent, and the Molarity of an Aqueous Solution?

Purpose

The purpose of this lab is to *introduce* students to the concepts of solutes, solvents, and molarity. This lab gives students an opportunity to use a computer simulation to explore the mathematical relationship between the moles of a solute, the volume of a solvent, and the molarity of an aqueous solution. Students will also learn about the wide range of methods that scientists can use during an investigation and the difference between data and evidence.

The Content

The properties and behavior of many solutions depend not only on the nature of the solute and the solvent but also on the concentration of the solute in the solution. Chemists use many different units when expressing concentration; however, one of the most common units is *molarity* (M), which is the number of moles of solute per liter of solution. The mathematical relationship between molarity, moles of solute, and volume of solution is

molarity = moles of solute / liters of solution

Timeline

The instructional time needed to complete this lab investigation is 130–200 minutes. Appendix 2 (p. 501) provides options for implementing this lab investigation over several class periods. Option C (200 minutes) should be used if students are unfamiliar with scientific writing because this option provides extra instructional time for scaffolding the writing process. You can scaffold the writing process by modeling, providing examples, and providing hints as students write each section of the report. Option D (130 minutes) should be used if students are familiar with scientific writing and have the skills needed to write an investigation report on their own. In option D, students complete stage 6 (writing the investigation report) and stage 8 (revising the investigation report) as homework.

Molarity

What Is the Mathematical Relationship Between the Moles of a Solute, the Volume of the Solvent, and the Molarity of an Aqueous Solution?

Materials and Preparation

The materials needed to implement this investigation are listed in Table 4.1. The *Molarity* simulation was developed by PhET Interactive Solutions, University of Colorado (*http://phet.colorado.edu*), and is available at *http://phet.colorado.edu/en/simulation/molarity*. It is free to use and can be run online using an internet browser. You should access the website and learn how the simulation works before beginning the lab investigation. In addition, it is important to check if students can access and use the simulation from a school computer because some schools have set up firewalls and other restrictions on web browsing.

TABLE 4.1

Materials list

Item	Quantity
Computer with internet access	1 per group
Whiteboard, 2' × 3' *	1 per group
Lab handout	1 per student
Peer-review guide and instructor scoring rubric	1 per student

* As an alternative, students can use computer and presentation software such as Microsoft PowerPoint or Apple Keynote to create their arguments.

Safety Precautions

Remind students to follow all normal lab safety rules.

Laboratory Waste Disposal

No waste disposal is needed in this lab investigation.

Topics for the Explicit and Reflective Discussion

Concepts That Can Be Used to Justify the Evidence

To provide an adequate justification of their evidence, students must explain why they included the evidence in their arguments and make the assumptions underlying their analysis and interpretation of the data explicit. In this investigation, students can use the following concepts to help justify their evidence:

- Moles
- Concentration
- The nature of aqueous solutions

LAB 4

We recommend that you discuss these fundamental concepts during the explicit and reflective discussion to help students make this connection.

How to Design Better Investigations

It is important for students to reflect on the strengths and weaknesses of the investigation they designed during the explicit and reflective discussion. Students should therefore be encouraged to discuss ways to eliminate potential flaws, measurement errors, or sources of bias in their investigations. To help students be more reflective about the design of their investigation, you can ask the following questions:

- What were some of the strengths of your investigation? What made it scientific?
- What were some of the weaknesses of your investigation? What made it less scientific?
- If you were to do this investigation again, what would you do to address the weaknesses in your investigation? What could you do to make it more scientific?

Crosscutting Concepts

This investigation is well aligned with two crosscutting concepts found in *A Framework for K–12 Science Education,* and you should review these concepts during the explicit and reflective discussion.

- *Scale, proportion, and quantity:* It is critical for scientists to be able to recognize what is relevant at different sizes, time frames, and scales. Scientists must also be able to recognize proportional relationships between categories or quantities. In this investigation, for example, students need to identify that the molarity of a solution is a ratio between the moles of solutes and the liters of solvent.
- *Energy and matter: Flows, cycles, and conservation:* Scientists must understand how energy and matter flow into, out of, and within an ecosystem in order to understand it and to understand how human activity can disrupt the natural balance of an ecosystem. This investigation is a perfect example of why this is important.

The Nature of Science and the Nature of Scientific Inquiry

This investigation is well aligned with two important concepts related to the *nature of science* (NOS) and the *nature of scientific inquiry* (NOSI), and you should review these concepts during the explicit and reflective discussion.

- *The difference between data and evidence in science:* Data are measurements, observations, and findings from other studies that are collected as part of an investigation. Evidence, in contrast, is analyzed data and an interpretation of the analysis.

- *Methods used in scientific investigations*: Examples of methods include experiments, systematic observations of a phenomenon, literature reviews, and analysis of existing data sets; the choice of method depends on the objectives of the research. There is no universal step-by-step scientific method that all scientists follow; rather, different scientific disciplines (e.g., chemistry vs. physics) and fields within a discipline (e.g., organic vs. physical chemistry) use different types of methods, use different core theories, and rely on different standards to develop scientific knowledge.

Hints for Implementing the Lab

- Learn how to use the online simulation before the lab begins. It is important for you to know how to use the simulation so you can help students when they get stuck or confused.

- A group of three students per computer tends to work well.

- Allow the students to play with the simulation as part of the tool talk before they begin to design their investigation. This gives students a chance to see what they can and cannot do with the simulation.

- Be sure that students record actual values (e.g., molarity, moles of solute, and volume) as they use the simulation.

Topic Connections

Table 4.2 (p. 74) provides an overview of the scientific practices, crosscutting concepts, disciplinary core ideas, and supporting ideas at the heart of this lab investigation. In addition, it lists NOS and NOSI concepts for the explicit and reflective discussion. Finally, it lists literacy and mathematics skills (*CCSS ELA* and *CCSS Mathematics*) that are addressed during the investigation.

TABLE 4.2 _____
Lab 4 alignment with standards

Scientific practices	• Asking questions and defining problems • Developing and using models • Planning and carrying out investigations • Analyzing and interpreting data • Using mathematics and computational thinking • Engaging in argument from evidence • Obtaining, evaluating, and communicating information
Crosscutting concepts	• Scale, proportion, and quantity • Energy and matter: Flows, cycles, and conservation
Core idea	• PS1.A: Structure and properties of matter
Supporting ideas	• Molarity • Solutes • Solvents • Solutions
NOS and NOSI concepts	• Difference between data and evidence • Methods used in scientific investigations
Literacy connections (*CCSS ELA*)	• *Reading:* Key ideas and details, craft and structure, integration of knowledge and ideas • *Writing:* Text types and purposes, production and distribution of writing, research to build and present knowledge, range of writing • *Speaking and listening:* Comprehension and collaboration, presentation of knowledge and ideas
Mathematics connections (*CCSS Mathematics*)	• Make sense of problems and persevere in solving them • Reason abstractly and quantitatively • Construct viable arguments and critique the reasoning of others • Model with mathematics • Use appropriate tools strategically • Attend to precision • Look for and make use of structure • Look for and express regularity in repeated reasoning

Lab Handout

Lab 4. Molarity: What Is the Mathematical Relationship Between the Moles of a Solute, the Volume of the Solvent, and the Molarity of an Aqueous Solution?

Introduction

Most of the matter around us is a mixture of pure substances. The main characteristic of a mixture is its variable composition. For example, a sports drink is a mixture of many substances, such as sugar and salt, with the proportions of substances varying depending on the type of sports drink. Mixtures can be classified as either homogeneous or heterogeneous. Homogeneous mixtures have parts that are not visually distinguishable, whereas heterogeneous mixtures have parts that can be distinguished visually. A homogeneous mixture is often called a solution. A sports drink therefore is a solution.

Much of the chemistry that affects us occurs among substances dissolved in water. It is therefore important to understand the nature of solutions in which water is the dissolving medium or the solvent. This type of solution is called an aqueous solution. An aqueous solution contains one or more chemicals (or *solutes*) dissolved in water (the *solvent*). The most common way to describe the concentration of a solute in an aqueous solution is to use a unit of measurement called *molarity*. In this lab investigation, you will explore the relationship between moles of solute, volume of solvent, and molarity.

Your Task

Use a computer simulation to determine the mathematical relationship between moles of solute, volume of solvent, and molarity. Once you have determined this relationship, you should be able to set up various functions that will allow you to accurately predict

- the molarity of a solution given the moles of solute and solvent volume,
- the moles of solute given the molarity of the solution and the volume of the solvent, and
- the volume of the solvent given the molarity of the solution and the moles of the solute.

The guiding question of this investigation is, **What is the mathematical relationship between the moles of a solute, the volume of the solvent, and the molarity of an aqueous solution?**

LAB 4

Materials

You will use an online simulation called *Molarity* to conduct your investigation. You can access the simulation by going to the following website: *http://phet.colorado.edu/en/ simulation/molarity*.

Safety Precautions

Follow all normal lab safety rules.

Investigation Proposal Required?   ☐ Yes   ☐ No

Getting Started

The first step in developing your mathematical function is to determine how moles of solute and volume of the solvent are related to the molarity of a solution. The *Molarity* simulation (see screenshot in Figure L4.1) allows you to mix different moles of solute in different volumes of water (the solvent). It then provides a measure of the molarity of the resulting aqueous solution.

FIGURE L4.1 _____

A screenshot from the *Molarity* simulation

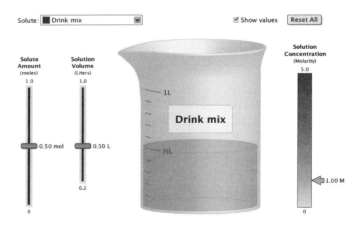

Before you start using the simulation, you must determine what type of data you will need to collect, how you will collect the data, and how you will analyze the data to answer the guiding question.

To determine *what type of data you need to collect*, think about the following questions:

- What type of observations will you need to record during your investigation?

Molarity

What Is the Mathematical Relationship Between the Moles of a Solute, the Volume of the Solvent, and the Molarity of an Aqueous Solution?

- When will you need to make these observations?

To determine *how you will collect the data*, think about the following questions:

- What types of comparisons will you need to make?
- How will you keep track of the data you collect and how will you organize it?

To determine *how you will analyze the data*, think about the following questions:

- What type of calculations will you need to make?
- What type of graph could you create to help make sense of your data?

Once you have collected and analyzed your data, your group will need to develop a function that can be used to predict (1) the molarity of a solution given the moles of solute and solvent volume, (2) the moles of solute given the molarity of the solution and the volume of the solvent, and (3) the volume of the solvent given the molarity of the solution and the moles of the solute. You will then need to test your function using the simulation. If you are able to use your function to make accurate predictions, then you will be able to generate the evidence you need to convince others that the function you developed is valid.

Connections to Crosscutting Concepts, the Nature of Science, and the Nature of Scientific Inquiry

As you work through your investigation, be sure to think about

- the importance of looking for proportional relationships between different quantities,
- why it is important to track what happens to matter within a system,
- the difference between data and evidence in science, and
- the wide range of methods that can be used during a scientific investigation.

Initial Argument

Once your group has finished collecting and analyzing your data, you will need to develop an initial argument. Your argument must include a *claim*, which is your answer to the guiding question. Your argument must also include *evidence* in support of your claim. The evidence is your analysis of the data and your interpretation of what the analysis means. Finally, you must include a *justification* of the evidence in your argument. You will therefore need to use a scientific concept or principle to explain why the evidence that you decided to use is relevant and important. You will create your initial argument on a whiteboard. Your whiteboard must include all the information shown in Figure L4.2.

FIGURE L4.2

Argument presentation on a whiteboard

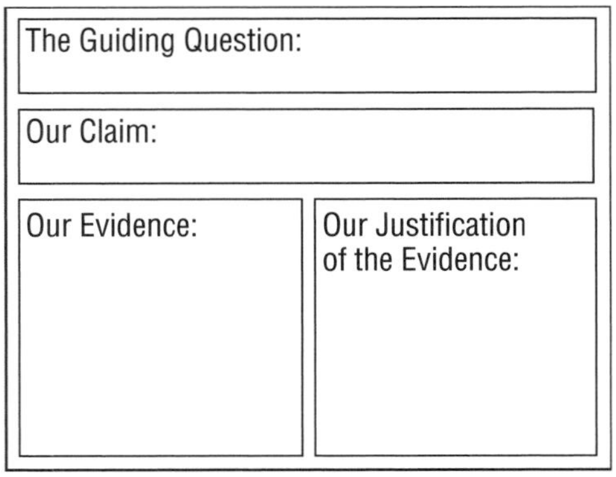

The Guiding Question:	
Our Claim:	
Our Evidence:	Our Justification of the Evidence:

Argumentation Session

The argumentation session allows all of the groups to share their arguments. One member of each group stays at the lab station to share that group's argument, while the other members of the group go to the other lab stations one at a time to listen to and critique the arguments developed by their classmates. The goal of the argumentation session is not to convince others that your argument is the best one; rather, the goal is to identify errors or instances of faulty reasoning in the initial arguments so these mistakes can be fixed. You will therefore need to evaluate the content of the claim, the quality of the evidence used to support the claim, and the strength of the justification of the evidence included in each argument that you see. To critique an argument, you might need more information than what is included on the whiteboard. You might therefore need to ask the presenter one or more follow-up questions, such as:

- What did your group do to analyze the data, and why did you decide to do it that way?
- Is that the only way to interpret the results of your group's analysis? How do you know that your interpretation of the analysis is appropriate?
- Why did your group decide to present your evidence in that manner?
- What other claims did your group discuss before deciding on that one? Why did you abandon those alternative ideas?
- How confident are you that your group's claim is valid? What could you do to increase your confidence?

Once the argumentation session is complete, you will have a chance to meet with your group and revise your original argument. Your group might need to gather more data or design a way to test one or more alternative claims as part of this process. Remember, your goal at this stage of the investigation is to develop the most valid or acceptable answer to the research question!

Report

Once you have completed your research, you will need to prepare an *investigation report* that consists of three sections that provide answers to the following questions:

1. What question were you trying to answer and why?
2. What did you do during your investigation and why did you conduct your investigation in this way?
3. What is your argument?

Molarity

What Is the Mathematical Relationship Between the Moles of a Solute, the Volume of the Solvent, and the Molarity of an Aqueous Solution?

Your report should answer these questions in two pages or less. The report must be typed and any diagrams, figures, or tables should be embedded into the document. Be sure to write in a persuasive style; you are trying to convince others that your claim is acceptable or valid!

LAB 4

Lab 4. Molarity: What Is the Mathematical Relationship Between the Moles of a Solute, the Volume of the Solvent, and the Molarity of an Aqueous Solution?

1. Describe the mathematical relationship between the moles of a solute, the volume of a solvent, and the molarity of an aqueous solution.

2. Some household cleaners come in concentrations stronger than necessary for basic cleaning jobs. Jeremy followed the instructions on a cleaning bottle and mixed enough cleaner with 4.0 L of water to form a 1.0 M cleaning solution. After testing his cleaning solution, he decided he should double the concentration for a tough stain. Jeremy added the same amount of cleaner and then another 4.0 L of water to his bucket.

   Using what you know about molarity, explain why Jeremy did not succeed in doubling the concentration of his cleaning solution.

3. All scientific investigations are experiments.

   a. I agree with this statement.
   b. I disagree with this statement.

   Explain your answer, using an example from your investigation about molarity.

Molarity

What Is the Mathematical Relationship Between the Moles of a Solute, the Volume of the Solvent, and the Molarity of an Aqueous Solution?

4. Evidence is data that support a claim.

    a. I agree with this statement.
    b. I disagree with this statement.

    Explain your answer, using an example from your investigation about molarity.

5. Scientists often need to look for proportional relationships between different quantities during an investigation. Explain what a proportional relationship is and why it is important, using an example from your investigation about molarity.

6. It is often important to track how matter flows into, out of, and within a system during an investigation. Explain why it is important to keep track of matter when studying a system, using an example from your investigation about molarity.

LAB 5

Teacher Notes

Lab 5. Temperature Changes Due to Evaporation: Which of the Available Substances Has the Strongest Intermolecular Forces?

Purpose

The purpose of this lab is to *introduce* students to intermolecular forces. This lab gives students an opportunity to determine the relative strength of the intermolecular forces in different substances based on the temperature change observed during evaporation of liquids. Students will also learn about the nature and role of experiments in science and the difference between observations and inferences in science.

The Content

Matter exists in three basic states: solid, liquid, and gas. Whether a substance is a solid, liquid, or gas at room temperature (20°C–25°C) depends on the properties of that specific substance. One of the main properties of matter that influences the state of a particular substance at room temperature is the strength of the intermolecular forces holding that substance together. *Intermolecular forces* are the attractive forces between different molecules within a substance. Intermolecular forces are considerably weaker than *intramolecular forces*, or the forces that hold atoms within a molecule together, such as covalent bonding. There are several types of intermolecular forces that vary in strength.

Dipole-dipole interactions occur in molecules with polar bonds. In polar molecules, one portion of the molecule has a more positive charge while another portion of the molecule has a more negative charge. In the case of electric charges, opposites attract; therefore the positive portion of one molecule is attracted to the negative portion of a neighboring molecule. Conversely, similar charges are repelled from each other. Therefore, within a polar substance the molecules arrange themselves so that the alignment of positive-negative interactions is maximized and the alignment of positive-positive or negative-negative interactions is minimized.

Hydrogen bonding is a special type of dipole-dipole interaction and is the strongest of the intermolecular forces. Strong dipole-dipole forces occur when hydrogen atoms are bonded to oxygen, nitrogen, or fluorine atoms within a molecule. Due to the highly electronegative properties of oxygen, nitrogen, and fluorine, the hydrogen atoms are held very close and result in a highly polar bond. Since these bonds hold the atoms close together, that allows the dipoles of neighboring molecules to come closer together than typical dipole-dipole interactions, which increases the strength of this particular type of intermolecular force.

London dispersion forces, a type of intermolecular force that occurs between molecules that do not demonstrate dipole moments, is the weakest of the intermolecular forces. London dispersion forces result from random and temporary dipole moments that are due to the distribution of electrons around the atoms within molecules. At any given moment, the distribution of electrons around an atom may become asymmetrical. When the electron distribution is asymmetrical, there is a momentary dipole, during which one area of the atom is slightly more or less negative than another area of the atom. When these brief dipole moments occur, they generate a force of attraction with similarly brief dipole moments in neighboring molecules. Given the random and temporary nature of these types of interactions, London dispersion forces are the weakest type of intermolecular force. These forces are very common in large nonpolar hydrocarbon molecules; the more atoms present in these types of molecules, the more opportunity for temporary dipoles to form.

For a phase change to occur, a substance must gain sufficient energy to overcome the intermolecular forces of attraction between the molecules. To overcome these forces, the individual molecules must gain energy. When a substance gains energy from its surroundings or from a heat source like a flame, the kinetic energy (the energy that an object possesses by virtue of being in motion) of the particles increases and they begin to move faster. The amount of kinetic energy a particle has depends not only on its velocity but also on its mass; *temperature* is a measure of the average kinetic energy of the particles of a substance. As the molecules gain kinetic energy (i.e., move faster), the intermolecular forces are not able to hold the molecules in place to maintain that state of matter. As the substance continues to gain energy, more and more of the molecules begin to overcome the intermolecular forces enough that a phase change occurs. This is not to say that the intermolecular forces disappear, only that they have less of an influence because of the increased kinetic energy of the molecules.

In the case of evaporation, some particles within a liquid have enough energy to escape the surface of a liquid and transition to the gas phase without reaching the boiling point of the substance. These relatively high-energy particles have enough kinetic energy to escape the hold of the intermolecular forces from neighboring molecules. All liquids are capable of evaporation, but the rate of evaporation is determined in part by how strong the intermolecular forces are. Therefore, substances with weak intermolecular forces would demonstrate high rates of evaporation because it is easier for the molecules to escape. The molecular mass of the substance is also an important factor. A particular substance may have relatively weak intermolecular forces, such as London dispersion forces, but when those molecules are large, they require a great amount of kinetic energy to escape the hold of those weak forces. Recall that kinetic energy is a function of mass; for a large molecule to move with a high velocity, it must gain a large amount of energy.

Finally, the process of evaporation is *endothermic* (absorbing energy). Particles of a substance must gain energy from their surroundings in order to evaporate. Those high-energy particles will be the first to evaporate from the surface of a liquid. As the liquid particles continue to gain heat energy from their surroundings, the average kinetic energy of the surroundings drops. Since temperature is a measure of average kinetic energy of a substance, when the kinetic energy of the surroundings drops, its temperature is reduced. This phenomenon is known as *evaporative cooling* and has important technological applications as well as serving as the mechanism that cools our bodies when we sweat. As the sweat evaporates from our skin, the sweat absorbs energy from our skin, which reduces the temperature of our skin.

Timeline

The instructional time needed to complete this lab investigation is 130–180 minutes. Appendix 2 (p. 501) provides options for implementing this lab investigation over several class periods. Option D (130 minutes) should be used if students are familiar with the lab equipment, particularly the temperature sensor and software, and the scientific writing process. Collecting the temperature change data takes a relatively short amount of time, and students familiar with the equipment could complete data collection on the first day of this investigation. Option F (180 minutes) incorporates an additional day for finishing data collection during stage 2, which makes this option appropriate for students who need more time to become familiar with the sensor equipment and software. Each of these options assumes that the students have experience with the scientific writing process and have the skills needed to write an investigation report on their own. In options D and F, students complete stage 6 (writing the investigation report) and stage 8 (revising the investigation report) as homework.

Materials and Preparation

The materials needed to implement this investigation are listed in Table 5.1. A temperature sensor kit includes an electronic temperature probe and any associated software or hardware such as a data collection interface or laptop computer. Temperature sensors are available for purchase from a variety of lab supply companies (e.g., Pasco or Vernier). It is recommended that the sensors and interface be positioned at lab stations prior to class to save time due to technical issues that may arise during setup. If possible, provide each station with two temperature probes (only one interface or computer is needed per station) in order to speed up the data collection process.

Figure 5.1 shows a suggested setup for adding a small amount of filter paper to the temperature probe to soak up each liquid. This step can be completed before students arrive in lab, or the students can complete this step. Additional filter paper and rubber bands should

FIGURE 5.1

Temperature probe with filter paper

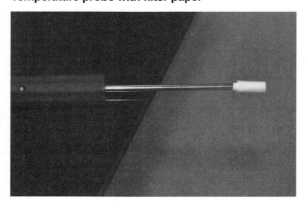

be available for each group so they may change the filter paper between tests. Stock containers for each test liquid should be kept in a fume hood. Students should bring their test tubes and test tube rack to the stock bottles and add 2–3 ml of each liquid to its own test tube. Students should stopper their test tubes when not actively using that liquid.

TABLE 5.1

Materials list

Item	Quantity
Consumables	
Filter paper	As needed
Rubber bands	As needed
Acetone, $(CH_3)_2CO$	2–3 ml per group
Ethyl alcohol, CH_3CH_2OH	2–3 ml per group
Heptane, C_7H_{16}	2–3 ml per group
Hexane, C_6H_{14}	2–3 ml per group
Isopropyl alcohol, $(CH_3)_2CHOH$	2–3 ml per group
Methyl alcohol, CH_3OH	2–3 ml per group
Equipment and other materials	
Temperature sensor kit with interface	1–2 per group
Test tubes with stoppers	6 per group
Test tube rack	1 per group
Pipettes	6 per group or 1 per stock solution bottle
Investigation Proposal C (optional)	1 per group
Whiteboard, 2' × 3' *	1 per group
Lab handout	1 per student
Peer-review guide and instructor scoring rubric	1 per student

* As an alternative, students can use computer and presentation software such as Microsoft PowerPoint or Apple Keynote to create their arguments.

Safety Precautions

Acetone, ethyl alcohol, heptane, hexane, isopropyl alcohol, and methyl alcohol are all flammable liquids. Acetone, ethyl alcohol, and heptane are each moderately toxic by ingestion and inhalation. You will therefore need to explain the potential hazards of working with these chemicals and how to work with hazardous chemicals. In addition, tell students to take the following safety precautions:

- Wear indirectly vented chemical-splash goggles and chemical-resistant gloves and aprons when they are collecting their data.
- Work under a fume hood when they are collecting their data.
- Handle all glassware with care.
- Wash their hands with soap and water when they are done collecting the data.

Laboratory Waste Disposal

We recommend following Flinn laboratory waste disposal method 26b to dispose of the liquids used in this lab. Isopropyl alcohol solutions can be disposed of down a drain if the drain is connected to a sanitation sewer system. Information about laboratory waste disposal methods is included in the Flinn Catalog and Reference Manual; you can request a free copy at *www.flinnsci.com*.

Topics for the Explicit and Reflective Discussion

Concepts That Can Be Used to Justify the Evidence

To provide an adequate justification of their evidence, students must explain why they included the evidence in their arguments and make the assumptions underlying their analysis and interpretation of the data explicit. In this investigation, students can use the following concepts to help justify their evidence:

- Intermolecular forces
- Molecular-kinetic theory of matter
- States of matter and phase change

We recommend that you discuss these fundamental concepts during the explicit and reflective discussion to help students make this connection.

How to Design Better Investigations

It is important for students to reflect on the strengths and weaknesses of the investigation they designed during the explicit and reflective discussion. Students should therefore be encouraged to discuss ways to eliminate potential flaws, measurement errors, or sources of bias in their investigations. To help students be more reflective about the design of their investigation, you can ask the following questions:

- What were some of the strengths of your investigation? What made it scientific?

- What were some of the weaknesses of your investigation? What made it less scientific?
- If you were to do this investigation again, what would you do to address the weaknesses in your investigation? What could you do to make it more scientific?

Crosscutting Concepts

This investigation is well aligned with two crosscutting concepts found in *A Framework for K–12 Science Education,* and you should review these concepts during the explicit and reflective discussion.

- *Energy and matter: Flows, cycles, and conservation:* In science it is important to track how energy and matter move into, out of, and within systems. For example, during a phase change a substance either gains energy from its surroundings or releases energy to its surroundings in predictable ways.
- *Structure and function:* The way an object is shaped or structured determines many of its properties and functions. The structure of a molecule, for example, determines how it can function and interact with its surroundings. In this investigation, polar molecules tend to exhibit hydrogen bonding and result in stronger intermolecular forces.

The Nature of Science and the Nature of Scientific Inquiry

This investigation is well aligned with two important concepts related to the *nature of science* (NOS) and the *nature of scientific inquiry* (NOSI), and you should review these concepts during the explicit and reflective discussion.

- *The difference between observations and inferences*: An observation is a descriptive statement about a natural phenomenon, whereas an inference is an interpretation of an observation. Students should also understand that current scientific knowledge and the perspectives of individual scientists guide both observations and inferences. Thus, different scientists can have different but equally valid interpretations of the same observations due to differences in their perspectives and background knowledge.
- *The nature and role of experiments:* Scientists use experiments to test the validity of a hypothesis (i.e., a tentative explanation) for an observed phenomenon. Experiments include a test and the formulation of predictions (expected results) if the test is conducted and the hypothesis is valid. The experiment is then carried out and the predictions are compared with the observed results of the experiment. If the predictions match the observed results, then the hypothesis is supported. If the observed results do not match the prediction, then the hypothesis is not supported. A signature feature of an experiment is the control of variables to help eliminate alternative explanations for observed results.

LAB 5

Hints for Implementing the Lab

- Learn how to use the temperature sensor kit and associated software before the lab begins. It is important for you to know how to use the equipment so you can help students when technical issues arise.

- Allow the students to become familiar with the temperature sensor kit and software as part of the tool talk before they begin to design their investigation. This gives students a chance to see what they can and cannot do with the equipment.

- Be sure that students record actual values (e.g., temperature readings and changes in temperature) or save any graphs generated by the sensor software, rather than just attempting to hand draw what they see on the computer screen.

- Data collection takes approximately three minutes for each liquid to obtain its greatest temperature drop.

- Temperature probes may be dipped directly into the test liquids without filter paper attached. However, using a bare temperature probe only allows about 30 seconds worth of data collection time.

Topic Connections

Table 5.2 provides an overview of the scientific practices, crosscutting concepts, disciplinary core ideas, and supporting ideas at the heart of this lab investigation. In addition, it lists NOS and NOSI concepts for the explicit and reflective discussion. Finally, it lists literacy and mathematics skills (*CCSS ELA* and *CCSS Mathematics*) that are addressed during the investigation.

TABLE 5.2

Lab 5 alignment with standards

Scientific practices	• Asking questions and defining problems • Planning and carrying out investigations • Analyzing and interpreting data • Constructing explanations and designing solutions • Engaging in argument from evidence • Obtaining, evaluating, and communicating information
Crosscutting concepts	• Energy and matter: Flows, cycles, and conservation • Structure and function
Core idea	• PS1.A: Structure and properties of matter
Supporting ideas	• Intermolecular forces • Molecular-kinetic theory of matter • States of matter and phase changes
NOS and NOSI concepts	• Observations and inferences • Nature and role of experiments
Literacy connections (*CCSS ELA*)	• *Reading:* Key ideas and details, craft and structure, integration of knowledge and ideas • *Writing:* Text types and purposes, production and distribution of writing, research to build and present knowledge, range of writing • *Speaking and listening:* Comprehension and collaboration, presentation of knowledge and ideas
Mathematics connections (*CCSS Mathematics*)	• Reason abstractly and quantitatively • Construct viable arguments and critique the reasoning of others • Look for and express regularity in repeated reasoning

Lab 5. Temperature Changes Due to Evaporation: Which of the Available Substances Has the Strongest Intermolecular Forces?

Introduction

Matter exists in three basic states: solid, liquid, and gas. Whether a substance is a solid, liquid, or gas at room temperature (20°C–25°C) depends on the properties of that specific substance. Oxygen, for instance, is a gas at room temperature, but water is a liquid. Substances also maintain their state over a broad range of conditions; for example, water is a liquid from 0°C all the way to 100°C. For a substance to transition from one phase to another, the conditions again must be just right. Transitioning from a solid to a liquid (i.e., melting) or transitioning from a liquid to a gas (i.e., boiling) requires a transfer of energy.

Boiling or vaporization is the process by which a substance changes from a liquid to a gas. Evaporation is vaporization that occurs at the surface of a liquid. Chemists explain the process of evaporation using the *molecular-kinetic theory of matter*. According to this theory, the molecules that make up a liquid are in constant motion but are attracted to other molecules by different types of *intermolecular forces*. The temperature of the liquid is proportional to the average *kinetic energy* of the molecules found within that liquid. Kinetic energy is the energy of motion. The kinetic energy of a molecule depends on its velocity and its mass. From this perspective, some of the molecules at the surface of a liquid will have greater or lower kinetic energy than the average. Some of the molecules will have enough kinetic energy to disrupt the intermolecular forces that hold it near the other molecules in the liquid. These molecules, as a result, will break away from the surface of the liquid. When this happens, the average kinetic energy of the remaining molecules in the liquid decreases. The temperature of a liquid therefore goes down as it evaporates.

There are several important factors that will influence the evaporation rate of a liquid. One of these factors is temperature. A liquid will evaporate faster at a higher temperature because more molecules at the surface will have enough kinetic energy to break free from the other molecules in that liquid. Another important factor is the type of intermolecular forces that exist between the molecules in that liquid. Intermolecular forces, as noted earlier, are attractive in nature. A liquid with weak intermolecular forces will evaporate quickly because it takes less kinetic energy for a molecule at the surface of the liquid to break away from the other molecules in the liquid. A liquid with strong intermolecular forces, in contrast, will evaporate slowly because the molecules that make up this type of liquid require more kinetic energy to break away from the other molecules at the surface of the liquid.

In this investigation, you will use what you have learned about evaporation, the molecular-kinetic theory of matter, and this brief introduction to intermolecular forces to determine the relative strength of the intermolecular forces that exist between the molecules in six different types of liquids. This is important for you to be able to do because chemists often need to rely on their understanding of intermolecular interactions to explain the bulk properties of matter. It is also important because it will help you understand the underlying reasons for several important natural phenomena, such as the water cycle or evaporative cooling, that have many practical applications for everyday life.

Your Task

Determine how temperature changes when six different liquids evaporate from a material and then rank the substances in terms of the relative strength of their intermolecular forces.

The guiding question of this investigation is, **Which of the available substances has the strongest intermolecular forces?**

Materials

You may use any of the following materials during your investigation:

Consumables	Equipment
• Filter paper	• Temperature sensor kit
• Rubber bands	• Test tubes w/ stoppers
• Acetone, $(CH_3)_2CO$	• Test tube rack
• Ethyl alcohol, CH_3CH_2OH	• Pipettes
• Heptane, C_7H_{16}	
• Hexane, C_6H_{14}	
• Isopropyl alcohol, $(CH3)_2CHOH$	
• Methyl alcohol, CH_3OH	

Safety Precautions

Follow all normal lab safety rules. Acetone, ethyl alcohol, heptane, hexane, isopropyl alcohol, and methyl alcohol are flammable liquids. Acetone, ethyl alcohol, and heptane are each moderately toxic by ingestion and inhalation. Given the fire and health hazards associated with these chemicals, all lab work must be done under a fume hood. In addition, take the following safety precautions:

- Wear indirectly vented chemical-splash goggles and chemical-resistant gloves and apron while in the laboratory.

- Handle all glassware with care.

- Wash your hands with soap and water before leaving the laboratory.

LAB 5

Getting Started

The first step in this investigation is to carry out a series of tests to determine how the temperature of each substance changes as it evaporates. One way to generate these data is by attaching a small piece of filter paper to a temperature probe using a rubber band (see Figure L5.1). Then briefly soak the filter paper in one of the substances, remove the apparatus from the liquid, and observe as the liquid evaporates from the filter paper.

FIGURE L5.1 _____

Temperature probe with filter paper

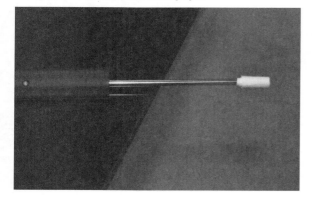

Before you begin your tests, however, you will need to determine what type of data you need to collect during each test, how you will collect the data, and how you will analyze the data.

To determine *what type of data you need to collect*, think about the following questions:

- What type of measurements or observations will you need to record during each test?
- How often will you need to make these measurements or observations?

To determine *how you will collect the data*, think about the following questions:

- How long will you need to conduct each test?
- What types of comparisons will you need to make?
- What will you do to reduce measurement error?
- How will you keep track of the data you collect and how will you organize it?

To determine *how you will analyze the data*, think about the following questions:

- What type of calculations will you need to make?
- What type of graph could you create to help make sense of your data?

The second step in your investigation will be to develop a way to rank order the substances in terms of the relative strength of their intermolecular forces. At this point in the investigation, it may also be useful to consider the molecular characteristics of each substance, such as the molecular structure or shape, molar mass, whether the molecule is polar, and so on. Table L5.1 provides the structural formula for each substance for comparison purposes.

TABLE L5.1

Molar mass and the structural formulas of the substances used in this investigation

Substance	Molar mass	Structural formula
Acetone	58.08 g/mol	
Ethyl alcohol	46.07 g/mol	
Heptane	100.21 g/mol	
Hexane	86.18 g/mol	
Isopropyl alcohol	60.10 g/mol	
Methyl alcohol	32.04 g/mol	

LAB 5

Connections to Crosscutting Concepts, the Nature of Science, and the Nature of Scientific Inquiry

As you work through your investigation, be sure to think about

- how energy and matter flow within systems,
- the importance of the structure of molecules in relation to the ways they function,
- the nature and role of experiments in science, and
- the difference between observations and inferences.

Initial Argument

Once your group has finished collecting and analyzing your data, you will need to develop an initial argument. Your argument must include a *claim*, which is your answer to the guiding question. Your argument must also include *evidence* in support of your claim. The evidence is your analysis of the data and your interpretation of what the analysis means. Finally, you must include a *justification* of the evidence in your argument. You will therefore need to use a scientific concept or principle to explain why the evidence that you decided to use is relevant and important. You will create your initial argument on a whiteboard. Your whiteboard must include all the information shown in Figure L5.2.

FIGURE L5.2

Argument presentation on a whiteboard

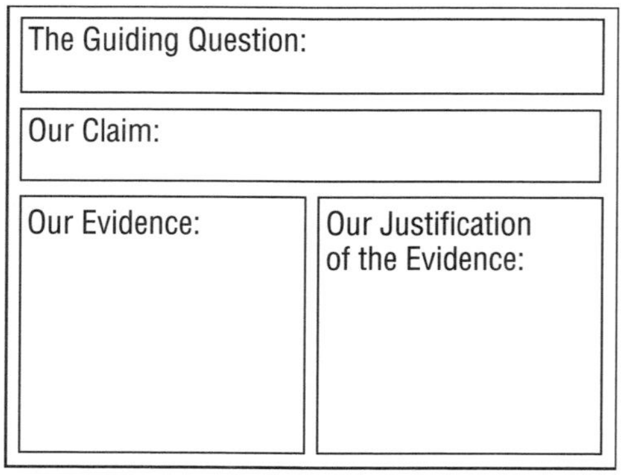

The Guiding Question:	
Our Claim:	
Our Evidence:	Our Justification of the Evidence:

Argumentation Session

The argumentation session allows all of the groups to share their arguments. One member of each group stays at the lab station to share that group's argument, while the other members of the group go to the other lab stations one at a time to listen to and critique the arguments developed by their classmates. The goal of the argumentation session is not to convince others that your argument is the best one; rather, the goal is to identify errors or instances of faulty reasoning in the initial arguments so these mistakes can be fixed. You will therefore need to evaluate the content of the claim, the quality of the evidence used to support the claim, and the strength of the justification of the evidence included in each argument that you see. To critique an argument, you might need more information than what is included on the whiteboard. You might therefore need to ask the presenter one or more follow-up questions, such as:

- What did your group do to analyze the data, and why did you decide to do it that way?
- Is that the only way to interpret the results of your group's analysis? How do you know that your interpretation of the analysis is appropriate?

- Why did your group decide to present your evidence in that manner?

- What other claims did your group discuss before deciding on that one? Why did you abandon those alternative ideas?

- How confident are you that your group's claim is valid? What could you do to increase your confidence?

Once the argumentation session is complete, you will have a chance to meet with your group and revise your original argument. Your group might need to gather more data or design a way to test one or more alternative claims as part of this process. Remember, your goal at this stage of the investigation is to develop the most valid or acceptable answer to the research question!

Report

Once you have completed your research, you will need to prepare an *investigation report* that consists of three sections that provide answers to the following questions:

1. What question were you trying to answer and why?

2. What did you do during your investigation and why did you conduct your investigation in this way?

3. What is your argument?

Your report should answer these questions in two pages or less. The report must be typed and any diagrams, figures, or tables should be embedded into the document. Be sure to write in a persuasive style; you are trying to convince others that your claim is acceptable or valid!

LAB 5

Lab 5. Temperature Changes Due to Evaporation: Which of the Available Substances Has the Strongest Intermolecular Forces?

1. Describe how the strength of intermolecular forces is related to the boiling point of a liquid.

2. Our bodies generate sweat to help regulate our temperature and stay cool on hot days or during exercise. Many people think that the sweat is cool and in turn cools our bodies, but the sweat our bodies produce is actually the same temperature as the inside of our body where the sweat was generated. Sweat actually cools our skin due to evaporation.

   Use what you know about the process of evaporation to explain how evaporative cooling works to keep our skin cool.

3. If two scientists observe the same event, it is likely that they will come to the same conclusions.

   a. I agree with this statement.
   b. I disagree with this statement.

   Explain your answer, using an example from your investigation about temperature changes due to evaporation.

4. Conducting an experiment is one way to investigate questions in science, but there are other ways to conduct scientific investigations.

    a. I agree with this statement.
    b. I disagree with this statement.

    Explain your answer, using an example from your investigation about temperature changes due to evaporation.

5. Understanding how matter and energy flow within and between systems is important in science. Explain why this is important, using an example from your investigation about temperature changes due to evaporation.

6. Some scientists devote their career to understanding the structure and function of just a handful of molecules. Explain why understanding the structure and function of a molecule is important, using an example from your investigation about temperature changes due to evaporation.

LAB 6

Lab 6. Pressure, Temperature, and Volume of Gases: How Does Changing the Volume or Temperature of a Gas Affect the Pressure of That Gas?

Purpose

The purpose of this lab is to *introduce* students to the gas laws related to pressure, temperature, and volume of gases. This lab gives students an opportunity to develop a mathematical model that can be used to describe the relationships between pressure, temperature, and volume of a gas based on the data they collect. Students will also learn about the difference between scientific laws and theories and between observations and inferences.

The Content

The *ideal gas law* describes the relationship between pressure, volume, temperature, and number of moles of a gas. Three laws form the foundation of the ideal gas law: Boyle's law, Charles' law, and Avogadro's law.

Boyle's law illustrates the inverse relationship between pressure and volume, such that as the pressure on a gas is increased its volume decreases. This relationship can be used to determine the change in volume or pressure to a gas after some event, using the equation $P_1V_1 = P_2V_2$. However, Boyle's law is only applicable if the temperature and amount of gas are held constant.

Charles' law illustrates the direct relationship between volume and temperature, such that as the temperature of a gas increases, so does its volume. This relationship can be used to determine the changes in volume or temperature to a gas after some event, using the equation $V_1/T_1 = V_2/T_2$. The temperature in this relationship is absolute temperature as measured on the Kelvin scale.

The third component to the ideal gas law is Avogadro's law, which illustrates that for a given mass of a gas the volume and amount of moles are directly proportional as long as the temperature and pressure are constant. The equation for Avogadro's law is $V_1/n_1 = V_2/n_2$, where n is the amount of substance of the gas as measured in moles.

The ideal gas law is summarized in the mathematical equation $PV = nRT$, where P is the pressure of the gas, V is the volume of the gas, n is the amount of gas in moles, R is the universal gas constant (0.0821 L•atm/mol•K), and T is the absolute temperature of the gas.

The molecular-kinetic theory of matter is often used to explain the gas laws. This theory suggests that all particles of matter are in constant motion (i.e., have kinetic energy) and that the temperature of a substance is a measure of the average kinetic energy of the particles in that substance. This theory is important in the context of gases because the constant random motion of the particles influences the volume a gas occupies and the amount of pressure exerted by the gas. The relationships described by Boyle's or Charles' law can be further explained through the lens of the molecular-kinetic theory of matter. As a gas gains heat energy the temperature of a gas increases, meaning the kinetic energy of the gas particles increases. As the particles move faster, the influence of the intermolecular forces of attraction are decreased and the particles are able to occupy more space, provided they are not confined to a rigid container. If they are confined to a rigid container, then a similar increase in kinetic energy would result in an increase of pressure. As the gas particles move faster, they collide with the container with greater force and more frequency, thus increasing the pressure.

Timeline

The instructional time needed to complete this lab investigation is 130–180 minutes. Appendix 2 (p. 501) provides options for implementing this lab investigation over several class periods. Option B (180 minutes) should be used if students are unfamiliar with the pressure sensor equipment and thus may need extra time to design their investigation and collect data. Option D (130 minutes) should be used if students are familiar with the equipment. Each of these options assumes that the students have experience with the scientific writing process and have the skills needed to write an investigation report on their own. In options B and D, students complete stage 6 (writing the investigation report) and stage 8 (revising the investigation report) as homework.

Materials and Preparation

The materials needed to implement this investigation are listed in Table 6.1 (p. 100). Pressure and temperature sensors as well as the sensor interface are available for purchase from a variety of lab supply companies (e.g., Pasco and Vernier). It is recommended that the sensors and interface be positioned at lab stations before class to save time due to technical issues that may arise during setup. Ice should be stored in a central location so that students have access for making ice baths and to minimize the risk of accidents involving ice/water and the sensor equipment.

LAB 6

TABLE 6.1
Materials list

Item	Quantity
Consumable	
Ice	As needed
Equipment and other materials	
Gas pressure sensor	1 per group
Temperature sensor	1 per group
Sensor interface	1 per group
Syringe	1 per group
Erlenmeyer flask	1 per group
Single-hole rubber stopper	1 per group
Rubber tubing	1 per group
Beaker, 500 ml	1 per group
Hot plate	1 per group
Investigation Proposal C (optional)	1 per group
Whiteboard, 2' × 3' *	1 per group
Lab handout	1 per student
Peer-review guide and instructor scoring rubric	1 per student

* As an alternative, students can use computer and presentation software such as Microsoft PowerPoint or Apple Keynote to create their arguments.

Safety Precautions

Remind students to follow all normal lab safety rules. In addition, tell students to take the following safety precautions:

- Wear indirectly vented chemical-splash goggles and chemical-resistant gloves and aprons when they are collecting their data.
- Use caution when working with hot plates, and keep them away from water and other liquids.
- Handle all glassware with care.
- Wash their hands with soap and water when they are done collecting the data.

Laboratory Waste Disposal

No waste disposal is needed in this lab investigation.

Topics for the Explicit and Reflective Discussion

Concepts That Can Be Used to Justify the Evidence

To provide an adequate justification of their evidence, students must explain why they included the evidence in their arguments and make the assumptions underlying their analysis and interpretation of the data explicit. In this investigation, students can use the following concepts to help justify their evidence:

- States of matter
- Molecular-kinetic theory of matter
- The differences between heat and temperature
- Interpreting graphs as mathematical relationships (i.e., inverse or direct)

We recommend that you discuss these fundamental concepts during the explicit and reflective discussion to help students make this connection.

How to Design Better Investigations

It is important for students to reflect on the strengths and weaknesses of the investigation they designed during the explicit and reflective discussion. Students should therefore be encouraged to discuss ways to eliminate potential flaws, measurement errors, or sources of bias in their investigations. To help students be more reflective about the design of their investigation, you can ask the following questions:

- What were some of the strengths of your investigation? What made it scientific?
- What were some of the weaknesses of your investigation? What made it less scientific?
- If you were to do this investigation again, what would you do to address the weaknesses in your investigation? What could you do to make it more scientific?

Crosscutting Concepts

This investigation is well aligned with two crosscutting concepts found in *A Framework for K–12 Science Education,* and you should review these concepts during the explicit and reflective discussion.

- *Cause and effect: Mechanism and explanation:* One of the main objectives of science is to identify and establish relationships between a cause and an effect. Scientists often attempt to determine how one variable influences another; for example,

LAB 6

in this investigation students manipulate variables to determine their impact on other variables in order to explain how gases behave under certain conditions.

- *Systems and system models:* Scientists often need to use models to understand complex phenomena. In this investigation, for example, students use empirical data to generate a mathematical model that describes the behavior of gases.

The Nature of Science and the Nature of Scientific Inquiry

This investigation is well aligned with two important concepts related to the *nature of science* (NOS) and the *nature of scientific inquiry* (NOSI), and you should review these concepts during the explicit and reflective discussion.

- *The difference between laws and theories in science:* A scientific law describes the behavior of a natural phenomenon or a generalized relationship under certain conditions; a scientific theory is a well-substantiated explanation of some aspect of the natural world. Theories do not become laws even with additional evidence; they explain laws. However, not all scientific laws have an accompanying explanatory theory. It is also important for students to understand that scientists do not discover laws or theories; the scientific community develops them over time.

- *The difference between data and evidence in science:* Data are measurements, observations, and findings from other studies that are collected as part of an investigation. Evidence, in contrast, is analyzed data and an interpretation of the analysis.

Hints for Implementing the Lab

- Learn how to use the pressure sensor kit and associated software before the lab begins. It is important for you to know how to use the equipment so you can help students when technical issues arise.

- Allow the students to become familiar with the sensors, the sensor interface, and the associated software as part of the tool talk before they begin to design their investigation. This gives students a chance to see what they can and cannot do with the equipment.

- Be sure that students record actual values (e.g., temperature readings, volume from syringe, or pressure) or save any graphs generated by the sensor software, rather than just attempting to hand draw what they see on the computer screen.

Topic Connections

Table 6.2 provides an overview of the scientific practices, crosscutting concepts, disciplinary core ideas, and supporting ideas at the heart of this lab investigation. In addition, it lists NOS and NOSI concepts for the explicit and reflective discussion. Finally, it lists literacy and mathematics skills (*CCSS ELA* and *CCSS Mathematics*) that are addressed during the investigation.

TABLE 6.2

Lab 6 alignment with standards

Scientific practices	• Asking questions and defining problems • Developing and using models • Planning and carrying out investigations • Analyzing and interpreting data • Using mathematics and computational thinking • Constructing explanations and designing solutions • Engaging in argument from evidence • Obtaining, evaluating, and communicating information
Crosscutting concepts	• Cause and effect: Mechanism and explanation • Systems and system models
Core idea	• PS1.A: Structure and properties of matter
Supporting ideas	• States of matter • Molecular-kinetic theory of matter • The connection and difference between heat and temperature • Graphs as representations of mathematical relationships
NOS and NOSI concepts	• Scientific laws and theories • Difference between data and evidence
Literacy connections (*CCSS ELA*)	• *Reading:* Key ideas and details, craft and structure, integration of knowledge and ideas • *Writing:* Text types and purposes, production and distribution of writing, research to build and present knowledge, range of writing • *Speaking and listening:* Comprehension and collaboration, presentation of knowledge and ideas
Mathematics connections (*CCSS Mathematics*)	• Reason abstractly and quantitatively • Construct viable arguments and critique the reasoning of others • Model with mathematics • Use appropriate tools strategically • Look for and make use of structure

LAB 6

Lab Handout

Lab 6. Pressure, Temperature, and Volume of Gases: How Does Changing the Volume or Temperature of a Gas Affect the Pressure of That Gas?

Introduction

There are three states of matter: solid, liquid, and gas. Each state of matter has physical properties that distinguish it from the other states; for example, matter in the solid phase has a definite shape, whereas matter in the liquid or gas phase will take on the shape of its container. The physical properties associated with the states of matter allow us to predict how different substances may react under various conditions. Particles in a gas move about more freely than those in a solid or liquid and therefore react to changes in temperature and pressure in a manner that is different than solids or liquids.

The *volume* of a sample of gas, or the amount of space that a sample of gas occupies, is particularly influenced by a variety of factors such as temperature or pressure. Just like the shape of a sample of gas or liquid is determined by its container, the volume of a sample of gas is influenced by its surroundings. A small sample of gas, like air, may be confined to a small container such as a balloon, or if the balloon pops the sample of gas can expand to occupy the entire volume of a classroom. Consider a tank of helium gas used to fill birthday balloons. There is a large amount of gas stored inside the tank, but several birthday balloons filled with a sample of the gas can easily expand to a size much larger than the tank. Understanding the physical properties of gases and how a gas interacts with its surroundings helps to explain this phenomenon. In this investigation you will explore the relationship between volume, temperature, and pressure for a gas within a closed system.

Your Task

Determine how changes to the volume and the temperature of a gas within a closed system affect the pressure of that gas. Then develop a general mathematical model that can be used to apply and describe these relationships with respect to all gases.

The guiding question for this lab is, **How does changing the volume or temperature of a gas affect the pressure of that gas?**

Materials

You may use any of the following materials during your investigation:

Consumable	Equipment
• Ice	• Gas pressure sensor
	• Temperature sensor
	• Sensor interface
	• Syringe
	• Erlenmeyer flask
	• Single-hole rubber stopper
	• Rubber tubing
	• Beaker (500 ml)
	• Hot plate

Safety Precautions

Follow all normal lab safety rules. In addition, take the following safety precautions:

- Wear indirectly vented chemical-splash goggles and chemical-resistant gloves and apron while in the laboratory.
- Handle all glassware with care.
- Use caution when working with hot plates because they can burn skin. Hot plates also need to be kept away from water and other liquids.
- Wash your hands with soap and water before leaving the laboratory.

Investigation Proposal Required?   ☐ Yes   ☐ No

Getting Started

To determine the relationship between the pressure, the volume, and the temperature of a gas, you will need to set up an apparatus that will allow you to first measure changes in gas pressure when the volume of gas changes. This can be accomplished with the apparatus shown in Figure L6.1 (p. 107). You will then need to be able to measure changes in gas pressure when the temperature of the gas changes. This can be accomplished with the apparatus shown in Figure L6.2 (p. 107). Once you have set up these apparatuses, you must determine what type of data you need to collect, how you will collect the data, and how you will analyze the data.

To determine *what type of data you need to collect*, think about the following questions:

- What type of measurements or observations will you need to record during your investigation?
- When will you need to make these measurements or observations?

To determine *how you will collect the data*, think about the following questions:

- What will serve as your dependent variable(s)?
- What will serve as a control (or comparison) condition?
- What types of treatment conditions will you need to set up and how will you do it?
- How will you make sure that your data are of high quality (i.e., how will you reduce error)?

To determine *how you will analyze the data*, think about the following questions:

- How will you determine if there is a difference between the treatment conditions and the control condition?
- What type of calculations will you need to make?
- What type of graph could you create to help make sense of your data?

Once you have finished collecting your data, your group will need to develop a mathematical model that describes how the pressure of a gas is affected by changes in the volume and the temperature. When developing a mathematical model, variables that are inversely related are multiplied and variables that are directly related are divided. Keep these mathematical relationships in mind as you develop your model.

The last step in this investigation is to test your model. To accomplish this goal, you can use your model to make predictions about the pressure of a gas in a closed system under different conditions. If you are able to make accurate predictions with your model, then you will be able to generate the evidence you need to convince others that your model is valid.

Connections to Crosscutting Concepts, the Nature of Science, and the Nature of Scientific Inquiry

As you work through your investigation, be sure to think about

- the importance of developing causal explanations for observations,
- how models are used to help understand natural phenomena,
- the difference between laws and theories in science, and
- the difference between data and evidence in science.

Initial Argument

Once your group has finished collecting and analyzing your data, you will need to develop an initial argument. Your argument must include a *claim*, which is your answer to the guiding question. Your argument must also include *evidence* in support of your claim. The evidence is your analysis of the data and your interpretation of what the analysis means. Finally, you must include a *justification* of the evidence in your argument. You will therefore need to use a scientific concept or principle to explain why the evidence that

FIGURE L6.1 _____

Apparatus used to measure changes in gas pressure in response to changes in the volume of the gas

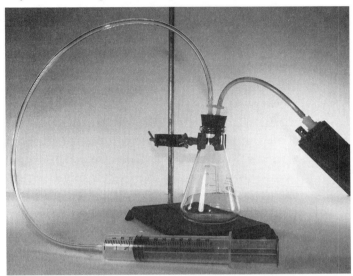

FIGURE L6.2 _____

Apparatus used to measure changes in gas pressure in response to changes in the temperature of the gas

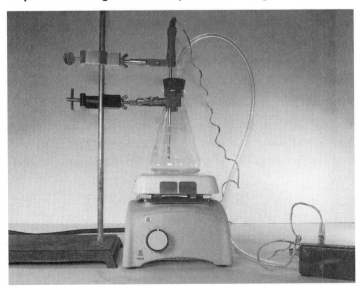

LAB 6

FIGURE L6.3 _____

Argument presentation on a whiteboard

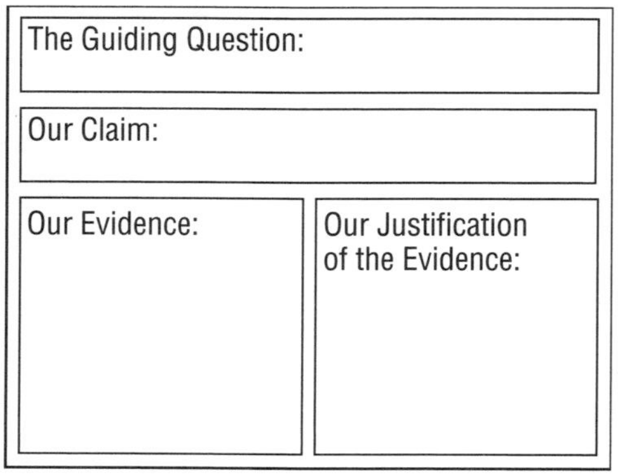

The Guiding Question:	
Our Claim:	
Our Evidence:	Our Justification of the Evidence:

you decided to use is relevant and important. You will create your initial argument on a whiteboard. Your whiteboard must include all the information shown in Figure L6.3.

Argumentation Session

The argumentation session allows all of the groups to share their arguments. One member of each group stays at the lab station to share that group's argument, while the other members of the group go to the other lab stations one at a time to listen to and critique the arguments developed by their classmates. The goal of the argumentation session is not to convince others that your argument is the best one; rather, the goal is to identify errors or instances of faulty reasoning in the initial arguments so these mistakes can be fixed. You will therefore need to evaluate the content of the claim, the quality of the evidence used to support the claim, and the strength of the justification of the evidence included in each argument that you see. To critique an argument, you might need more information than what is included on the whiteboard. You might therefore need to ask the presenter one or more follow-up questions, such as:

- What did your group do to analyze the data, and why did you decide to do it that way?
- Is that the only way to interpret the results of your group's analysis? How do you know that your interpretation of the analysis is appropriate?
- Why did your group decide to present your evidence in that manner?
- What other claims did your group discuss before deciding on that one? Why did you abandon those alternative ideas?
- How confident are you that your group's claim is valid? What could you do to increase your confidence?

Once the argumentation session is complete, you will have a chance to meet with your group and revise your original argument. Your group might need to gather more data or design a way to test one or more alternative claims as part of this process. Remember, your goal at this stage of the investigation is to develop the most valid or acceptable answer to the research question!

Report

Once you have completed your research, you will need to prepare an *investigation report* that consists of three sections that provide answers to the following questions:

1. What question were you trying to answer and why?

2. What did you do during your investigation and why did you conduct your investigation in this way?

3. What is your argument?

Your report should answer these questions in two pages or less. The report must be typed and any diagrams, figures, or tables should be embedded into the document. Be sure to write in a persuasive style; you are trying to convince others that your claim is acceptable or valid!

LAB 6

Lab 6. Pressure, Temperature, and Volume of Gases: How Does Changing the Volume or Temperature of a Gas Affect the Pressure of That Gas?

1. Describe the relationship between pressure and volume for a gas and volume and temperature for a gas. What are the mathematical equations that describe these relationships?

2. Susan releases a birthday balloon filled with helium into the air. The balloon quickly rises into the sky until it is out of sight. Gradually as the balloon rises, it begins to expand. Eventually, the balloon gets so large that it pops.

   Use what you know about the relationships between the pressure, volume, and temperature of a gas to explain what conditions must have been present for the balloon to expand.

3. Data and evidence are the same thing in science.

   a. I agree with this statement.
   b. I disagree with this statement.

   Explain your answer, using an example from your investigation about pressure, temperature, and volume of gases.

4. It is common for a theory to become a law in science.

    a. I agree with this statement.
    b. I disagree with this statement.

    Explain your answer, using an example from your investigation about pressure, temperature, and volume of gases.

5. Scientists often generate causal explanations for the observations they make. Explain why this is important, using an example from your investigation about pressure, temperature, and volume of gases.

6. Scientists create models to help understand natural phenomena. Explain what a model is and why models are useful, using an example from your investigation about pressure, temperature, and volume of gases.

LAB 7

Teacher Notes

Lab 7. Periodic Trends: Which Properties of the Elements Follow a Periodic Trend?

Purpose

The purpose of this lab is to *introduce* students to periodic trends. This lab gives students an opportunity to explore the values of different properties for all elements in the periodic table to determine if they follow trends. Students will also learn about the importance of identifying patterns in science; the importance of scale, proportion, and quantity; the difference between observations and inferences; and how science is influenced by social and cultural values.

The Content

The development of the periodic table of elements has served as one of the foundational activities for the discipline throughout the history of chemistry. Not only does it serve as one of the central, organizing graphic models for the entire field but, through that organization, it also provides a vast amount of information about the elements that allows chemists to predict the behavior and tendencies of substances before testing them. Most periodic tables are printed with basic information about each element, including its symbol, atomic number, and molar mass.

The periodic table also includes other useful information. Chemists, for example, can use *periodic trends* to determine the nature of several properties of each element based on its location in the table. Periodic trends are the tendencies of certain properties of the elements to increase or decrease as you progress along a row or column of the periodic table. These trends are related to the physical structure of the atoms of each element (i.e., the number of protons and electrons present) and the chemical properties of those atoms (i.e., how those atoms interact with electrons from other sources). Some of these trends are termed *periodic* because they occur across periods (a row in the periodic table is called a period) and within groups (a column in the periodic table is called a group). Other trends are termed *quasi-periodic* because they occur within groups of elements but not across periods. Figure 7.1 provides a graphic representation of some periodic trends.

Students explore several different properties of the elements during this investigation, including both periodic and quasi-periodic trends. Atomic mass is a periodic trend. Atomic mass increases as you move along a period from left to right, and down a group from top to bottom. This is simply due to the fact that as you move in those directions, the number of protons (and thus neutrons and electrons) continues to increase, meaning those atoms

FIGURE 7.1

Examples of periodic trends

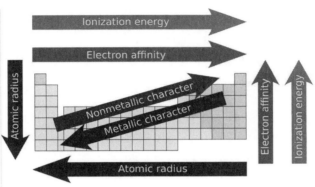

have greater mass. In a related but inverse manner, atomic radius is also a periodic trend; atomic radius, however, decreases across a period from left to right.

Electronegativity is the tendency to draw shared electrons in a covalent bond. When considering groups in the periodic table, smaller atoms toward the top have greater affinity for electrons than larger atoms. Thus, electronegativity generally increases as you go from left to right in a period and from the bottom to the top of a group of elements. In a related sense, ionization energy, which refers to the amount of energy needed to remove an electron from an atom, follows the same trend as electronegativity.

Melting point can be considered a quasi-periodic trend because patterns are not consistent across periods but are consistent within specific groups. Figure 7.2 shows the melting points of the first 36 elements in the periodic table. It is easy to see from this chart that a

FIGURE 7.2

Melting points for several elements

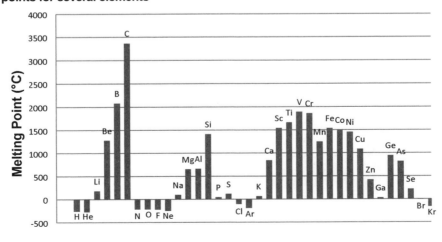

trend across a period does not exist. Melting point seems to increase going from left to right, but then there are some dramatic drops in melting point when considering nonmetals. However, when observing elements located in the same group on the periodic table, trends emerge. Density is another quasi-periodic trend. In general, density increases as you move down a group, but there is no trend across periods. Specific heat capacity is also a quasi-periodic trend. Specific heat capacity tends to decrease as you move down a group, but there is no trend across periods.

Timeline

The instructional time needed to complete this lab investigation is 150–200 minutes. Appendix 2 (p. 501) provides options for implementing this lab investigation over several class periods. Option C (200 minutes) should be used if students are unfamiliar with scientific writing because this option provides extra instructional time for scaffolding the writing process. You can scaffold the writing process by modeling, providing examples, and providing hints as students write each section of the report. Option D (150 minutes) should be used if students are familiar with scientific writing and have the skills needed to write an investigation report on their own. In option D, students complete stage 6 (writing the investigation report) and stage 8 (revising the investigation report) of the investigation outside of class, which reduces the amount of time needed to complete the lab.

Materials and Preparation

The materials needed to implement this investigation are listed in Table 7.1. The "Properties of Elements" Excel file (available at *www.nsta.org/publications/press/extras/adi-chem.aspx*) can be loaded onto student computers before the investigation, e-mailed to students, or uploaded to a class website that students can access.

TABLE 7.1

Materials list

Item	Quantity
Computer with Excel application	1 per group
"Properties of Elements" Excel file	1 per group
Periodic table	1 per student
Whiteboard, 2' × 3' *	1 per group
Lab handout	1 per student
Peer-review guide and instructor scoring rubric	1 per student

* As an alternative, students can use computer and presentation software such as Microsoft PowerPoint or Apple Keynote to create their arguments.

Safety Precautions

Remind students to follow all normal lab safety rules.

Laboratory Waste Disposal

No waste disposal is needed in this lab investigation.

Topics for the Explicit and Reflective Discussion

Concepts That Can Be Used to Justify the Evidence

To provide an adequate justification of their evidence, students must explain why they included the evidence in their arguments and make the assumptions underlying their analysis and interpretation of the data explicit. In this investigation, students can use the following concepts to help justify their evidence:

- Atomic structure
- Physical and chemical properties

We recommend that you discuss these fundamental concepts during the explicit and reflective discussion to help students make this connection.

How to Design Better Investigations

It is important for students to reflect on the strengths and weaknesses of the investigation they designed during the explicit and reflective discussion. Students should therefore be encouraged to discuss ways to eliminate potential flaws, measurement errors, or sources of bias in their investigations. To help students be more reflective about the design of their investigation, you can ask the following questions:

- What were some of the strengths of your investigation? What made it scientific?
- What were some of the weaknesses of your investigation? What made it less scientific?
- If you were to do this investigation again, what would you do to address the weaknesses in your investigation? What could you do to make it more scientific?

Crosscutting Concepts

This investigation is well aligned with two crosscutting concepts found in *A Framework for K–12 Science Education,* and you should review these concepts during the explicit and reflective discussion.

- *Patterns:* Observed patterns guide the way scientists organize and classify substances and interactions. Scientists also explore the relationships between and the underlying causes of the patterns they observe in nature. Chemists, for example, can predict characteristics of different elements based on patterns found in the periodic table.

- *Scale, proportion, and quantity:* It is critical for scientists to be able to recognize what is relevant at different sizes, time frames, and scales. Scientists must also be able to recognize proportional relationships between categories or quantities. The periodic trends studied in this activity are related to atomic size and proportional relationships among their structures.

The Nature of Science and the Nature of Scientific Inquiry

This investigation is well aligned with two important concepts related to the *nature of science* (NOS) and the *nature of scientific inquiry* (NOSI), and you should review these concepts during the explicit and reflective discussion.

- *The difference between observations and inferences:* An observation is a descriptive statement about a natural phenomenon, whereas an inference is an interpretation of an observation. Students should also understand that current scientific knowledge and the perspectives of individual scientists guide both observations and inferences. Thus, different scientists can have different but equally valid interpretations of the same observations due to differences in their perspectives and background knowledge.

- *The influence of society and culture on science:* Science is influenced by the society and culture in which it is practiced because science is a human endeavor. Cultural values and expectations determine what scientists choose to investigate, how investigations are conducted, how research findings are interpreted, and what people see as implications. People also view some research as being more important than others because of cultural values and current events.

Hints for Implementing the Lab

- This lab works best if students are not already familiar with common periodic trends. Therefore, it should be used at the beginning of the unit on the periodic table.

- Different groups can be assigned sets of properties to investigate, and then they can see how other groups investigated other trends. This can enhance the amount of discussion during the argumentation session.

- Emphasize the need for students to develop high-quality evidence. Information regarding different periodic trends is easy to find with access to the internet. Stress that they need to use information from the spreadsheet to support their claims.

Topic Connections

Table 7.2 provides an overview of the scientific practices, crosscutting concepts, disciplinary core ideas, and supporting ideas at the heart of this lab investigation. In addition, it lists NOS and NOSI concepts for the explicit and reflective discussion. Finally, it lists literacy and mathematics skills (*CCSS ELA* and *CCSS Mathematics*) that are addressed during the investigation.

TABLE 7.2

Lab 7 alignment with standards

Scientific practices	• Asking questions and defining problems • Planning and carrying out investigations • Analyzing and interpreting data • Using mathematics and computational thinking • Constructing explanations and designing solutions • Engaging in argument from evidence • Obtaining, evaluating, and communicating information
Crosscutting concepts	• Patterns • Scale, proportion, and quantity
Core idea	• PS1.A: Structure and properties of matter
Supporting ideas	• Atomic structure • Physical and chemical properties
NOS and NOSI concepts	• Observations and inferences • Social and cultural influences
Literacy connections (*CCSS ELA*)	• *Reading*: Key ideas and details, craft and structure, integration of knowledge and ideas • *Writing*: Text types and purposes, production and distribution of writing, research to build and present knowledge, range of writing • *Speaking and listening*: Comprehension and collaboration, presentation of knowledge and ideas
Mathematics connection (*CCSS Mathematics*)	• Look for and express regularity in repeated reasoning

LAB 7

Lab Handout

Lab 7. Periodic Trends: Which Properties of the Elements Follow a Periodic Trend?

Introduction

Periodic trends are the tendencies of certain properties of the elements to increase or decrease as you progress along a row or a column of the periodic table. A row in the periodic table is called a *period*, and a column in the periodic table is called a *group*. These trends can occur in both physical and atomic properties of the elements. The periodic table is organized in a way that makes these trends relatively easy to determine. Scientists can use these trends to help them predict an element's properties, which can determine how it will react in certain situations. The similar atomic structure among elements in a group or period helps explain how these trends occur.

FIGURE L7.1

Atomic radius is an example of a periodic trend.

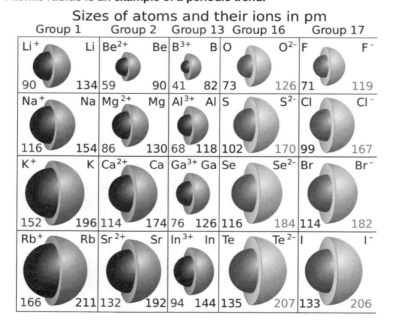

Atomic radius is an example of a property that has a periodic trend; as illustrated in Figure L7.1, atomic radius increases as you move down the periodic table regardless of group. Some properties, however, only change in a uniform manner within a specific group. These properties are often described as having a *quasi-periodic trend*. Boiling point is an example of a property that has a quasi-periodic trend; boiling point only changes in a uniform manner within a group, but it does not follow a similar pattern when you look across a period. In this investigation, you will explore how the physical and atomic properties of the element change across periods and groups in order to identify the other periodic trends.

Your Task

You will be given an Excel file that includes a list of the known elements and information about the atomic mass, density, melting point, specific heat capacity, and electronegativity of the elements. You must use this information along with the definitions provided in the "Introduction" to determine which of these properties follow a periodic trend and which ones do not.

The guiding question of this investigation is, **Which properties of the elements follow a periodic trend?**

Materials

You may use any of the following materials during your investigation:

- Computer
- "Properties of Elements" Excel file
- Periodic table

Safety Precautions

Follow all normal lab safety rules.

Investigation Proposal Required?   ☐ Yes   ☐ No

Getting Started

To answer the guiding question, you will need to analyze an existing data set. To accomplish this task, you must determine what type of data you need to examine and how you will analyze the data.

To determine *what type of data you will need to examine*, think about the following questions:

- What makes something a periodic trend or a quasi-period trend?
- Which elements will you need to look up to establish a trend?

To determine *how you will analyze the data*, think about the following questions:

- What type of graph could you create to help make sense of your data?
- What types of calculations will you need to make?

Connections to Crosscutting Concepts, the Nature of Science, and the Nature of Scientific Inquiry

As you work through your investigation, be sure to think about

- the importance of identifying patterns in science,
- the importance of proportional relationships,

LAB 7

- the difference between observations and inferences, and
- how science is influenced by social and cultural values.

Initial Argument

Once your group has finished collecting and analyzing your data, you will need to develop an initial argument. Your argument must include a *claim*, which is your answer to the guiding question. Your argument must also include *evidence* in support of your claim. The evidence is your analysis of the data and your interpretation of what the analysis means. Finally, you must include a *justification* of the evidence in your argument. You will therefore need to use a scientific concept or principle to explain why the evidence that you decided to use is relevant and important. You will create your initial argument on a whiteboard. Your whiteboard must include all the information shown in Figure L7.2.

FIGURE L7.2

Argument presentation on a whiteboard

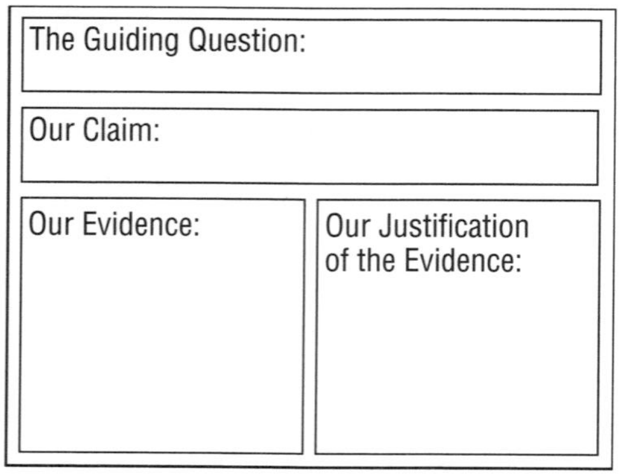

The Guiding Question:	
Our Claim:	
Our Evidence:	Our Justification of the Evidence:

Argumentation Session

The argumentation session allows all of the groups to share their arguments. One member of each group stays at the lab station to share that group's argument, while the other members of the group go to the other lab stations one at a time to listen to and critique the arguments developed by their classmates. The goal of the argumentation session is not to convince others that your argument is the best one; rather, the goal is to identify errors or instances of faulty reasoning in the initial arguments so these mistakes can be fixed. You will therefore need to evaluate the content of the claim, the quality of the evidence used to support the claim, and the strength of the justification of the evidence included in each argument that you see. To critique an argument, you might need more information than what is included on the whiteboard. You might therefore need to ask the presenter one or more follow-up questions, such as:

- What did your group do to analyze the data, and why did you decide to do it that way?
- Is that the only way to interpret the results of your group's analysis? How do you know that your interpretation of the analysis is appropriate?
- Why did your group decide to present your evidence in that manner?
- What other claims did your group discuss before deciding on that one? Why did you abandon those alternative ideas?
- How confident are you that your group's claim is valid? What could you do to increase your confidence?

Once the argumentation session is complete, you will have a chance to meet with your group and revise your original argument. Your group might need to gather more data or design a way to test one or more alternative claims as part of this process. Remember, your goal at this stage of the investigation is to develop the most valid or acceptable answer to the research question!

Report

Once you have completed your research, you will need to prepare an *investigation report* that consists of three sections that provide answers to the following questions:

1. What question were you trying to answer and why?

2. What did you do during your investigation and why did you conduct your investigation in this way?

3. What is your argument?

Your report should answer these questions in two pages or less. The report must be typed and any diagrams, figures, or tables should be embedded into the document. Be sure to write in a persuasive style; you are trying to convince others that your claim is acceptable or valid!

LAB 7

Lab 7. Periodic Trends: Which Properties of the Elements Follow a Periodic Trend?

1. Describe the difference between a periodic and a quasi-periodic trend.

2. Fluorine (F), boron (B), and lithium (Li) are all in the same period, but are in different groups. Fluorine has a higher electronegativity than boron, which has a higher electronegativity than lithium (F = 3.98, B = 2.04, Li = 0.98; all using the Pauling scale). However, boron has a higher melting point than both fluorine and lithium, and lithium's melting point is higher than that of fluorine (F = −220°C, B = 2300°C, Li = 180°C).

   Using your knowledge of periodic trends and the information above, describe the nature of the periodic trends for electronegativity and melting point.

3. "The element has high ionization energy" is an example of an observation.

   a. I agree with this statement.
   b. I disagree with this statement.

   Explain your answer, using an example from your investigation about periodic trends.

4. Cultural values and expectations affect how scientists conduct an investigation.

    a. I agree with this statement.

    b. I disagree with this statement.

    Explain your answer, using an example from your investigation about periodic trends.

5. Scientists often look for and attempt to explain patterns in nature. Explain why patterns are important, using an example from your investigation about periodic trends.

6. Scientists often need to look for proportional relationships. Explain why looking for a proportional relationship is often useful in science, using an example from your investigation about periodic trends.

LAB 8

Teacher Notes

Lab 8. Solutes and the Freezing Point of Water: How Does the Addition of Different Types of Solutes Affect the Freezing Point of Water?

Purpose

The purpose of this lab is to *introduce* students to *colligative properties*, which are the physical changes that result from adding solute to a solvent. Colligative properties depend on how many solute particles are present as well as the solvent amount, but they do *not* depend on the type of solute particles. During this investigation, students explore how the addition of different types of solutes can affect the freezing point of water. Students will also explore the need for imagination and creativity in conducting scientific investigations and the nature and role of experiments in science.

The Content

The homogeneous combination of two or more substances results in the formation of a solution. Solutions can exist in any phase (solid, liquid, or gas), although the most common solutions that students are familiar with are aqueous solutions, in which water is the primary solvent. The making of a solution can result in either a physical change or a chemical change. A physical change will result from a solution that involves the even mixing of the solute and solvent chemical components. Kool-Aid is an example of a solution that results in a physical change; the sugar molecules and dye molecules are physically distributed throughout the water used to dissolve them, and no chemical changes occur in any of the molecules. A solution that results in a chemical change involves a change in the chemical structure of the original solute and/or solvent; sodium chloride (NaCl) dissolved in water is an example of a solution resulting in a chemical change. When NaCl is dissolved in water, it chemically changes into the ions Na^+ and Cl^-. These ions are different from the original solute, NaCl. The charged nature of these ions allows the solution to conduct electricity. The original solute, NaCl, is not capable of conducting electricity. Ionic compounds, such as NaCl, are called *electrolytes*. Molecular compounds, such as sugar, are called *non-electrolytes*.

With respect to this investigation, the chemicals suggested for students to use represent both electrolytes and non-electrolytes. Students are tasked with investigating how dissolving these substances in water will affect the freezing point of the solution. In a broad sense, this investigation concerns the impact of physical or chemical changes on the physical properties (specifically freezing point) of the product. All of the product solutions formed by dissolving any of the suggested solutes in water will have a freezing point lower than that of water (0°C). This phenomenon is due to the increased number of molecules coming from the solutes that will be present in all of the resulting solutions.

The freezing point of an aqueous solution is affected by the overall amount of energy present in the system. As energy is removed from the system, all the water molecules will begin to lose kinetic energy and slow down. Once water molecules have slowed down enough, they begin to interact with each other to form the crystalline structures present in ice. However, the addition of a solute to the water will disrupt that process, making it more difficult for those water molecule interactions to occur. Thus, more energy must be removed from the aqueous solution to slow the water molecules down enough to allow the formation of crystals. The amount of energy that has to be removed for this interaction to occur depends on the amount of solute present in the solution.

The differences in freezing point between the electrolyte and non-electrolyte solutes relate to the total amount of solute particles in the solution stemming from a physical or chemical change. Non-electrolytes undergo a physical change when dissolved in water, remaining in their original chemical structure. However, electrolytes dissociate into their component ions when dissolved in water, representing a chemical change since these ions are different chemical structures. The differing effects on freezing temperature are due to differences in the type of change occurring.

Both types of solutes add particles that will require more energy to be removed from the solution to reach the freezing point, but the total amount of particles added will be different. Non-electrolytes, which only undergo a physical change, will add the same number of moles, and thus molecules, as the original amount of solute dissolved. Electrolytes, which undergo a chemical change and disassociate into ions when dissolved, will add an increased amount of total molecules based on molar ratios in the reaction. That is, one mole of sugar will only add one mole of sugar molecules to the solution, but one mole of NaCl will add one mole of Na^+ ions and one mole of Cl^- ions to the solution. Thus, a NaCl solution will have twice the number of particles affecting the freezing point when compared with a sugar solution. Because of the increased number of solute particles, more energy needs to be removed to allow the water to begin crystallizing, thus lowering the freezing point of the solution.

Timeline

The instructional time needed to complete this lab investigation is 200–300 minutes. Appendix 2 (p. 501) provides options for implementing this lab investigation over several class periods. Option A (250 minutes) should be used if students are unfamiliar with scientific writing because this option provides extra instructional time for scaffolding the writing process. You can scaffold the writing process by modeling, providing examples, and providing hints as students write each section of the report. Option B (200 minutes) should be used if students are familiar with scientific writing and have the skills needed to

LAB 8

write an investigation report on their own. In option B, students complete stage 6 (writing the investigation report) and stage 8 (revising the investigation report) as homework. It is assumed that stage 7 (peer review session) will take an entire 50 minutes, rather than the 30 minutes listed in the appendix. Also, stage 2 (design a method and collect data) may take longer than two class sessions, which could require another 50 minutes, leading to a total of 300 minutes. However, this extra time will allow for students to deeply engage in experimental design and evaluation.

Materials and Preparation

The materials needed to implement this investigation are listed in Table 8.1. The consumables and equipment can be purchased from a science supply company such as Carolina, Flinn Scientific, or Ward's Science. We recommend that you use a set routine for distributing and collecting the materials during the lab investigation. For example, the consumables and equipment for each group can be set up at each group's lab station before class begins, or one member from each group can collect them from a table or a cart when needed during class.

Safety Precautions

Remind students to follow all normal lab safety rules. Because 2-propanol is flammable, you will need to explain the potential hazards of working with this chemical and how to work with flammable chemicals. In addition, tell students to take the following safety precautions:

- Wear indirectly vented chemical-splash goggles and chemical-resistant gloves and aprons when they are collecting their data.
- Keep flammable chemicals away from open flame.
- Handle all glassware with care.
- Wash their hands with soap and water when they are done collecting the data.

Laboratory Waste Disposal

The aqueous solutions of glycerin, potassium chloride, 2-propanol, sodium chloride, sodium nitrate, sucrose, and rock salt can be disposed of down a drain if it is connected to a sanitation sewer system. We recommend following Flinn laboratory waste disposal method 26b for these liquids. Information about laboratory waste disposal methods is included in the Flinn Catalog and Reference Manual; you can request a free copy at *www.flinnsci.com*.

TABLE 8.1

Materials list

Item	Quantity
Consumables	
Glycerin, $C_3H_8O_3$	1–2 g per group
Potassium chloride, KCl	1–2 g per group
2-Propanol, C_3H_8O	1–2 g per group
Sodium chloride, NaCl	1–2 g per group
Sodium nitrate, $NaNO_3$	1–2 g per group
Sucrose, $C_{12}H_{22}O_{11}$	1–2 g per group
Rock salt	As needed
Ice	As needed
Distilled water	As needed
Equipment and other materials	
Temperature probe	1 per group
Styrofoam cup	1 per group
Test tube, medium (20 × 150 mm)	4–6 per group
Erlenmeyer flasks	2–3 per group
Graduated cylinder, 100 ml	1 per group
Parafilm, 1" squares	4–6 per group
Investigation Proposal C (optional)	1 per group
Whiteboard, 2' × 3' *	1 per group
Lab handout	1 per student
Peer-review guide and instructor scoring rubric	1 per student

*As an alternative, students can use computer and presentation software such as Microsoft PowerPoint or Apple Keynote to create their arguments.

LAB 8

Topics for the Explicit and Reflective Discussion

Concepts That Can Be Used to Justify the Evidence

To provide an adequate justification of their evidence, students must explain why they included the evidence in their arguments and make the assumptions underlying their analysis and interpretation of the data explicit. In this investigation, students can use the following concepts to help justify their evidence:

- Solutes, solvents, and solutions
- Physical and chemical properties
- Physical and chemical changes
- Phase change and freezing point

We recommend that you discuss these fundamental concepts during the explicit and reflective discussion to help students make this connection.

How to Design Better Investigations

It is important for students to reflect on the strengths and weaknesses of the investigation they designed during the explicit and reflective discussion. Students should therefore be encouraged to discuss ways to eliminate potential flaws, measurement errors, or sources of bias in their investigations. To help students be more reflective about the design of their investigation, you can ask the following questions:

- What were some of the strengths of your investigation? What made it scientific?
- What were some of the weaknesses of your investigation? What made it less scientific?
- If you were to do this investigation again, what would you do to address the weaknesses in your investigation? What could you do to make it more scientific?

Crosscutting Concepts

This investigation is well aligned with two crosscutting concepts found in *A Framework for K–12 Science Education*, and you should review these concepts during the explicit and reflective discussion.

- *Patterns:* Scientists look for patterns in nature and attempt to understand the underlying cause of these patterns. Chemists, for example, look for patterns in the ways matter interacts with other matter.
- *Energy and matter: Flows, cycles, and conservation:* In science it is important to track how energy and matter move into, out of, and within systems. For example,

during a phase change a substance either gains energy from its surroundings or releases energy to its surroundings in predictable ways.

The Nature of Science and the Nature of Scientific Inquiry

This investigation is well aligned with two important concepts related to the *nature of science* (NOS) and the *nature of scientific inquiry* (NOSI), and you should review these concepts during the explicit and reflective discussion.

- *The importance of imagination and creativity in science:* Students should learn that developing explanations for or models of natural phenomena and then figuring out how they can be put to the test of reality is as creative as writing poetry, composing music, or designing skyscrapers. Scientists must also use their imagination and creativity to figure out new ways to test ideas and collect or analyze data.

- *The nature and role of experiments:* Scientists use experiments to test the validity of a hypothesis (i.e., a tentative explanation) for an observed phenomenon. Experiments include a test and the formulation of predictions (expected results) if the test is conducted and the hypothesis is valid. The experiment is then carried out and the predictions are compared with the observed results of the experiment. If the predictions match the observed results, then the hypothesis is supported. If the observed results do not match the prediction, then the hypothesis is not supported. A signature feature of an experiment is the control of variables to help eliminate alternative explanations for observed results.

Hints for Implementing the Lab

- Distilled water should be used to make all solutions. Tap water will likely contain added minerals and other compounds that will affect the freezing point of the water.

- Students should be allowed to design their own investigations. However, students may not be familiar with using rock salt and ice to create a supercool environment. You can demonstrate the procedure as part of the tool talk.

- Depending on the temperature probe used, students may be able to instantly visualize a freezing curve for each solution using the data collection software that works with the probe. Students should be advised to look for an extended temperature plateau in the readout that will indicate the phase transition point.

- Students should investigate at least two different electrolytes and two different non-electrolytes to develop a more robust argument in response to the guiding question. However, depending on the amount of time available, you can also have student groups conduct investigations into various pairs of solutes, both similar and different, to set the stage for different results that can foster good discussions during the argumentation session.

Topic Connections

Table 8.2 provides an overview of the scientific practices, crosscutting concepts, disciplinary core ideas, and supporting ideas at the heart of this lab investigation. In addition, it lists the NOS and NOSI concepts for the explicit and reflective discussion. Finally, it lists literacy and mathematics skills (*CCSS ELA* and *CCSS Mathematics*) that are addressed during the investigation.

TABLE 8.2

Lab 8 alignment with standards

Scientific practices	• Asking questions and defining problems • Developing and using models • Planning and carrying out investigations • Analyzing and interpreting data • Using mathematics and computational thinking • Constructing explanations and designing solutions • Engaging in argument from evidence • Obtaining, evaluating, and communicating information
Crosscutting concepts	• Patterns • Energy and matter: Flows, cycles, and conservation
Core idea	• PS1.A: Structure and properties of matter
Supporting ideas	• Colligative properties • Solutes, solvents, solutions • Physical and chemical properties • Physical and chemical changes • Phase change and freezing point
NOS and NOSI concepts	• Imagination and creativity in science • Nature and role of experiments
Literacy connections (*CCSS ELA*)	• *Reading*: Key ideas and details, craft and structure, integration of knowledge and ideas • *Writing*: Text types and purposes, production and distribution of writing, research to build and present knowledge, range of writing • *Speaking and listening*: Comprehension and collaboration, presentation of knowledge and ideas
Mathematics connections (*CCSS Mathematics*)	• Model with mathematics • Attend to precision • Look for and express regularity in repeated reasoning

Lab Handout

Lab 8. Solutes and the Freezing Point of Water: How Does the Addition of Different Types of Solutes Affect the Freezing Point of Water?

Introduction

When substances undergo a change in phase, whether it is solid to liquid, liquid to gas, or the reverse, a transfer of energy must occur. Phase changes can be described as *endothermic* (absorbing energy) or *exothermic* (releasing energy) processes. Consider what happens when water freezes. When water changes from a liquid to a solid, the molecules lose energy and become more ordered; thus, freezing is an exothermic process because energy is released. Alternatively, when ice melts or water boils, the ice or liquid water gains or absorbs energy from its surroundings; thus, melting and boiling are endothermic processes.

A graph of a cooling curve provides a visual representation of what happens during phase changes. Figure L8.1 shows a cooling curve for a sample of water being cooled at a constant rate. The sample of water starts at 120°C in the gas phase. As it is cooled, the temperature of the sample decreases because the average kinetic energy of the water molecules in the sample is decreasing over time. At 100°C, the cooling curve levels off and the temperature remains constant until all of the steam has condensed into a liquid. This period of time represents the *condensation point*. At this point, the water molecules no longer have enough kinetic energy to remain in the gas phase. The liquid water then begins to decrease in temperature again as it loses more kinetic energy. At the freezing point, 0°C, the temperature of the sample

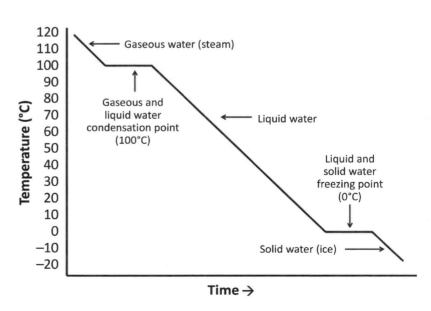

FIGURE L8.1

Cooling curve for water

levels off again and the liquid water transitions to a solid state. Once all the liquid water has become ice, the sample will once again decrease in temperature as it continues to lose energy.

When a solute is added to water, some of its physical properties may change. An example of a physical property of water that can change with the addition of a solute is its ability to conduct electricity. Pure water does not conduct electricity. However, when an *electrolyte* is added to water, the resulting solution is able to conduct electricity. Water, however, will not conduct electricity when a *non-electrolyte* is added to it. Electrolytes are ionic compounds such as sodium chloride and potassium chloride. Ionic compounds consist of ions joined together by ionic bonds, but these compounds dissociate into individual ions when mixed with water. Non-electrolytes, in contrast, are molecular compounds such as sucrose and propanol. Molecular compounds are made up of molecules that consist of atoms joined together by covalent bonds. The molecules found within a molecular compound do not break down when the compound is added to water; the molecules simply dissociate from each other as a result of the dissolution process. Other physical properties of water, such as its freezing point or boiling point, may also change in different ways when different types of solutes are added to it. In this investigation, you will determine how the addition of different types of solutes affects the freezing point of water.

Your Task

Determine how the addition of electrolytes and non-electrolytes affects the freezing point of water. Your group may test any of the following solutes:

Non-electrolytes	Electrolytes
• Glycerin ($C_3H_8O_3$) • 2-Propanol (C_3H_8O) • Sucrose ($C_{12}H_{22}O_{11}$)	• Potassium chloride (KCl) • Sodium chloride (NaCl) • Sodium nitrate ($NaNO_3$)

The guiding question of this investigation is, **How does the addition of different types of solutes affect the freezing point of water?**

Materials

You may use any of the following materials during your investigation:

Consumables	Equipment
• $C_3H_8O_3$ • C_3H_8O • $C_{12}H_{22}O_{11}$ • KCl • NaCl • $NaNO_3$ • Rock salt • Ice • Distilled water	• Temperature probe • Styrofoam cup • Test tube (medium: 20 × 150 mm) • Erlenmeyer flasks • Graduated cylinder (100 ml) • Parafilm

Safety Precautions

Follow all normal lab safety rules. 2-Propanol is flammable so be sure to keep it away from flames. Your teacher will explain relevant and important information about working with the chemicals associated with this investigation. In addition, take the following safety precautions:

- Wear indirectly vented chemical-splash goggles and chemical-resistant gloves and apron while in the laboratory.

- Handle all glassware with care.

- Wash your hands with soap and water before leaving the laboratory.

Investigation Proposal Required? ☐ Yes ☐ No

Getting Started

To measure the freezing point of distilled water or a solution made from one of the solutes, you can use rock salt and ice to create a supercool environment (less than 0°C). The basic setup is illustrated in Figure L8.2. The aqueous solutions that you test should have a concentration of at least 0.5 M and no more than 1.0 M. When you make your solutions, use 100 ml of H_2O and prepare them in an Erlenmeyer flask. Be sure to swirl the mixture until all the solute dissolves.

Now that you know how to measure the freezing point of distilled water or an aqueous solution using this equipment, you will need to design an investigation to answer the guiding question. You will therefore need to think about what type of data you need to collect, how you will collect the data, and how you will analyze the data.

FIGURE L8.2

Equipment used to measure the freezing point of water or an aqueous solution

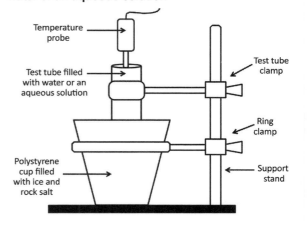

To determine *what type of data you need to collect*, think about the following questions:

- How will you know when an aqueous solution has reached its freezing point?
- What type of measurements will you need to record during your investigation?

To determine *how you will collect the data*, think about the following questions:

- What will serve as a control (or comparison) condition?
- What types of treatment conditions will you need to set up and how will you do it?
- What will you need to keep constant across comparisons?

LAB 8

- How will you make sure that your data are of high quality (i.e., how will you reduce error)?

To determine *how you will analyze the data*, think about the following questions:

- How will you determine if there is a difference between the treatment conditions and the control condition?
- What type of table could you create to help make sense of your data?

Connections to Crosscutting Concepts, the Nature of Science, and the Nature of Scientific Inquiry

As you work through your investigation, be sure to think about

- the importance of identifying patterns,
- how energy and matter are related to each other,
- how imagination and creativity are used during your investigation, and
- the nature and role of experiments.

FIGURE L8.3

Argument presentation on a whiteboard

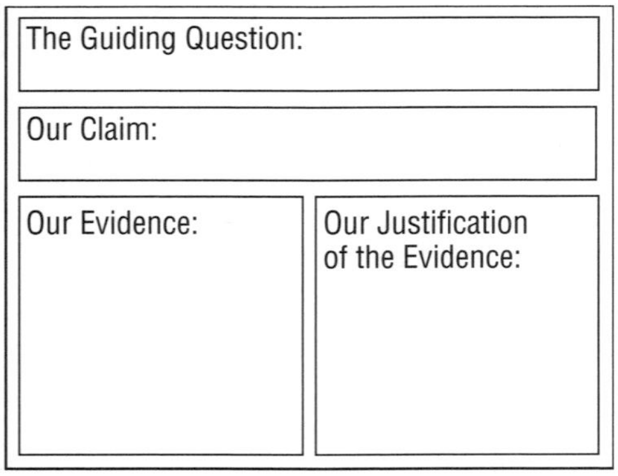

The Guiding Question:	
Our Claim:	
Our Evidence:	Our Justification of the Evidence:

Initial Argument

Once your group has finished collecting and analyzing your data, you will need to develop an initial argument. Your argument must include a *claim*, which is your answer to the guiding question. Your argument must also include *evidence* in support of your claim. The evidence is your analysis of the data and your interpretation of what the analysis means. Finally, you must include a *justification* of the evidence in your argument. You will therefore need to use a scientific concept or principle to explain why the evidence that you decided to use is relevant and important. You will create your initial argument on a whiteboard. Your whiteboard must include all the information shown in Figure L8.3.

Argumentation Session

The argumentation session allows all of the groups to share their arguments. One member of each group stays at the lab station to share that group's argument, while the other members of the group go to the other lab stations one at a time to listen to and critique the arguments developed by their classmates. The goal of the argumentation session is not to convince others that your argument is the best one; rather, the goal is to identify errors or instances of faulty reasoning in the initial arguments so these mistakes can be fixed. You will therefore need to evaluate the content of the claim, the quality of the evidence used to support the claim, and the strength of the justification of the evidence included in each argument that you see. To critique an argument, you might need more information than what is included on the whiteboard. You might therefore need to ask the presenter one or more follow-up questions, such as:

- How did your group collect the data? Why did you use that method?
- What did your group do to make sure the data you collected are reliable? What did you do to decrease measurement error?
- What did your group do to analyze the data, and why did you decide to do it that way? Did you check your calculations?
- Is that the only way to interpret the results of your group's analysis? How do you know that your interpretation of the analysis is appropriate?
- Why did your group decide to present your evidence in that manner?
- What other claims did your group discuss before deciding on that one? Why did you abandon those alternative ideas?
- How confident are you that your group's claim is valid? What could you do to increase your confidence?

Once the argumentation session is complete, you will have a chance to meet with your group and revise your original argument. Your group might need to gather more data or design a way to test one or more alternative claims as part of this process. Remember, your goal at this stage of the investigation is to develop the most valid or acceptable answer to the research question!

Report

Once you have completed your research, you will need to prepare an *investigation report* that consists of three sections that provide answers to the following questions:

1. What were you trying to do and why?
2. What did you do during your investigation and why did you conduct your investigation in this way?
3. What is your argument?

Your report should answer these questions in two pages or less. The report must be typed and any diagrams, figures, or tables should be embedded into the document. Be sure to write in a persuasive style; you are trying to convince others that your claim is acceptable or valid!

LAB 8

Lab 8. Solutes and the Freezing Point of Water: How Does the Addition of Different Types of Solutes Affect the Freezing Point of Water?

1. Describe the differences between electrolytes and non-electrolytes.

2. Cities and towns that frequently experience snow in the winter will usually keep warehouses full of salt. When the public roads become covered in snow and ice, they send out trucks to pour the salt over those roads.

   Using what you know about electrolytes and physical properties, explain why it is beneficial to put salt on icy roads.

3. Measuring the freezing point of a solution is an example of an experiment.

   a. I agree with this statement.
   b. I disagree with this statement.

   Explain your answer, using an example from your investigation about solutes and the freezing point of water.

4. Scientists need to be creative or have a good imagination to excel in science.

   a. I agree with this statement.
   b. I disagree with this statement.

   Explain your answer, using an example from your investigation about solutes and the freezing point of water.

5. Scientists often need to track how matter moves into and within a system. Explain why this is important, using an example from your investigation about solutes and the freezing point of water.

6. Scientists often look for and attempt to explain patterns in nature. Explain why patterns are important, using an example from your investigation about solutes and the freezing point of water.

Application Labs

LAB 9

Teacher Notes

Lab 9. Melting and Freezing Points: Why Do Substances Have Specific Melting and Freezing Points?

Purpose

The purpose of this lab is for students to *apply* their understanding of intermolecular forces to explain observed differences in the melting point of four substances. This lab gives students an opportunity to develop a conceptual model that accounts for the different freezing points of water, lauric acid, oleic acid, palmitic acid, and stearic acid. The students' conceptual models should be able to explain the shape of the heating or cooling curve for each of these substances. Students will also learn about the different types of investigations conducted in science and the role imagination and creativity play in science.

The Content

Matter exists in three basic states: solid, liquid, and gas. Whether a substance is a solid, liquid, or gas at room temperature (20°C–25°C) depends on the properties of that specific substance. One of the main properties of matter that influences the state of a particular substance at room temperature is the strength of the intermolecular forces holding that substance together. *Intermolecular forces* are the attractive forces between different molecules within a substance. Intermolecular forces are considerably weaker than *intramolecular forces*, or the forces that hold atoms within a molecule together, such as covalent bonding. There are several types of intermolecular forces that vary in strength.

Dipole-dipole interactions occur in molecules with polar bonds. In polar molecules, one portion of the molecule has a more positive charge while another portion of the molecule has a more negative charge. In the case of electric charges, opposites attract, therefore the positive portion of one molecule is attracted to the negative portion of a neighboring molecule. Conversely, similar charges are repelled from each other. Therefore, within a polar substance the molecules arrange themselves so that the alignment of positive-negative interactions is maximized and the alignment of positive-positive or negative-negative interactions is minimized.

Hydrogen bonding is a special type of dipole-dipole interaction and is the strongest of the intermolecular forces. Strong dipole-dipole forces occur when hydrogen atoms are bonded to oxygen, nitrogen, or fluorine atoms within a molecule. Due to the highly electronegative properties of oxygen, nitrogen, and fluorine, the hydrogen atoms are held very close and result in a highly polar bond. Since these bonds hold the atoms close together, that allows the dipoles of neighboring molecules to come closer together than typical dipole-dipole interactions, which increases the strength of this particular type of intermolecular force.

London dispersion forces, a type of intermolecular force that occurs between molecules that do not demonstrate dipole moments, is the weakest of the intermolecular forces. London dispersion forces result from random and temporary dipole moments that are due to the distribution of electrons around the atoms within molecules. At any given moment, the distribution of electrons around an atom may become asymmetrical. When the electron distribution is asymmetrical, there is a momentary dipole, during which one area of the atom is slightly more or less negative than another area of the atom. When these brief dipole moments occur, they generate a force of attraction with similarly brief dipole moments in neighboring molecules. Given the random and temporary nature of these types of interactions, London dispersion forces are the weakest type of intermolecular force. These forces are very common in large nonpolar hydrocarbon molecules; the more atoms present in these types of molecules, the more opportunity for temporary dipoles to form.

For a phase change to occur, a substance must gain sufficient energy to overcome the intermolecular forces of attraction between the molecules. To overcome these forces the individual molecules must gain energy. When a substance gains energy from its surroundings or from a heat source like a flame, the *kinetic energy* (the energy that an object possesses by virtue of being in motion) of the particles increases and they begin to move faster. The amount of kinetic energy a particle has depends not only on its velocity but also on its mass; *temperature* is a measure of the average kinetic energy of the particles of a substance. In the case of melting, or transitioning from a solid to a liquid, as the molecules gain kinetic energy (i.e., move faster) the intermolecular forces are not able to hold the molecules in place to maintain that state of matter. As the substance continues to gain energy, more and more of the molecules begin to overcome the intermolecular forces enough that a phase change occurs. This is not to say that the intermolecular forces disappear, only that they have less of an influence because of the increased kinetic energy of the molecules. In the case of freezing (transitioning from a liquid to a solid), the opposite occurs. During the freezing process, molecules lose kinetic energy or begin to move slower. As the molecules slow, they are less able to overcome intermolecular forces of attraction with neighboring molecules.

When substances undergo a change in phase, whether it is solid to liquid, liquid to gas, or the reverse of either phase change, a transfer of energy must occur. Therefore, phase changes can be described as *endothermic* or *exothermic* processes. In the case of freezing, molecules lose energy and become more ordered, and as the molecules lose energy to their surroundings the substance transitions from a liquid to a solid; thus, the process of freezing is exothermic because energy is released. Alternatively, when substances melt or even boil, they must gain or absorb energy from their surroundings to overcome the intermolecular forces; thus, melting and boiling are endothermic processes.

Graphs of heating or cooling curves provide a visual representation of what is happening during phase changes. Figure 9.1 (p. 142) shows a heating curve for a sample of

FIGURE 9.1

Heating curve for water

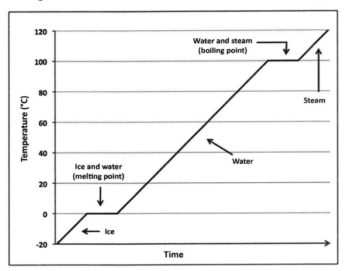

water being heated at a constant rate. The water starts at –20°C in the solid phase; as it is heated, it gradually increases in temperature. At 0°C the heating curve plateaus and the temperature remains constant for some time before increasing again. This plateau represents the melting point of water. Prior to the plateau, the water was solid ice and any energy the water gained caused the water molecules to move and vibrate more, which increased their average kinetic energy and the temperature. At the melting point, energy gained by the water molecules was used to overcome the intermolecular forces of attraction between the molecules; therefore, the temperature did not change even though the sample was still gaining energy. The energy gained during melting increased the potential energy between the molecules. Once the intermolecular forces are overcome, the sample transitions to the liquid phase and again increases in temperature as it gains energy. At the boiling point, 100°C, the temperature levels off again as the molecules gain potential energy and overcome the intermolecular forces to transition to the gas phase.

Lauric acid, oleic acid, palmitic acid, and stearic acid are all fatty acids, which are carboxylic acids with long hydrocarbon chains. The hydrocarbon chain length may vary from 10 to 30 carbons. Fatty acids tend to be nonpolar molecules. Lauric acid, palmitic acid, and stearic acid are saturated fatty acids, and oleic acid is an unsaturated fatty acid. A saturated fatty acid has all the bonding positions between the carbons in the hydrocarbon chain occupied by hydrogen. Unsaturated fatty acids, in contrast, have one or more double bonds present somewhere in the hydrocarbon chain.

The melting point of saturated fatty acids tends to increase as the molar mass of the molecule increases (see Table 9.1). Unsaturated fatty acids tend to have lower melting points than saturated fatty acids because of their differences in the shape of these groups of molecules. Saturated fatty acids have a relatively linear shape. This shape allows many fatty acid molecules to be "stacked" together. The alignment of the molecules in this manner results in greater intermolecular interactions and a relatively high melting point. In contrast, unsaturated fatty acids have one or more "bends" in the molecule and, as a result, these molecules do not "stack" together very well. The intermolecular interactions are therefore much weaker than the ones in saturated molecules. The melting points of unsaturated fatty acids, as a result, tend to be much lower than the melting points of saturated fatty acids.

TABLE 9.1

Information about the fatty acids used in this investigation

Substance	Formula	Molar mass	Structure	Melting point
Lauric acid	$C_{12}H_{24}O_2$	200.32 g/mol	Saturated	43°C
Oleic acid	$C_{18}H_{34}O_2$	282.46 g/mol	Unsaturated	14°C
Palmitic acid	$C_{16}H_{32}O_2$	256.42 g/mol	Saturated	63°C
Stearic acid	$C_{18}H_{36}O_2$	284.48 g/mol	Saturated	70°C

Timeline

The instructional time needed to complete this lab investigation is 180 minutes. Appendix 2 (p. 501) provides options for implementing this lab investigation over several class periods. Option B (180 minutes) should be used if students are not familiar with the lab equipment, particularly the temperature sensor and software. Collecting the temperature change data and generating the heating (or cooling) curves takes the better portion of a 50-minute class period. Option H (also 180 minutes) incorporates only a portion of day 2 for additional data collection during stage 2, which makes this option appropriate for students who are more familiar with the sensor equipment and software. Each of these options assumes that the students have experience with the scientific writing process and have the skills needed to write an investigation report on their own. In options B and H, students complete stage 6 (writing the investigation report) and stage 8 (revising the investigation report) as homework.

Materials and Preparation

The materials needed to implement this investigation are listed in Table 9.2 (p. 144). A temperature sensor kit includes an electronic temperature probe and any associated software or hardware such as a data collection interface or laptop computer. Temperature sensors are available for purchase from a variety of lab supply companies (e.g., Pasco or Vernier). It is recommended that the sensors and interface be positioned at lab stations prior to class to save time due to technical issues that may arise during setup.

To complete this investigation, students will need to generate either a cooling curve or a heating curve as each of the three substances transition between the solid and liquid phases. It is recommended that you prepare four test tubes for each group prior to the lab, containing one each of the following samples: lauric acid, stearic acid, oleic acid, and palmitic acid. All of these fatty acids, except oleic acid, will be solid at room temperature.

LAB 9

TABLE 9.2

Materials list

Item	Quantity
Consumables	
$C_{12}H_{24}O_2$	75–100 g per group
$C_{18}H_{34}O_2$	75–100 g per group
$C_{16}H_{32}O_2$	75–100 g per group
$C_{18}H_{36}O_2$	75–100 g per group
Rock salt	As needed
Ice	As needed
Distilled water	20 ml per group
Equipment and other materials	
Temperature sensor kit with interface	1 per group
Hot plate	1 per group
Test tube with stopper, prefilled with $C_{12}H_{24}O_2$	1 per group
Test tube with stopper, prefilled with $C_{18}H_{34}O_2$	1 per group
Test tube with stopper, prefilled with $C_{16}H_{32}O_2$	1 per group
Test tube with stopper, prefilled with $C_{18}H_{36}O_2$	1 per group
Styrofoam cup	1 per group
Beaker, 500 ml	1 per group
Investigation Proposal C (optional)	1 per group
Whiteboard, 2' × 3' *	1 per group
Lab handout	1 per student
Peer-review guide and instructor scoring rubric	1 per student

* As an alternative, students can use computer and presentation software such as Microsoft PowerPoint or Apple Keynote to create their arguments.

Some student groups may wish to investigate the freezing point of each substance and others the melting point (they should obtain similar results either way). To facilitate both modes of investigating, keep half of the test tubes of lauric acid, palmitic acid, and stearic acid at room temperature so those substances are in their solid form, and keep half of the test tubes of all four fatty acids in a warm-water bath so those substances are in their liquid form. Students can use those test tubes to generate cooling curves by placing them in a cool (or room-temperature water bath) to cause the acids to freeze. When investigating water, the students may melt a sample of ice using a hot plate and an ice bath. To freeze a sample of water, the students will need to place a small amount of water in a test tube. The test tube should then be placed inside a Styrofoam cup and surrounded with an ice/rock salt mixture to obtain temperatures sufficient to freeze the sample.

Safety Precautions

Remind students to follow all normal lab safety rules. Lauric acid, oleic acid, and rock salt can irritate eyes and skin. Palmitic acid and stearic acid can cause eye damage, and the fumes from these fatty acids can also irritate the respiratory tract. Therefore, you will need to explain the potential hazards of working with these chemicals and how to work with hazardous chemicals. In addition, tell students to take the following safety precautions:

- Wear indirectly vented chemical-splash goggles and chemical-resistant gloves and aprons when they are collecting their data.
- Use caution when working with hot plates, and keep them away from water and other liquids.
- Handle all glassware with care.
- Wash their hands with soap and water when they are done collecting the data.

Laboratory Waste Disposal

We recommend following Flinn laboratory waste disposal method 24a to dispose of the fatty acid samples used in this lab. Information about laboratory waste disposal methods is included in the Flinn Catalog and Reference Manual; you can request a free copy at *www.flinnsci.com*. Alternatively, the samples may be covered, stored, and used again when this investigation is repeated. If you choose to reuse the materials, be aware that impurities can be introduced into the samples over time. These impurities will depress the melting points and complicate their freezing or melting behaviors.

Topics for the Explicit and Reflective Discussion

Concepts That Can Be Used to Justify the Evidence
To provide an adequate justification of their evidence, students must explain why they included the evidence in their arguments and make the assumptions underlying their analysis and interpretation of the data explicit. In this investigation, students can use the following concepts to help justify their evidence:

- Molar mass
- Molecular geometry
- Intermolecular forces

LAB 9

- Kinetic molecular theory of matter
- States of matter and phase changes

We recommend that you discuss these fundamental concepts during the explicit and reflective discussion to help students make this connection.

How to Design Better Investigations

It is important for students to reflect on the strengths and weaknesses of the investigation they designed during the explicit and reflective discussion. Students should therefore be encouraged to discuss ways to eliminate potential flaws, measurement errors, or sources of bias in their investigations. To help students be more reflective about the design of their investigation, you can ask the following questions:

- What were some of the strengths of your investigation? What made it scientific?
- What were some of the weaknesses of your investigation? What made it less scientific?
- If you were to do this investigation again, what would you do to address the weaknesses in your investigation? What could you do to make it more scientific?

Crosscutting Concepts

This investigation is well aligned with two crosscutting concepts found in *A Framework for K–12 Science Education,* and you should review these concepts during the explicit and reflective discussion.

- *Systems and system models:* It is critical for scientists to be able to define the system under study and then make a model of it to understand it. Models can be physical, conceptual, or mathematical.
- *Energy and matter: Flows, cycles, and conservation:* In science it is important to track how energy and matter move into, out of, and within systems. For example, during a phase change a substance either gains energy from its surroundings or releases energy to its surroundings in predictable ways.

The Nature of Science and the Nature of Scientific Inquiry

This investigation is well aligned with two important concepts related to the *nature of science* (NOS) and the *nature of scientific inquiry* (NOSI), and you should review these concepts during the explicit and reflective discussion.

- *Methods used in scientific investigations*: Examples of methods include experiments, systematic observations of a phenomenon, literature reviews, and analysis of existing data sets; the choice of method depends on the objectives of the research.

There is no universal step-by-step scientific method that all scientists follow; rather, different scientific disciplines (e.g., chemistry vs. physics) and fields within a discipline (e.g., organic vs. physical chemistry) use different types of methods, use different core theories, and rely on different standards to develop scientific knowledge.

- *The importance of imagination and creativity in science:* Students should learn that developing explanations for or models of natural phenomena and then figuring out how they can be put to the test of reality is as creative as writing poetry, composing music, or designing skyscrapers. Scientists must also use their imagination and creativity to figure out new ways to test ideas and collect or analyze data.

Hints for Implementing the Lab

- Learn how to use the temperature sensor kit and associated software before the lab begins. It is important for you to know how to use the equipment so you can help students when technical issues arise.

- Allow the students to become familiar with the temperature sensor kit and software as part of the tool talk before they begin to design their investigation. This gives students a chance to see what they can and cannot do with the equipment.

- Be sure that students record actual values (e.g., temperature readings and changes in temperature) or save any graphs generated by the sensor software, rather than just attempting to hand draw what they see on the computer screen.

- Data collection can take several minutes as each liquid undergoes its phase change. You and the students should keep a close eye on time to ensure that data collection is not rushed due to limited class time.

- When freezing water using the ice/rock salt method, it is possible for the sample of water to supercool before freezing. Be mindful of this phenomenon, and be prepared to follow up with questions to help students understand what has happened.

Topic Connections

Table 9.3 (p. 148) provides an overview of the scientific practices, crosscutting concepts, disciplinary core ideas, and supporting ideas at the heart of this lab investigation. In addition, it lists NOS and NOSI concepts for the explicit and reflective discussion. Finally, it lists literacy and mathematics skills (*CCSS ELA* and *CCSS Mathematics*) that are addressed during the investigation.

LAB 9

TABLE 9.3

Lab 9 alignment with standards

Scientific practices	• Asking questions and defining problems • Developing and using models • Planning and carrying out investigations • Analyzing and interpreting data • Constructing explanations and designing solutions • Engaging in argument from evidence • Obtaining, evaluating, and communicating information
Crosscutting concepts	• Systems and system models • Energy and matter: Flows, cycles, and conservation
Core idea	• PS1.A: Structure and properties of matter
Supporting ideas	• Molar mass • Molecular geometry • Intermolecular forces • Kinetic molecular theory of matter • States of matter and phase changes
NOS and NOSI concepts	• Methods used in scientific investigations • Imagination and creativity in science
Literacy connections (CCSS ELA)	• *Reading:* Key ideas and details, craft and structure, integration of knowledge and ideas • *Writing:* Text types and purposes, production and distribution of writing, research to build and present knowledge, range of writing • *Speaking and listening:* Comprehension and collaboration, presentation of knowledge and ideas
Mathematics connections (CCSS Mathematics)	• Reason abstractly and quantitatively • Construct viable arguments and critique the reasoning of others • Look for and express regularity in repeated reasoning

Lab Handout

Lab 9. Melting and Freezing Points: Why Do Substances Have Specific Melting and Freezing Points?

Introduction

All molecules are constantly in motion, so they have *kinetic energy*. The amount of kinetic energy a molecule has depends on its velocity and its mass. *Temperature* is a measure of the average kinetic energy of all the molecules found within a substance. The temperature of a substance, as a result, will go up when the average kinetic energy of the molecules found within that substance increases, and the temperature of a substance will go down when the average kinetic energy of these molecules decreases. Temperature is therefore a useful way to measure how energy moves into, through, or out of a substance.

A substance will change state when enough energy transfers into or out of it. For example, when a solid gains enough energy from its surroundings, it will change into a liquid. This process is called *melting*. Melting is an endothermic process because the substance absorbs energy. The opposite occurs when a substance changes from a liquid to a solid (i.e., *freezing*). When a liquid substance releases enough energy into its surroundings, it will turn into a solid. Freezing is therefore an *exothermic* process because the substance releases energy. All substances have a specific melting point and a specific *freezing point*. The melting point is the temperature at which a substance transitions from a solid to a liquid. The freezing point is the temperature at which a substance transitions from a liquid to a solid.

There are three important properties of a substance that will affect its melting or freezing point. The first property is the molar mass of the molecules that make up that substance. The second property is the nature of the intermolecular forces that hold the molecules of a substance together. The third property is the shape of the molecules that make up that substance.

There are various types of intermolecular forces with different amounts of strength. One type of intermolecular force is caused by *dipole-dipole* interactions. This type of intermolecular force occurs when slightly negative atoms within a molecule are attracted to the slightly positive atoms in nearby molecules. *Hydrogen bonding* is a type of dipole-dipole interaction, in which oxygen, nitrogen, or fluorine atoms in a molecule are strongly attracted to hydrogen atoms in neighboring molecules. Another type of intermolecular force is caused when the arrangement of electrons within an atom has a momentary distribution that is not symmetrical. An asymmetrical distribution of electrons results in the molecule having a temporary dipole. This type of intermolecular interaction, called *London dispersion forces*, tends to occur in nonpolar molecules. London dispersion forces are much weaker than other intermolecular forces such as hydrogen bonding and other dipole-dipole interactions. Even though the relative strength of London dispersion forces is low, the influence of this type of intermolecular force can be magnified when large molecules interact with each other because the opportunity for these types of forces to occur increases.

In this investigation, you will explore how the three properties listed earlier in this section—molar mass, intermolecular forces, and shape of molecules—are related to the specific melting or freezing point of a substance.

Your Task

Determine the melting or freezing point of water, lauric acid, oleic acid, and stearic acid. Then develop a conceptual model that can be used to explain the observed differences in the melting or freezing points of these four substances. Your model should also be able to explain the shape of a heating or cooling curve for each of these substances. Once you have developed your conceptual model, you will need to test it to determine if it allows you to predict the melting point of palmitic acid.

The guiding question of this investigation is, **Why do substances have specific melting and freezing points?**

Materials

You may use any of the following materials during your investigation:

Consumables	Equipment
• Ice	• Temperature sensor kit
• Rock salt	• Hot plate
• Lauric acid, $C_{12}H_{24}O_2$	• Test tubes
• Oleic acid, $C_{18}H_{34}O_2$	• Beaker (500 ml)
• Palmitic acid, $C_{16}H_{32}O_2$	• Styrofoam cup
• Stearic acid, $C_{18}H_{36}O_2$	
• Distilled water	

Safety Precautions

Follow all normal lab safety rules. Your teacher will explain relevant and important information about working with the chemicals associated with this investigation. In addition, take the following safety precautions:

- Wear indirectly vented chemical-splash goggles and chemical-resistant gloves and apron while in the laboratory.

- Handle all glassware with care.

- Use caution when working with hot plates because they can burn skin. Hot plates also need to be kept away from water and other liquids.

- Wash your hands with soap and water before leaving the laboratory.

Investigation Proposal Required? ☐ Yes ☐ No

Getting Started

The first step in developing your model is to determine the melting point or the freezing point of each substance using the temperature sensor kit. To conduct these tests, you must determine what type of data you need to collect, how you will collect the data, and how you will analyze the data.

To determine *what type of data you need to collect*, think about the following questions:

- What type of measurements or observations will you need to record during each test?
- When will you need to make these measurements or observations?

To determine *how you will collect the data*, think about the following questions:

- What factors will you control during your different tests?
- What types of comparisons will you need to make?
- What will you do to reduce measurement error?
- How will you keep track of the data you collect and how will you organize it?

To determine *how you will analyze the data*, think about the following questions:

- What type of calculations will you need to make?
- What type of graph could you create to help make sense of your data?
- How will you determine the melting or freezing point of the different substances?
- How many different factors do you need to include in your explanatory model?

Once you have determined the melting or freezing points of each substance, your group will need to develop your conceptual model. The model must be able to explain the differences in melting point that you observe in each substance and the shape of the heating or cooling curves you generated. Your model also must include information about what is taking place between molecules on the submicroscopic level. To develop your model, you will need to consider how the molecular characteristics of each substance, such as its molecular structure or shape and its molar mass, may or may not affect the melting or freezing point of each substance. The table below lists the molar mass and structural formula of each substance for comparison purposes. You can also go to the Chemical Education Digital Library (*www.chemeddl.org*) to learn more about these substances and to view 3-D interactive virtual models of each one.

The last step in this investigation is to test your model. To accomplish this goal, you can use palmitic acid to determine if your model enables you to make accurate predictions

LAB 9

about the melting or freezing point of a different substance. The molar mass and structural formula of palmitic acid is provided in Table L9.1. If you are able to use your model to make accurate predictions, then you will be able to generate the evidence you need to convince others that your model is valid.

TABLE L9.1

Molar mass and structural formulas for the substances used in this investigation

Substance	Molar mass	Structural formula
Water	18.02 g/mol	
Lauric acid	200.32 g/mol	
Oleic acid	282.46 g/mol	
Palmitic acid	256.42 g/mol	
Stearic acid	284.48 g/mol	

Connections to Crosscutting Concepts, the Nature of Science, and the Nature of Scientific Inquiry

As you work through your investigation, be sure to think about

- how models are used to help understand natural phenomena,
- how energy and matter flow within systems,
- the different types of investigations conducted in science, and
- the role imagination and creativity play in science.

Initial Argument

Once your group has finished collecting and analyzing your data, you will need to develop an initial argument. Your argument must include a *claim*, which is your answer to the guiding question. Your argument must also include *evidence* in support of your claim. The evidence is your analysis of the data and your interpretation of what the analysis means. Finally, you must include a *justification* of the evidence in your argument. You

National Science Teachers Association

will therefore need to use a scientific concept or principle to explain why the evidence that you decided to use is relevant and important. You will create your initial argument on a whiteboard. Your whiteboard must include all the information shown in Figure L9.1.

Argumentation Session

The argumentation session allows all of the groups to share their arguments. One member of each group stays at the lab station to share that group's argument, while the other members of the group go to the other lab stations one at a time to listen to and critique the arguments developed by their classmates. The goal of the argumentation session is not to convince others that your argument is the best one; rather, the goal is to identify errors or instances of faulty reasoning in the initial arguments so these mistakes can be fixed. You will therefore need to evaluate the content of the claim, the quality of the evidence used to support the claim, and the strength of the justification of the evidence included in each argument that you see. To critique an argument, you might need more information than what is included on the whiteboard. You might therefore need to ask the presenter one or more follow-up questions, such as:

FIGURE L9.1

Argument presentation on a whiteboard

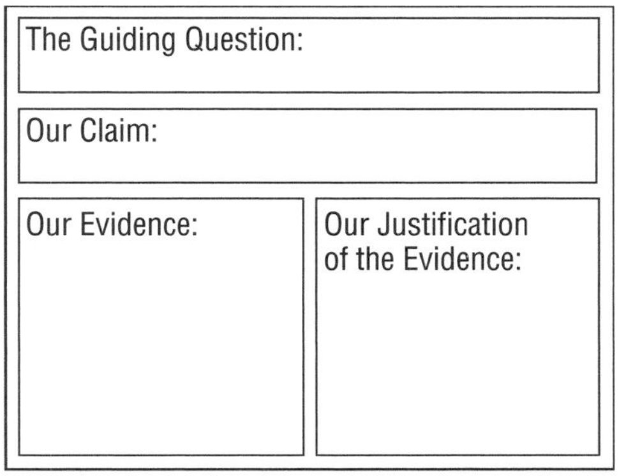

The Guiding Question:	
Our Claim:	
Our Evidence:	Our Justification of the Evidence:

- What did your group do to analyze the data, and why did you decide to do it that way?

- Is that the only way to interpret the results of your group's analysis? How do you know that your interpretation of the analysis is appropriate?

- Why did your group decide to present your evidence in that manner?

- What other claims did your group discuss before deciding on that one? Why did you abandon those alternative ideas?

- How confident are you that your group's claim is valid? What could you do to increase your confidence?

Once the argumentation session is complete, you will have a chance to meet with your group and revise your original argument. Your group might need to gather more data or design a way to test one or more alternative claims as part of this process. Remember, your goal at this stage of the investigation is to develop the most valid or acceptable answer to the research question!

Report

Once you have completed your research, you will need to prepare an *investigation report* that consists of three sections that provide answers to the following questions:

LAB 9

1. What question were you trying to answer and why?

2. What did you do during your investigation and why did you conduct your investigation in this way?

3. What is your argument?

Your report should answer these questions in two pages or less. The report must be typed and any diagrams, figures, or tables should be embedded into the document. Be sure to write in a persuasive style; you are trying to convince others that your claim is acceptable or valid!

Checkout Questions

Lab 9. Melting and Freezing Points: Why Do Substances Have Specific Melting and Freezing Points?

1. Describe the differences between dipole-dipole interactions and London dispersion forces.

2. Jake is cooking dinner for his family. In order to cook the different parts of his meal he needs to heat one pot of water to boiling (100°C) and another pot of vegetable oil until it reaches about 150°C.

   Use what you know about dipole-dipole interactions, London dispersion forces, molecular mass, and energy transfer to explain how the vegetable oil can reach a higher temperature than the water before it begins to boil.

3. In science, experiments are the best way to investigate any question.

   a. I agree with this statement.
   b. I disagree with this statement.

   Explain your answer, using an example from your investigation about melting and freezing points.

4. A painter can be creative during her work, but there is no room for creativity when conducting scientific investigations.

   a. I agree with this statement.
   b. I disagree with this statement.

   Explain your answer, using an example from your investigation about melting points and freezing points.

5. Understanding how matter and energy flow within and between systems is important in science. Explain why this is important, using an example from your investigation about melting and freezing points.

6. Scientists rely on models when they are trying to explain or understand complex phenomena. Explain what a model is and why models are important, using an example from your investigation about melting and freezing points.

LAB 10

Lab 10. Identification of an Unknown Based on Physical Properties: What Type of Solution Is the Unknown Liquid?

Purpose

The purpose of this lab is for students to *apply* what they have learned about physical properties and the nature of solutes, solvents, and solutions to identify two unknown solutions. Through this activity, students will have an opportunity to explore the importance of scale, proportion, and quantity as a way to understand the world and how matter and energy relate to each other. Students will also learn about the differences between data and evidence and between observations and inferences.

The Content

Physical changes to matter occur when the physical appearance and structure of a sample of matter is changed but the chemical structure of that sample is not changed. When physical changes occur to matter, change to a substance's physical properties is apparent. Physical properties include, among many other qualities, boiling point, freezing point, color, luster, and electrical conductivity. For example, when a liquid element reaches its boiling point, the atoms making up that matter have enough energy to move fast enough to change their physical state, going from liquid to gas. Although a physical change has occurred, with liquid transitioning to gas, the chemical structure of the element has not changed at all. The same idea holds for molecules as well as atoms of elements.

Chemical changes to matter involve the reorganization of matter into one or more completely different substances. The chemical changes that occur to a substance when it interacts with another substance are influenced by its chemical properties. Examples of chemical properties include flammability, oxidation states, toxicity, and the types of bonds a molecule or atom is capable of forming. Chemical properties are only observable and measurable when a substance interacts with another substance. The flammability of a substance, for example, can only be determined by attempting to combust that matter in the presence of oxygen and a flame. If the material "catches fire" and combusts, then much of the reactant matter is chemically changed into carbon dioxide and water, and the reactant matter is considered flammable.

The homogeneous combination of two or more substances results in the formation of a solution. Solutions can exist in any phase (solid, liquid, or gas), although the most common solutions students are familiar with involve aqueous solutions, in which water is the primary solvent. The making of a solution can result in either a physical change or a chemical change. A physical change will result from a solution that involves the even mixing of the solute and solvent chemical components. Kool-Aid is an example of a solution that results in a physical change; the sugar molecules and dye molecules are physically distributed throughout the water used to dissolve them, and no chemical changes occur in any of the molecules. A solution that results in a chemical change involves a change in the chemical structure of the original solute and/or solvent; potassium chloride (KCl) dissolved in water is an example of a solution resulting in a chemical change. This solution is formed when solid KCl is added to water. KCl chemically changes into the ions K^+ and Cl^- when it dissolves into water. The chemical change to the KCl results in a solution with different physical properties. The aqueous solution of KCl, for example, is able to conduct electricity whereas solid KCl and pure water cannot. The dissociation of KCl into two ions also increases the total number of particles present in the solution. The addition of these particles will also change the boiling point and freezing point of the water.

Colligative properties are physical changes that result from adding a solute to a solvent. Colligative properties depend on how many solute particles are present as well as the solvent amount but they do not depend on the type of solute particle. Boiling point elevation and freezing point depression are two examples of colligative properties. When the number of particles is increased in a solution, it increases the amount of energy that is required for that solution to transition from a liquid to a gas. The boiling point of the solution, as a result, elevates with the addition of more solute. The addition of a greater number of particles to a solution also lowers its freezing point because more energy must be removed from the solution before it transitions from a liquid to a solid. Thus, differences in the amount of particles present will alter the temperatures that are considered the boiling point and freezing point of a solution, changing the physical properties of that solution.

Students are tasked with identifying an unknown liquid sample by comparing the physical properties of that sample with those of several known solutions. The materials provided include known solutions with the same solute but different concentrations. Further, the solutes used represent ionizing and nondissociating solutes, which results in differences in the number of total particles present in each solution. The addition of different concentrations adds complexity by adding another factor influencing the amount of particles present in the known solutions.

With the materials provided, the students should develop investigations that attempt to determine the boiling points of both the known and unknown solutions. However, with the variation in concentrations of solutions of the same solute, the boiling point data alone will not provide a robust enough set of evidence to support a final claim as to the identity of the unknown. To develop further evidence, student groups should work to measure other physical or chemical properties of the solutions, including density, pH, and conductivity. You should encourage groups to collect multiple types of data to develop a more complete set of evidence. Such work offers the opportunity to discuss the difference between data and evidence explicitly with students. Furthermore, using the type of analysis and comparison also provides opportunities to distinguish between observations (in this activity,

LAB 10

the various physical property data collected) and inferences (the identity of the unknown is inferred through comparison of characteristics).

Timeline

The instructional time needed to complete this lab investigation is 200–300 minutes. Appendix 2 (p. 501) provides options for implementing this lab investigation over several class periods. Option A (250 minutes) should be used if students are unfamiliar with scientific writing, because this option provides extra instructional time for scaffolding the writing process. You can scaffold the writing process by modeling, providing examples, and providing hints as students write each section of the report. Option B (200 minutes) should be used if students are familiar with scientific writing and have the skills needed to write an investigation report on their own. In option B, students complete stage 6 (writing the investigation report) and stage 8 (revising the investigation report) as homework. It is assumed that stage 7 (peer review session) will take an entire 50 minutes, rather than the 30 minutes listed in the appendix. Also, stage 2 (design a method and collect data) may take longer than two class sessions, which could require another 50 minutes, leading to a total of 300 minutes. However, this extra time will allow for students to deeply engage in experimental design and evaluation.

Materials and Preparation

The materials needed to implement this investigation are listed in Table 10.1. The consumables and equipment can be purchased from a science supply company such as Carolina, Flinn Scientific, or Ward's Science. You will need to make about 1 L of each solution for a class of 30. For best results, prepare all the solutions with analytical precision using an analytical balance and volumetric flasks. Prepare the solutions needed for this investigation as follows:

- *1 M solution of acetic acid, CH_3COOH*: Add 57.1 ml of 17.4 M solution of CH_3CO_2H to 500 ml of distilled water in a 1 L volumetric flask. Mix well. Dilute to the 1 L mark with distilled water. *Caution: Be sure to add the acid to the water (**DO NOT** add the water to the acid).*

- *2 M solution of CH_3COOH*: Add 114.3 ml of 17.4 M solution of CH_3CO_2H to 500 ml of distilled water in a 1 L volumetric flask. Mix well. Dilute to the 1 L mark with distilled water. *Caution: Be sure to add the acid to the water (do NOT add the water to the acid).*

- *1 M solution of sodium chloride, NaCl*: Add 58.44 g of NaCl to 500 ml of distilled water in a 1 L volumetric flask. Mix well. Dilute to the 1 L mark with distilled water.

- *2 M solution of NaCl*: Add 116.88 g of NaCl to 500 ml of distilled water in a 1 L volumetric flask. Mix well. Dilute to the 1 L mark with distilled water.

- *1 M solution of sucrose, $C_{12}H_{22}O_{11}$*: Add 342.30 g of $C_{12}H_{22}O_{11}$ to 500 ml of distilled water in a 1 L volumetric flask. Mix well. Dilute to the 1 L mark with distilled water.

- *2 M solution of $C_{12}H_{22}O_{11}$*: Add 684.60 g of $C_{12}H_{22}O_{11}$ to 500 ml of distilled water in a 1 L volumetric flask. Mix well. Dilute to the 1 L mark with distilled water.

We recommend that you use a set routine for distributing and collecting the materials during the lab investigation. For example, the consumables and equipment for each group can be set up at each group's lab station before class begins, or one member from each group can collect them from a table or a cart when needed during class.

TABLE 10.1

Materials list

Item	Quantity
Consumables	
1 M solution of CH_3COOH	100 ml per group
2 M solution of CH_3COOH	100 ml per group
1 M solution of NaCl	100 ml per group
2 M solution of NaCl	100 ml per group
1 M solution of $C_{12}H_{22}O_{11}$	100 ml per group
2 M solution of $C_{12}H_{22}O_{11}$	100 ml per group
Unknown solution A (this should be one of the known solutions but labeled as unknown)	100 ml per group
Unknown solution B (this should be one of the known solutions but labeled as unknown)	100 ml per group
Equipment and other materials	
Erlenmeyer flasks or other sample containers	8 per group
Temperature probe	1 per group
Hot plate	1 per group
Electronic or triple beam balance	1 per group
Graduated cylinder, 50 ml	2 per group
Conductivity meter	1 per group
pH paper	Several strips per group
Investigation Proposal C (optional)	1 per group
Whiteboard, 2' × 3' *	1 per group
Lab handout	1 per student
Peer-review guide and instructor scoring rubric	1 per student

*As an alternative, students can use computer and presentation software such as Microsoft PowerPoint or Apple Keynote to create their arguments.

LAB 10

Safety Precautions

Remind students to follow all normal lab safety rules. In addition, tell students to take the following safety precautions:

- Wear indirectly vented chemical-splash goggles and chemical-resistant gloves and aprons when they are collecting their data.
- Use caution when working with hot plates, and keep them away from water and other liquids.
- Handle all glassware with care.
- Wash their hands with soap and water when they are done collecting the data.

Laboratory Waste Disposal

The aqueous solutions of acetic acid, sodium chloride, and sucrose can be disposed of down a drain if it is connected to a sanitation sewer system. We recommend following Flinn laboratory waste disposal method 26b for these liquids. Information about laboratory waste disposal methods is included in the Flinn Catalog and Reference Manual; you can request a free copy at *www.flinnsci.com.*

Topics for the Explicit and Reflective Discussion

Concepts That Can Be Used to Justify the Evidence

To provide an adequate justification of their evidence, students must explain why they included the evidence in their arguments and make the assumptions underlying their analysis and interpretation of the data explicit. In this investigation, students can use the following concepts to help justify their evidence:

- Solutions, solvents, and solutes
- Physical and chemical properties
- The nature of ionic compounds
- Colligative properties

We recommend that you discuss these fundamental concepts during the explicit and reflective discussion to help students make this connection.

How to Design Better Investigations

It is important for students to reflect on the strengths and weaknesses of the investigation they designed during the explicit and reflective discussion. Students should therefore be encouraged to discuss ways to eliminate potential flaws, measurement errors, or sources of bias in their investigations. To help students be more reflective about the design of their investigation, you can ask the following questions:

- What were some of the strengths of your investigation? What made it scientific?

- What were some of the weaknesses of your investigation? What made it less scientific?

- If you were to do this investigation again, what would you do to address the weaknesses in your investigation? What could you do to make it more scientific?

Crosscutting Concepts

This investigation is well aligned with two crosscutting concepts found in *A Framework for K–12 Science Education*, and you should review these concepts during the explicit and reflective discussion.

- *Scale, proportion, and quantity*. It is critical for scientists to be able to recognize what is relevant at different sizes, time frames, and scales. Scientists must also be able to recognize proportional relationships between categories or quantities.

- *Energy and matter: Flows, cycles, and conservation*. In science it is important to track how energy and matter move into, out of, and within systems. In this investigation, for example, it is important to know the concentration of each solution in order to identify it.

The Nature of Science and the Nature of Scientific Inquiry

This investigation is well aligned with two important concepts related to the *nature of science* (NOS) and the *nature of scientific inquiry* (NOSI), and you should review these concepts during the explicit and reflective discussion.

- *The difference between observations and inferences:* An observation is a descriptive statement about a natural phenomenon, whereas an inference is an interpretation of an observation. Students should also understand that current scientific knowledge and the perspectives of individual scientists guide both observations and inferences. Thus, different scientists can have different but equally valid interpretations of the same observations due to differences in their perspectives and background knowledge.

- *The difference between data and evidence in science*: Data are measurements, observations, and findings from other studies that are collected as part of an investigation. Evidence, in contrast, is analyzed data and an interpretation of the analysis.

Hints for Implementing the Lab

- Distilled water should be used to make all solutions. Tap water will likely contain added minerals and other compounds that will affect the boiling point of the water.

LAB 10

- We recommend making one of the unknowns a 2 M $C_{12}H_{22}O_{11}$ solution and the other unknown a 1 M NaCl solution.

- Two unknown solutions have been included in this activity to foster higher-quality argumentation. The more identities that student groups have to determine, the more opportunities there are for variation among groups that can lead to critical questioning and discussion during the argumentation session. However, if necessary for time or scheduling issues, the groups can be divided to determine the different identities of the two unknown solutions, and there will still be some groups that may have different results for the same unknown solution.

- The pH paper is more of a methodology distractor in this activity. It will identify acetic acid quickly, but it will not really be useful in distinguishing the NaCl and $C_{12}H_{22}O_{11}$ solutions. If acid/base topics have not been covered at this point, the pH paper can be left out. However, high school students will have most likely interacted with pH paper by this time in their schooling, so they may have some prior understanding about its use.

- The students will be able to determine the boiling point and the density of the solutions using the materials provided. These properties should be enough to determine the identity of the unknown solutions. However, the conductivity meters can also be easily used to make some distinguishing observations. Ideally, students would use all three of these physical properties to create a robust body of evidence to support their claim.

Topic Connections

Table 10.2 provides an overview of the scientific practices, crosscutting concepts, disciplinary core ideas, and supporting ideas at the heart of this lab investigation. In addition, it lists the NOS and NOSI concepts for the explicit and reflective discussion. Finally, it lists literacy and mathematics skills (*CCSS ELA* and *CCSS Mathematics*) that are addressed during the investigation.

TABLE 10.2

Lab 10 alignment with standards

Scientific practices	• Asking questions and defining problems • Planning and carrying out investigations • Analyzing and interpreting data • Using mathematics and computational thinking • Constructing explanations and designing solutions • Engaging in argument from evidence • Obtaining, evaluating, and communicating information
Crosscutting concepts	• Scale, proportion, and quantity • Energy and matter: Flows, cycles, and conservation
Core idea	• PS1.A: Structure and properties of matter
Supporting ideas	• Solutions, solvents, and solutes • Physical and chemical properties • The nature of ionic compounds • Colligative properties
NOS and NOSI concepts	• Observations and inferences • Difference between data and evidence
Literacy connections (CCSS ELA)	• *Reading*: Key ideas and details, craft and structure, integration of knowledge and ideas • *Writing*: Text types and purposes, production and distribution of writing, research to build and present knowledge, range of writing • *Speaking and listening*: Comprehension and collaboration, presentation of knowledge and ideas
Mathematics connections (CCSS Mathematics)	• Reason abstractly and quantitatively • Construct viable arguments and critique the reasoning of others • Use appropriate tools strategically • Attend to precision • Look for and express regularity in repeated reasoning

LAB 10

Lab Handout

Lab 10. Identification of an Unknown Based on Physical Properties: What Type of Solution Is the Unknown Liquid?

Introduction

Physical changes to chemical substances occur when the physical appearance and structure of a substance change but the chemical structure does not. *Chemical changes* happen when the chemical structure of matter is altered into a different structure that is a completely different substance. Chemists can understand which kind of changes have occurred by examining their relevant physical and chemical properties. *Physical properties* of matter are characteristics of substances that relate to the structure of matter and can be measured without changing them. These properties can be observed or measured in a variety of ways. Physical properties can include boiling point, color, conductivity, density, freezing point, melting point, viscosity, and many more. *Chemical properties* refer to characteristics of matter that can only be observed by changing the structure of a substance through a chemical change. Examples of chemical properties include reactivity, flammability, and oxidation states.

Many of the liquids that you interact with daily are aqueous solutions (see Figure L10.1). Dissolving a solute in water can result in either a physical change or a chemical change, depending on the chemical properties of the solute. If a solute can separate into different chemical substances when mixed in water, then a chemical change has occurred. But if the solute stays in the same chemical structure when mixed with water, then only a physical change has occurred because the solute has not changed its chemical structure; it has only changed its physical distribution in the water. The differences between these changes affect the *colligative properties* of the solutions. Colligative properties of solutions are characteristics that are dependent on the number of particles present and not on the specific type of particles present in a solution.

Your Task

Determine the identity of an unknown solution by comparing its physical and chemical properties with the same properties of known solutions.

The guiding question of this investigation is, **What type of solution is the unknown liquid?**

FIGURE L10.1 _____

A sample of methyl blue (left) and an aqueous solution of methyl blue (right)

Materials

You may use any of the following materials during your investigation:

Consumables	Equipment
• 1 M solution of acetic acid, CH_3COOH • 2 M solution of CH_3COOH • 1 M solution of sodium chloride, NaCl • 2 M solution of NaCl • 1 M solution of sucrose, $C_{12}H_{22}O_{11}$ • 2 M solution of $C_{12}H_{22}O_{11}$ • Unknown solution A • Unknown solution B	• Erlenmeyer flasks • Temperature probe • Hot plate • Electronic or triple beam balance • Graduated cylinder (50 ml) • Conductivity meter • pH paper

Safety Precautions

Follow all normal lab safety rules. Your teacher will explain relevant and important information about working with the chemicals associated with this investigation. In addition, take the following safety precautions:

- Wear indirectly vented chemical-splash goggles and chemical-resistant gloves and apron while in the laboratory.

LAB 10

- Use caution when working with hot plates because they can burn skin. Hot plates also need to be kept away from water and other liquids.
- Handle all glassware with care.
- Wash your hands with soap and water before leaving the laboratory.

Investigation Proposal Required?   ☐ Yes   ☐ No

Getting Started

The first step in this investigation is to identify all the various physical and chemical properties that are possible to measure using the available materials. Once you have determined which physical and chemical properties you can measure, you can then design your investigation. To do this, you will need to think about what type of data you need to collect, how you will collect the data, and how you will analyze the data. To determine *what type of data you need to collect*, think about what type of measurements you will need to record during your investigation.

To determine *how you will collect the data*, think about the following questions:

- What will serve as a control (or comparison) condition?
- What types of treatment conditions will you need to set up and how will you do it?
- How will you make sure that your data are of high quality (i.e., how will you reduce error)?

To determine *how you will analyze the data*, think about the following questions:

- How will you determine if there is a difference between the treatment and control conditions?
- What type of table could you create to help make sense of your data?

Connections to Crosscutting Concepts, the Nature of Science, and the Nature of Scientific Inquiry

As you work through your investigation, be sure to think about

- how scale, proportion, and quantity are important in understanding the natural world;
- how energy and matter are related to each other;
- the difference between observations and inferences; and
- the difference between data and evidence.

Initial Argument

Once your group has finished collecting and analyzing your data, you will need to develop an initial argument. Your argument must include a *claim*, which is your answer to the guiding question. Your argument must also include *evidence* in support of your claim. The evidence is your analysis of the data and your interpretation of what the analysis means. Finally, you must include a *justification* of the evidence in your argument. You will therefore need to use a scientific concept or principle to explain why the evidence that you decided to use is relevant and important. You will create your initial argument on a whiteboard. Your whiteboard must include all the information shown in Figure L10.2.

FIGURE L10.2

Argument presentation on a whiteboard

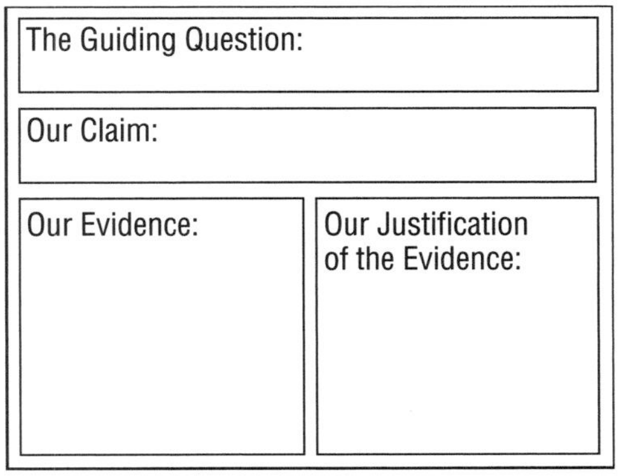

The Guiding Question:	
Our Claim:	
Our Evidence:	Our Justification of the Evidence:

Argumentation Session

The argumentation session allows all of the groups to share their arguments. One member of each group stays at the lab station to share that group's argument, while the other members of the group go to the other lab stations one at a time to listen to and critique the arguments developed by their classmates. The goal of the argumentation session is not to convince others that your argument is the best one; rather, the goal is to identify errors or instances of faulty reasoning in the initial arguments so these mistakes can be fixed. You will therefore need to evaluate the content of the claim, the quality of the evidence used to support the claim, and the strength of the justification of the evidence included in each argument that you see. To critique an argument, you might need more information than what is included on the whiteboard. You might therefore need to ask the presenter one or more follow-up questions, such as:

- What did your group do to make sure the data you collected are reliable? What did you do to decrease measurement error?

- What did your group do to analyze the data, and why did you decide to do it that way? Did you check your calculations?

- Is that the only way to interpret the results of your group's analysis? How do you know that your interpretation of the analysis is appropriate?

- Why did your group decide to present your evidence in that manner?

- What other claims did your group discuss before deciding on that one? Why did you abandon those alternative ideas?

- How confident are you that your group's claim is valid? What could you do to increase your confidence?

LAB 10

Once the argumentation session is complete, you will have a chance to meet with your group and revise your original argument. Your group might need to gather more data or design a way to test one or more alternative claims as part of this process. Remember, your goal at this stage of the investigation is to develop the most valid or acceptable answer to the research question!

Report

Once you have completed your research, you will need to prepare an *investigation report* that consists of three sections that provide answers to the following questions:

1. What question were you trying to answer and why?

2. What did you do during your investigation and why did you conduct your investigation in this way?

3. What is your argument?

Your report should answer these questions in two pages or less. The report must be typed and any diagrams, figures, or tables should be embedded into the document. Be sure to write in a persuasive style; you are trying to convince others that your claim is acceptable or valid!

Checkout Questions

Lab 10. Identification of an Unknown Based on Physical Properties: What Type of Solution Is the Unknown Liquid?

1. What are colligative properties?

2. Tree sap is the liquid that flows within trees to carry nutrients to different parts of the tree. Tree sap is primarily water mixed with other sugars from the tree; this mixture forms a thick syrup-like substance. For trees that grow in very cold climates, where winter temperatures drop below 0°C, it is important that their sap contain a lot of dissolved sugars; otherwise, the trees could die because their sap would freeze.

   Use what you know about colligative properties to explain why tree sap with a lot of dissolved sugars is beneficial to trees living in cold climates.

LAB 10

3. "The freezing point of the solution is –3°C" is an example of an observation.

    a. I agree with this statement.

    b. I disagree with this statement.

Explain your answer, using an example from your investigation about identification of an unknown based on physical properties.

4. "The freezing points of solutions A and B are both –1°C" is an example of evidence.

    a. I agree with this statement.

    b. I disagree with this statement.

Explain your answer, using an example from your investigation about identification of an unknown based on physical properties.

5. Scientists often need to look for proportional relationships between different quantities during an investigation. Explain what a proportional relationship is and why these relationships are important, using an example from your investigation about identification of an unknown based on physical properties.

6. It is often important to track how matter flows into, out of, and within system during an investigation. Explain why it is important to keep track of matter when studying a system, using an example from your investigation about identification of an unknown based on physical properties.

LAB 11

Teacher Notes

Lab 11. Atomic Structure and Electromagnetic Radiation: What Are the Identities of the Unknown Powders?

Purpose

The purpose of this lab is for students to *apply* what they know about atomic structure to identify four unknown powders using a flame test and spectroscopy. Through this activity, students will learn about the importance of identifying patterns in science, how models are used to make sense of scientific ideas, the difference between scientific laws and theories, and the role that imagination and creativity play in science.

The Content

The structure of atoms is directly related to the "size" of the atom, particularly concerning the number of protons and electrons the atom contains. The positively charged protons located in the nucleus of the atom provide the pull to keep the negatively charged electrons within the atom. Electrons are found around the nucleus in regions called *orbitals*. Orbitals represent the potential position of an electron at any given point in time. Figure 11.1 illustrates several different types of orbitals (s, p, and d).

FIGURE 11.1 _____

Electron orbitals

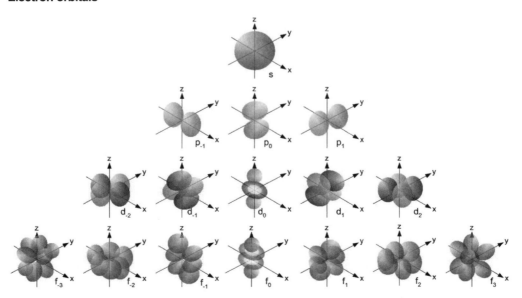

Orbitals are located at different distances from the nucleus and have different energy levels associated with them. Each orbital, however, can only hold two electrons. Electrons fill orbitals at lower energy levels before they fill orbitals at higher energy levels. Low-energy orbitals are the ones that are closer to the nucleus. Electrons will also fill orbitals one at a time when there is more than one orbital at the same energy level. For example, at the 2p energy level there are three p orbitals, so an electron will fill each of the p orbitals before two electrons will share the same p orbital. This behavior helps to minimize the repulsions between electrons and makes the atom more stable. Figure 11.2 summarizes the energies of the orbitals up to the 4d level.

FIGURE 11.2

Atomic energy levels with their sublevels

Notice that the s orbital from the next higher energy level has slightly lower energy than the d orbitals in the lower energy level.

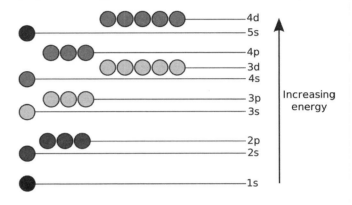

The electrons found in different orbitals, as noted earlier, have different energy levels. When an atom absorbs energy from an outside source, its electrons gain energy and jump to a higher energy level. The electrons are said to be an *excited* state when this happens. When those electrons return to a ground state by releasing that energy, it is emitted in the form of electromagnetic radiation. If that electromagnetic radiation falls between 400 and 700 nanometers (nm) in wavelength, it is given off in the form of visible light. Since each element has its own characteristic electron configuration, these light emissions are indicative of the electron configuration for the element. The electron configuration of an atom is an intrinsic quality of the atomic structure. Thus, the different colors of light produced by different elements during a flame test can be used to identify a compound.

Timeline

The instructional time needed to complete this lab investigation is 130–200 minutes. Appendix 2 (p. 501) provides options for implementing this lab investigation over several class periods. Option C (200 minutes) should be used if students are unfamiliar with scientific writing because this option provides extra instructional time for scaffolding the writing process. You can scaffold the writing process by modeling, providing examples, and providing hints as students write each section of the report. Option D (130 minutes) should be used if students are familiar with scientific writing and have the skills needed to write an investigation report on their own. In option D, students complete stage 6 (writing the investigation report) and stage 8 (revising the investigation report) as homework.

LAB 11

Materials and Preparation

The materials needed to implement this investigation are listed in Table 11.1. The different compounds listed should be in powder form. Students can place a small amount of each powder on the end of a wire loop or a wooden splint that has been soaked in distilled water and hold the setup over the flame. Samples of each powder can be prepared for each group if time allows; this will reduce a lot of classroom traffic and prevent contamination. However, central collection of samples can also be used, with groups sending students to collect enough to conduct their tests. The unknown samples will be unlabeled samples of the known compounds used in the activity. Select any combination of powders that seems appropriate and manageable with the available resources.

Safety Precautions

Remind students to follow all normal lab safety rules. Calcium chloride, potassium chloride, and sodium chloride are mild irritants to skin, eyes, and mucous membranes. Copper(II) chloride is corrosive to skin and eyes and is highly toxic by ingestion or inhalation. Lithium chloride causes serious eye damage and is moderately toxic by ingestion. Strontium chloride causes serious eye damage. Therefore, you will need to explain the potential hazards of working with these chemicals and how to work with hazardous chemicals. In addition, tell students to take the following safety precautions:

- Wear indirectly vented chemical-splash goggles and chemical-resistant gloves and aprons when they are collecting their data.
- Use caution when working with Bunsen burners. They can burn skin, and combustibles and flammables must be kept away from the open flame. Students with long hair should tie it back behind their heads.
- Remind students to handle all glassware with care.
- Wash their hands with soap and water when they are done collecting the data.

Laboratory Waste Disposal

We recommend following Flinn laboratory waste disposal method 26a to dispose of the chlorides used in this lab. Information about laboratory waste disposal methods is included in the Flinn Catalog and Reference Manual; you can request a free copy at *www.flinnsci.com*.

Topics for the Explicit and Reflective Discussion

Concepts That Can Be Used to Justify the Evidence

To provide an adequate justification of their evidence, students must explain why they included the evidence in their arguments and make the assumptions underlying their

TABLE 11.1

Materials list

Item	Quantity
Consumables	
Calcium chloride, $CaCl_2$	5 g per group
Copper(II) chloride, $CuCl_2$	5 g per group
Lithium chloride, LiCl	5 g per group
Potassium chloride, KCl	5 g per group
Sodium chloride, NaCl	5 g per group
Strontium chloride, $SrCl_2$	5 g per group
Unknown powder A (this should be one of the known powders but labeled as unknown)	5 g per group
Unknown powder B (this should be one of the known powders but labeled as unknown)	5 g per group
Unknown powder C (this should be one of the known powders but labeled as unknown)	5 g per group
Unknown powder D (this should be one of the known powders but labeled as unknown)	5 g per group
Equipment and other materials	
Beakers	1 per group
Bunsen burner	1 per group
Wire loop*	1 per group
Wooden splints*	13 per group
Spectroscope	1 per group
Whiteboard, 2' × 3' †	1 per group
Lab handout	1 per student
Peer-review guide and instructor scoring rubric	1 per student

* Students can use either a wire loop or wooden splints.

† As an alternative, students can use computer and presentation software such as Microsoft PowerPoint or Apple Keynote to create their arguments.

analysis and interpretation of the data explicit. In this investigation, students can use the following concepts to help justify their evidence:

- Atomic structure
- Electromagnetic radiation

We recommend that you discuss these fundamental concepts during the explicit and reflective discussion to help students make this connection.

How to Design Better Investigations

It is important for students to reflect on the strengths and weaknesses of the investigation they designed during the explicit and reflective discussion. Students should therefore be encouraged to discuss ways to eliminate potential flaws, measurement errors, or sources of bias in their investigations. To help students be more reflective about the design of their investigation, you can ask the following questions:

- What were some of the strengths of your investigation? What made it scientific?
- What were some of the weaknesses of your investigation? What made it less scientific?
- If you were to do this investigation again, what would you do to address the weaknesses in your investigation? What could you do to make it more scientific?

Crosscutting Concepts

This investigation is well aligned with three crosscutting concepts found in *A Framework for K–12 Science Education,* and you should review these concepts during the explicit and reflective discussion.

- *Patterns:* Observed patterns guide the way scientists organize and classify substances and interactions. Scientists also explore the relationships between and the underlying causes of the patterns they observe in nature.
- *Systems and system models:* Defining a system under study (such as an atom) and making a model (such as the one developed by Bohr or the one developed by Heisenberg and Schrödinger) of it are tools for developing a better understanding of natural phenomena in science. Models can be physical, conceptual, or mathematical.

The Nature of Science and the Nature of Scientific Inquiry

This investigation is well aligned with two important concepts related to the *nature of science* (NOS) and the *nature of scientific inquiry* (NOSI), and you should review these concepts during the explicit and reflective discussion.

- *The difference between laws and theories in science:* A scientific law describes the behavior of a natural phenomenon or a generalized relationship under certain conditions; a scientific theory is a well-substantiated explanation of some aspect of the natural world. Theories do not become laws even with additional evidence; they explain laws. However, not all scientific laws have an accompanying explanatory theory. It is also important for students to understand that scientists do not discover laws or theories; the scientific community develops them over time.

- *The importance of imagination and creativity in science:* Students should learn that developing explanations for or models of natural phenomena and then figuring out how they can be put to the test of reality is as creative as writing poetry, composing music, or designing skyscrapers. Scientists must also use their imagination and creativity to figure out new ways to test ideas and collect or analyze data.

Hints for Implementing the Lab

- Be sure to demonstrate the proper way to conduct a flame test as part of the tool talk.

- Depending on the time and resources available, you can adjust the number of unknown powders that students are asked to identify to add richness to their arguments or allow for completion in a smaller amount of time. Be sure to have multiple unknown powders, because this enhances the argumentation that can occur during the argumentation session.

- Consider changing the identities of the unknown powders tested for different classes, to avoid the possibility of students from different classes sharing results. Even within the same class, have different student groups test different unknown powders, with some overlap of samples occurring among all groups. Again, the similarities and differences among the groups can contribute to a more engaging argumentation session.

- Tell students to avoid dropping any powder onto the burner. The dropped powder will remain on the burner and continue to produce colored flames, skewing any further flame test results.

- You can also make 1 M solutions of each known and unknown compound. Students can then dip the wire loops or wooden splints into the solutions to conduct the flame tests. Remind students to avoid contamination of the burner with solution that may drop from the end of the loop or splint.

- Show students how to use a spectroscope as part of the tool talk.

LAB 11

Topic Connections

Table 11.2 provides an overview of the scientific practices, crosscutting concepts, disciplinary core ideas, and supporting ideas at the heart of this lab investigation. In addition, it lists the NOS and NOSI concepts for the explicit and reflective discussion. Finally, it lists literacy and mathematics skills (*CCSS ELA* and *CCSS Mathematics*) that are addressed during the investigation.

TABLE 11.2

Lab 11 alignment with standards

Scientific practices	• Asking questions and defining problems • Developing and using models • Planning and carrying out investigations • Analyzing and interpreting data • Constructing explanations and designing solutions • Engaging in argument from evidence • Obtaining, evaluating, and communicating information
Crosscutting concepts	• Patterns • Systems and system models
Core idea	• PS1.A: Structure and properties of matter
Supporting ideas	• Atomic structure • Electromagnetic radiation
NOS and NOSI concepts	• Scientific laws and theories • Imagination and creativity in science
Literacy connections (*CCSS ELA*)	• *Reading*: Key ideas and details, craft and structure, integration of knowledge and ideas • *Writing*: Text types and purposes, production and distribution of writing, research to build and present knowledge, range of writing • *Speaking and listening*: Comprehension and collaboration, presentation of knowledge and ideas
Mathematics connection (*CCSS Mathematics*)	• Look for and express regularity in repeated reasoning

Lab Handout

Lab 11. Atomic Structure and Electromagnetic Radiation: What Are the Identities of the Unknown Powders?

Introduction

According to our current theory about the structure of atoms, electrons are found around the nucleus in regions called orbitals (see Figure L11.1). Orbitals represent the potential position of an electron at any given point in time. Orbitals are located at different distances from the nucleus and have different energy levels associated with them. Each orbital, however, can only hold two electrons. The electrons of an atom fill low-energy orbitals, which are the ones closer to the nucleus, before they fill higher-energy ones.

Electrons are in a ground state when under stable conditions. When the electrons in an atom are bombarded with energy from an outside source, however, they absorb that energy and jump temporarily to a higher energy level. The electrons are said to be in an excited state when this happens. When those electrons release that energy, it is emitted in the form of electromagnetic radiation. If that electromagnetic radiation falls between 400 and 700 nanometers (nm) in wavelength, it is given off in the form of visible light.

Many common metal ions, such as Li^+, Na^+, K^+, Ca^{2+}, Ba^{2+}, Sr^{2+}, and Cu^{2+}, produce a distinct color of visible light when they are heated. These ions emit a unique color of light because they consist of atoms that have a unique electron configuration. Chemists can therefore identify these elements with a flame test. To conduct a flame test, a clean wire loop or a wooden splint that has been soaked in distilled water is dipped into a powder or solution and then placed into the hottest portion of a flame (see Figure Lll.2).

FIGURE L11.1

Each of the three p orbitals (top row) and all three together on the same atom (bottom)

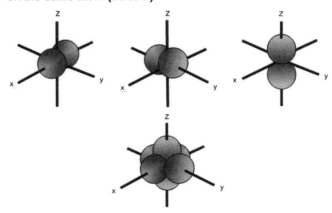

FIGURE L11.2

Flame test

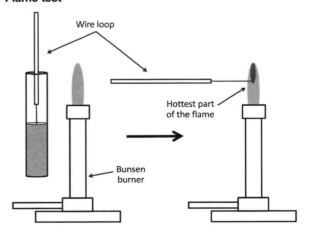

LAB 11

The unique color that we observe during a flame test is actually a mixture of several different wavelengths of visible light. Chemists can use a spectroscope to identify these various wavelengths. This technique is known as spectroscopy. A spectroscope splits light to form an emission line spectrum. The emission line spectrum for hydrogen is provided in Figure L11.3. The emission line spectrum for hydrogen consists of four different wavelengths of light (410 nm, 434 nm, 486 nm, and 656 nm). In this investigation, you will have an opportunity to conduct a flame test and use a spectroscope to identify four unknown powders.

FIGURE L11.3

The hydrogen emission spectrum with wavelength labels

Your Task

Use a flame test and a spectroscope to determine the emission line spectrum of six different powders. Then determine the identity of four unknown powders using a flame test, a spectroscope, and the emission line spectra from the six known powders.

The guiding question of this investigation is, **What are the identities of the unknown powders?**

Materials

You may use any of the following materials during your investigation:

Consumables	Equipment
• Calcium chloride, $CaCl_2$ • Copper(II) chloride, $CuCl_2$ • Lithium chloride, LiCl • Potassium chloride, KCl • Sodium chloride, NaCl • Strontium chloride, $SrCl_2$ • 4 unknown powders	• Beakers • Bunsen burner • Wooden splints • Spectroscope

Safety Precautions

Follow all normal lab safety rules. Your teacher will explain relevant and important information about working with the chemicals associated with this investigation. In addition, take the following safety precautions:

- Wear indirectly vented chemical-splash goggles and chemical-resistant gloves and apron while in the laboratory.
- Use caution when working with Bunsen burners. They can burn skin, and combustibles and flammables must be kept away from the open flame. If you have long hair, tie it back behind your head.
- Handle all glassware with care.
- Wash your hands with soap and water before leaving the laboratory.

Investigation Proposal Required?   ☐ Yes   ☐ No

Getting Started

To answer the guiding question, you will need to design and conduct an investigation. To accomplish this task, you must determine what type of data you need to collect, how you will collect the data, and how you will analyze the data.

To determine *what type of data you need to collect*, think about the following questions:

- How will you be able to identify a substance based on a flame test?
- What type of measurements or observations will you need to record during your investigation?

To determine *how you will collect the data*, think about the following questions:

- How often will you collect data and when will you do it?
- How will you make sure that your data are of high quality (i.e., how will you reduce error)?
- How will you keep track of the data you collect and how will you organize it?

To determine *how you will analyze the data*, think about the following questions:

- What type of data table could you create to help make sense of your data?
- What types of calculations will you need to make?

Connections to Crosscutting Concepts, the Nature of Science, and the Nature of Scientific Inquiry

As you work through your investigation, be sure to think about

- the importance of identifying patterns,
- how system models contribute to understanding science,
- the difference between laws and theories in science, and
- the importance of imagination and creativity in your investigation.

LAB 11

Initial Argument

Once your group has finished collecting and analyzing your data, you will need to develop an initial argument. Your argument must include a *claim*, which is your answer to the guiding question. Your argument must also include *evidence* in support of your claim. The evidence is your analysis of the data and your interpretation of what the analysis means. Finally, you must include a *justification* of the evidence in your argument. You will therefore need to use a scientific concept or principle to explain why the evidence that you decided to use is relevant and important. You will create your initial argument on a whiteboard. Your whiteboard must include all the information shown in Figure L11.4.

FIGURE L11.4

Argument presentation on a whiteboard

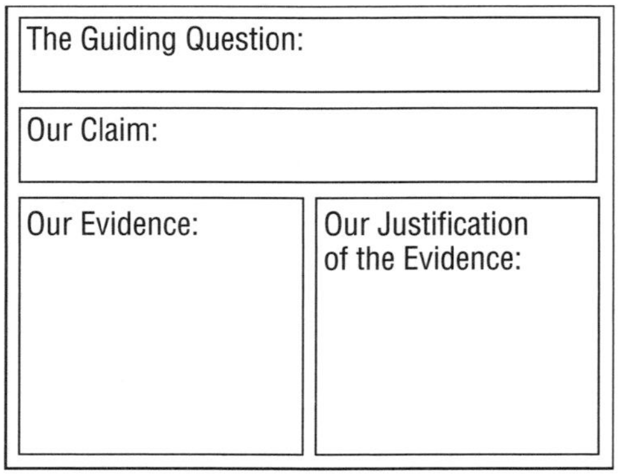

The Guiding Question:	
Our Claim:	
Our Evidence:	Our Justification of the Evidence:

Argumentation Session

The argumentation session allows all of the groups to share their arguments. One member of each group stays at the lab station to share that group's argument, while the other members of the group go to the other lab stations one at a time to listen to and critique the arguments developed by their classmates. The goal of the argumentation session is not to convince others that your argument is the best one; rather, the goal is to identify errors or instances of faulty reasoning in the initial arguments so these mistakes can be fixed. You will therefore need to evaluate the content of the claim, the quality of the evidence used to support the claim, and the strength of the justification of the evidence included in each argument that you see. To critique an argument, you might need more information than what is included on the whiteboard. You might therefore need to ask the presenter one or more follow-up questions, such as:

- How did your group collect the data? Why did you use that method?
- What did your group do to make sure the data you collected are reliable? What did you do to decrease measurement error?
- What did your group do to analyze the data, and why did you decide to do it that way?
- Is that the only way to interpret the results of your group's analysis? How do you know that your interpretation of the analysis is appropriate?
- Why did your group decide to present your evidence in that manner?
- What other claims did your group discuss before deciding on that one? Why did you abandon those alternative ideas?
- How confident are you that your group's claim is valid? What could you do to increase your confidence?

Once the argumentation session is complete, you will have a chance to meet with your group and revise your original argument. Your group might need to gather more data or design a way to test one or more alternative claims as part of this process. Remember, your goal at this stage of the investigation is to develop the most valid or acceptable answer to the research question!

Report

Once you have completed your research, you will need to prepare an *investigation report* that consists of three sections that provide answers to the following questions:

1. What question were you trying to answer and why?

2. What did you do during your investigation and why did you conduct your investigation in this way?

3. What is your argument?

Your report should answer these questions in two pages or less. The report must be typed and any diagrams, figures, or tables should be embedded into the document. Be sure to write in a persuasive style; you are trying to convince others that your claim is acceptable or valid!

LAB 11

Lab 11. Atomic Structure and Electromagnetic Radiation: What Are the Identities of the Unknown Powders?

1. Describe how photons can be emitted from an atom.

2. Neon lights used in many signs and commercial displays give off a unique red light. These lights work by passing an electric current through a glass tube filled with neon gas (a noble gas). Neon gas produces a red color, however "neon" lights can be found in a variety of other colors too.

   Use what you know about photons and atomic structure to explain how it is possible to produce other colors of "neon" light.

3. Theories and laws are different kinds of scientific knowledge.

   a. I agree with this statement.

   b. I disagree with this statement.

   Explain your answer, using an example from your investigation about atomic structure and electromagnetic radiation.

4. Scientists need to be creative or have a good imagination to excel in science.

    a. I agree with this statement.

    b. I disagree with this statement.

    Explain your answer, using an example from your investigation about atomic structure and electromagnetic radiation.

5. Scientists often use models to help them understand natural phenomena. Explain what a model is and why models are important, using an example from your investigation about atomic structure and electromagnetic radiation.

6. Scientists often look for and attempt to explain patterns in nature. Explain why patterns are important, using an example from your investigation about atomic structure and electromagnetic radiation.

LAB 12

Lab 12. Magnetism and Atomic Structure: What Relationships Exist Between the Electrons in a Substance and the Strength of Magnetic Attraction?

Purpose

The purpose of this lab is for students to *apply* what they have learned about atomic structure to explore the relationship between electron configuration and the strength of an element's magnetic field. Through this activity, students will have the opportunity to explore how the presence of paired and unpaired electrons in the orbitals of atoms influences the magnetic properties of a substance. Students will also learn about the importance of identifying patterns in science, how models are used to make sense of scientific ideas, the difference between scientific laws and theories, and how scientific knowledge changes over time in light of new evidence.

The Content

The structure of atoms is directly related to the "size" of the atom, particularly concerning the number of protons and electrons the atom contains. The positively charged protons located in the nucleus of the atom provide the pull to keep the negatively charged electrons within the atom. Electrons are found around the nucleus in regions called *orbitals*. Orbitals represent the probable position of an electron at any given point in time. Figure 12.1 illustrates several different types of orbitals (s, p, and d).

Orbitals are located at different distances from the nucleus and have different energy levels associated with them. At the first energy level is the 1s orbital, and at the second energy level is the 2s orbital. The 2s orbital is similar to the 1s orbital except that the region where there is the greatest chance of finding the electron is farther out from the nucleus. At the second energy level, in addition to the 2s orbital, there are three different p orbitals. These p orbitals are called $2p_x$, $2p_y$, and $2p_z$ (the 2 indicates that they are at the second energy level, and the letters indicate their relative orientation to each other). There are also p orbitals at higher energy levels (e.g., $3p_x$, $3p_y$, $3p_z$, $4p_x$, $4p_y$, and $4p_z$). In addition to s and p orbitals, there are two other sets of orbitals that can be found at higher energy levels. At the third level, for example, there is a set of five d orbitals (with complicated shapes and names) in addition to the one 3s and the 3p orbitals, for a total of nine orbitals all together. At the fourth level, there are seven f orbitals in addition to the 4s, 4p, and 4d orbitals.

FIGURE 12.1

Electron orbitals

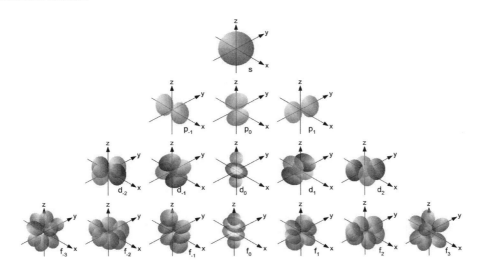

The electron configuration of any atom can be determined using the *Aufbau principle.* According to this principle, electrons fill low-energy orbitals before they fill higher-energy ones. Electrons, however, are mutually repulsive, so individual electrons will occupy different orbitals at the same energy level before sharing the same orbital at a specific energy level. This tendency to occupy different orbitals at the same energy level, which is called Hund's rule helps to minimize the repulsions between electrons and makes the atom more stable. As an example, consider the electron configuration of chlorine and iron. A diagram of the electron configuration for chlorine and for iron is provided in Figure 12.2 (p. 190). In this diagram, orbitals are represented as horizontal lines with the electrons in them depicted as arrows. Up-and-down arrows are used to indicate that the electrons have different spins. The 17 electrons found in an atom of chlorine are distributed across 9 different orbitals according to the Aufbau principle and Hund's rule. The 1s, 2s, 2p, and 3s orbitals are filled. The 3p orbitals are each occupied by at least one electron, with the $3p_x$ and $3p_y$ orbitals being completely filled. An atom of iron, in contrast, has 26 electrons distributed across 15 orbitals according to the Aufbau principle and Hund's rule. The 1s, 2s, 2p, 3s, 3p, and 4s orbitals are completely filled. The five 3d orbitals are each occupied by at least one electron, and only one of these orbitals contains two electrons.

LAB 12

FIGURE 12.2 _____

Electron configuration diagram for chlorine and iron

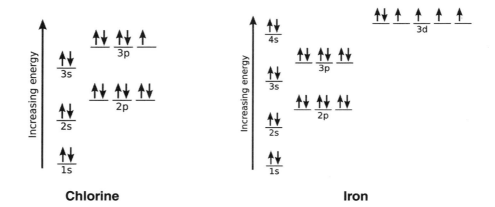

Chlorine   Iron

The electron configuration of the atoms in a substance also influences the magnetic properties of the substance. Many pure substances, besides iron metal, are attracted to a strong magnetic field. These substances are said to be *paramagnetic.* Substances that are not affected by strong magnetic fields are said to be *diamagnetic.* Past experiments have demonstrated a connection between the magnetic properties of a substance and the movement of electrons. Indeed, the magnetic field created by the movement of electrons is how scientists actually detect their presence, since it is difficult to actually pinpoint an electron. Electrons move very fast, so if one was detected by an instrument, that electron would already be in a different place by the time the detection registered. Thus, scientists focus on their magnetic fields to understand their movement and interactions.

In the quantum mechanical model, one of the energy states of electrons is related to the electron's spin ($m_s = -1/2$ or $+1/2$). This electron spin produces a magnetic field with electrons spinning in opposite directions that have reverse magnetic poles. Paired electrons have the opposite spin from their partner, which means their magnetic fields cancel out each other's effect. Unpaired electrons can spin in either direction and can change that direction when interacting with the magnetic fields of other unpaired electrons. This change in spin direction allows for unpaired electrons in different atoms to pair up. The spin of the electrons is important in shaping many properties of elements. The number of orbitals present in an atom (and the varying amounts of energy they have) will also influence the strength of the attraction between the outer electrons and the nucleus, which can affect the strength of the magnetic field they produce. That relationship is the focus of this investigation.

Timeline

The instructional time needed to complete this lab investigation is 130–200 minutes. Appendix 2 (p. 501) provides options for implementing this lab investigation over several class periods. Option C (200 minutes) should be used if students are unfamiliar with scientific writing, because this option provides extra instructional time for scaffolding the writing process. You can scaffold the writing process by modeling, providing examples, and providing hints as students write each section of the report. Option D (130 minutes) should be used if students are familiar with scientific writing and have the skills needed to write an investigation report on their own. In option D, students complete stage 6 (writing the investigation report) and stage 8 (revising the investigation report) of the investigation outside of class, which reduces the amount of time needed to complete the lab.

Materials and Preparation

The materials needed to implement this investigation are listed in Table 12.1. Be sure to put 1 mole of each compound into their respective vials. You will also want to inform your students that you have put 1 mole (or equal molar amounts) in each vial so they know that variable is controlled during their analysis.

TABLE 12.1

Materials list

Item	Quantity
Consumables	
Vial of copper(II) chloride, Cu^{2+}	1 per group
Vial of iron(II) chloride, Fe^{2+}	1 per group
Vial of manganese(II) chloride, Mn^{2+}	1 per group
Vial of zinc chloride, Zn^{2+}	1 per group
Equipment and other materials	
Empty vial	1 per group
Neodymium-iron-boron (NdFeB) magnet	1 per group
Electronic balance	1 per group
Plastic collar for balance plate	1 per group
Splints/Rods (vial supports)	2 per group
Whiteboard, 2' × 3' *	1 per group
Lab handout	1 per student
Peer-review guide and instructor scoring rubric	1 per student

*As an alternative, students can use computer and presentation software such as Microsoft PowerPoint or Apple Keynote to create their arguments.

LAB 12

Safety Precautions

Remind students to follow all normal lab safety rules. Copper(II) chloride, iron(II) chloride, manganese(II) chloride, and zinc chloride can cause skin irritation. You will therefore need to explain the potential hazards of working with these chemicals and how to work with hazardous chemicals. You also need to caution students about working with the NdFeB magnets because they are powerful but extremely brittle. In addition, tell students to take the following safety precautions:

- Wear indirectly vented chemical-splash goggles and chemical-resistant gloves and aprons when they are collecting their data.
- Wash their hands with soap and water when they are done collecting the data.

Laboratory Waste Disposal

The chlorides can be disposed of in a landfill. We recommend following Flinn laboratory waste disposal method 26a to dispose of these solids. Information about laboratory waste disposal methods is included in the Flinn Catalog and Reference Manual; you can request a free copy at *www.flinnsci.com*.

Topics for the Explicit and Reflective Discussion

Concepts That Can Be Used to Justify the Evidence

To provide an adequate justification of their evidence, students must explain why they included the evidence in their arguments and make the assumptions underlying their analysis and interpretation of the data explicit. In this investigation, students can use the following concepts to help justify their evidence:

- Atomic structure
- Electron configuration
- Electron spin
- Magnetism

We recommend that you discuss these fundamental concepts during the explicit and reflective discussion to help students make this connection.

How to Design Better Investigations

It is important for students to reflect on the strengths and weaknesses of the investigation they designed during the explicit and reflective discussion. Students should therefore be encouraged to discuss ways to eliminate potential flaws, measurement errors, or sources of bias in their investigations. To help students be more reflective about the design of their investigation, you can ask the following questions:

- What were some of the strengths of your investigation? What made it scientific?
- What were some of the weaknesses of your investigation? What made it less scientific?
- If you were to do this investigation again, what would you do to address the weaknesses in your investigation? What could you do to make it more scientific?

Crosscutting Concepts

This investigation is well aligned with two crosscutting concepts found in *A Framework for K–12 Science Education,* and you should review these concepts during the explicit and reflective discussion.

- *Systems and system models:* Defining a system under study (such as an atom) and making a model (such as the one developed by Bohr or the one developed by Heisenberg and Schrödinger) of it are tools for developing a better understanding of natural phenomena in science. Models can be physical, conceptual, or mathematical.
- *Structure and function:* The way an object (such as an atom or a molecule) is shaped or structured determines many of its properties and functions.

The Nature of Science and the Nature of Scientific Inquiry

This investigation is well aligned with two important concepts related to the *nature of science* (NOS) and the *nature of scientific inquiry* (NOSI), and you should review these concepts during the explicit and reflective discussion.

- *Changes in scientific knowledge over time:* A person can have confidence in the validity of scientific knowledge but must also accept that scientific knowledge may be abandoned or modified in light of new evidence or because existing evidence has been reconceptualized by scientists. There are many examples in the history of science of both evolutionary changes (i.e., the slow or gradual refinement of ideas) and revolutionary changes (i.e., the rapid abandonment of a well-established idea) in scientific knowledge.
- *The difference between laws and theories in science:* A scientific law describes the behavior of a natural phenomenon or a generalized relationship under certain conditions; a scientific theory is a well-substantiated explanation of some aspect of the natural world. Theories do not become laws even with additional evidence; they explain laws. However, not all scientific laws have an accompanying explanatory theory. It is also important for students to understand that scientists do not discover laws or theories; the scientific community develops them over time.

LAB 12

Hints for Implementing the Lab

- Have students write out the electron configuration for each of the elements being tested in this investigation. This can help them visually connect the idea of electron configuration with the differences seen in their data.

- Have a sample setup for the magnetic attraction measuring device prepared so you can demonstrate its use during the tool talk. Keep the setup available so students have a model to follow as they construct their own.

- To make the investigation more complex for the whole class, include several more vials of other similar, paramagnetic compounds, and let different groups of students work on different combinations of vials. This can lead to a more robust argumentation session.

Topic Connections

Table 12.2 provides an overview of the scientific practices, crosscutting concepts, disciplinary core ideas, and supporting ideas at the heart of this lab investigation. In addition, it lists the NOS and NOSI concepts for the explicit and reflective discussion. Finally, it lists literacy and mathematics skills (*CCSS ELA* and *CCSS Mathematics*) that are addressed during the investigation.

TABLE 12.2

Lab 12 alignment with standards

Scientific practices	• Asking questions and defining problems • Developing and using models • Planning and carrying out investigations • Analyzing and interpreting data • Constructing explanations and designing solutions • Engaging in argument from evidence • Obtaining, evaluating, and communicating information
Crosscutting concepts	• Systems and system models • Structure and function
Core idea	• PS1.A: Structure and properties of matter
Supporting ideas	• Atomic structure • Electron configuration • Electron spin • Magnetism
NOS and NOSI concepts	• Changes in scientific knowledge over time • Scientific laws and theories
Literacy connections (CCSS ELA)	• *Reading*: Key ideas and details, craft and structure, integration of knowledge and ideas • *Writing*: Text types and purposes, production and distribution of writing, research to build and present knowledge, range of writing • *Speaking and listening*: Comprehension and collaboration, presentation of knowledge and ideas
Mathematics connection (CCSS Mathematics)	• Look for and express regularity in repeated reasoning

LAB 12

Lab Handout

Lab 12. Magnetism and Atomic Structure: What Relationships Exist Between the Electrons in a Substance and the Strength of Magnetic Attraction?

Introduction

The structure of atoms is both simple and complex at the same time. The bulk of the mass of an atom is located in the nucleus, where protons and neutrons are relatively evenly distributed. Outside an atom's nucleus are electrons, which are negatively charged particles that are attracted to the positively charged nucleus. Electrons inhabit regions of space known as orbitals. These regions represent the probable position of an electron at any given point in time. Each orbital has a unique shape and can hold only two electrons. Each orbital also has a defined energy level.

At the first energy level is the 1s orbital. The "1" indicates that the orbital is in the energy level closest to the nucleus, and the "s" describes the shape of the orbital; s orbitals are spheres. At the second energy level, there is a second s orbital called the 2s orbital. This orbital is similar to the 1s orbital except that the region where there is the greatest chance of finding the electron is farther out from the nucleus. There are similar orbitals at higher energy levels (e.g., 3s, 4s, and 5s). At the second energy level there are also p orbitals; a p orbital looks like two balloons tied together. There are three different p orbitals that point at right angles to each other. These orbitals are called p_x, p_y and p_z. The p orbitals at the second energy level are called $2p_x$, $2p_y$ and $2p_z$ (the 2 indicates that they are at the second energy level). There are also p orbitals at higher energy levels (e.g., $3p_x$, $3p_y$, $3p_z$, $4p_x$, $4p_y$ and $4p_z$). In addition to s and p orbitals, there are two other sets of orbitals at higher energy levels. At the third energy level, for example, there is a set of five d orbitals (with complicated shapes and names) as well as the 3s and 3p orbitals. At the third energy level there are a total of nine orbitals. At the fourth energy level, there are seven f orbitals as well the 4s, 4p, and 4d orbitals. Figure L12.1 illustrates the 1s, 2s, and 2p orbitals.

The electron configuration of any atom can be determined using the *Aufbau principle*. According to this principle, electrons fill low-energy orbitals before they fill higher-energy ones. Electrons, however, are mutually repulsive, so individual electrons will occupy different orbitals at the same energy level before sharing the same orbital at a specific energy level. This tendency to occupy different orbitals at the same energy level, which is called Hund's rule, helps to minimize the repulsions between electrons and makes the atom more stable. As an example, consider the electron configuration of iron as illustrated in Figure L12.2. In this figure orbitals are represented as horizontal lines with the electrons in them

FIGURE L12.1

3-D representation of s and p electron orbitals

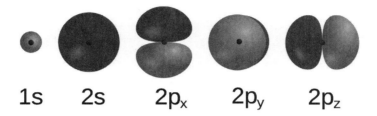

1s   2s   2p$_x$ 2p$_y$ 2p$_z$

FIGURE L12.2

Electron configuration diagram for iron

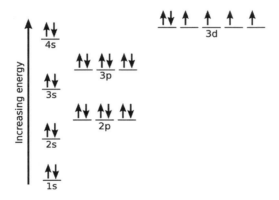

depicted as arrows; up and down arrows are used to indicate that the electrons have different spins. The 26 electrons that are present in an iron atom are distributed across 15 orbitals according to the Aufbau principle and Hund's rule. The 1s, 2s, 2p, 3s, and 4s orbitals are completely filled. The five 3d orbitals are each occupied by at least one electron, and only one of these orbitals is completely filled.

The electron configuration of the atoms found within a substance will affect its magnetic properties. Many pure substances, besides iron metal, are attracted to a strong magnetic field. These substances are said to be paramagnetic. Substances that are not affected by strong magnetic fields are said to be diamagnetic. Past experiments with electricity and magnetism have demonstrated a connection between the magnetic properties of a substance and the movement of electrons. In the quantum mechanical model, one of the energy states of electrons is related to the electron's spin ($m_s = -1/2$ or $+1/2$). This electron spin produces a magnetic field with electrons spinning in opposite directions that have reverse magnetic poles. Paired electrons always have the opposite spin from their partner, which means their magnetic fields cancel out each other's effect. Unpaired electrons can spin in either direction and can change that direction when interacting with the magnetic

LAB 12

fields of other unpaired electrons. In this investigation you will explore the relationship between electron configuration and magnetism in ionic compounds.

Your Task

Determine how electron configuration affects the magnetic properties of a substance.

The guiding question of this investigation is, **What relationships exist between the electrons in a substance and the strength of magnetic attraction?**

Materials

You may use any of the following materials during your investigation:

Consumables	Equipment
Vials containing copper(II) chloride (Cu^{2+}), iron(II) chloride (Fe^{2+}), manganese(II) chloride (Mn^{2+}), and zinc chloride (Zn^{2+})	• Empty vial • Neodymium-iron-boron magnet • Electronic balance • Plastic collar for electronic balance plate • Splints or rods

Safety Precautions

Follow all normal lab safety rules. Your teacher will explain relevant and important information about working with the chemicals associated with this investigation. *Caution: The neodymium-iron-boron magnet that you will use in this investigation is very powerful but extremely brittle. **DO NOT** test the magnet on the exposed metal at your laboratory station or bring anything metallic near your magnet.* In addition, take the following safety precautions:

- Wear indirectly vented chemical-splash goggles and chemical-resistant gloves and apron while in the laboratory.
- Wash your hands with soap and water before leaving the laboratory.

Investigation Proposal Required? ☐ Yes ☐ No

Getting Started

The first step in your investigation will be to create a device that you can use to measure the magnetic properties of different substances. Figure L12.3 shows how you can use an electronic balance to measure magnetic attraction. Once you have created your magnetic attraction measuring device, you will need to develop a procedure to determine how electron configuration affects the magnetic properties of the different chlorides that are available. Before you can design your procedure, however, your group will need to determine what type of data you need to collect, how you will collect the data, and how you will analyze the data.

To determine *what type of data you need to collect,* think about the following questions:

- What type of measurements will you need to record during your investigation?
- When will you need to make these measurements?

To determine *how you will collect the data,* think about the following questions:

- What comparisons will you need to make?
- How will you hold other variables constant?
- How will you use reduce measurement error?
- How will you keep track of the data you collect and how you will organize it?

To determine *how you will analyze the data,* think about the following questions:

- What are the similarities and differences in the electron configurations for each of the metal ions you used?
- What type of graph or table could you create to help make sense of your data?

FIGURE L12.3

Magnetic attraction measuring device

Place the vial on the support
directly over the magnet

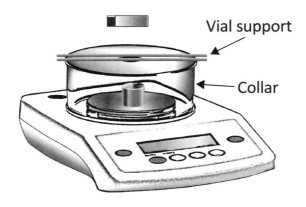

Vial support

Collar

LAB 12

Connections to Crosscutting Concepts, the Nature of Science, and the Nature of Scientific Inquiry

As you work through your investigation, be sure to think about

- the relationship between structure and function,
- how system models contribute to understanding science,
- the difference between laws and theories in science, and
- how scientific knowledge can change over time in light of new evidence.

Initial Argument

Once your group has finished collecting and analyzing your data, you will need to develop an initial argument. Your argument must include a *claim*, which is your answer to the guiding question. Your argument must also include *evidence* in support of your claim. The evidence is your analysis of the data and your interpretation of what the analysis means. Finally, you must include a *justification* of the evidence in your argument. You will therefore need to use a scientific concept or principle to explain why the evidence that you decided to use is relevant and important. You will create your initial argument on a whiteboard. Your whiteboard must include all the information shown in Figure L12.4.

FIGURE L12.4 _____

Argument presentation on a whiteboard

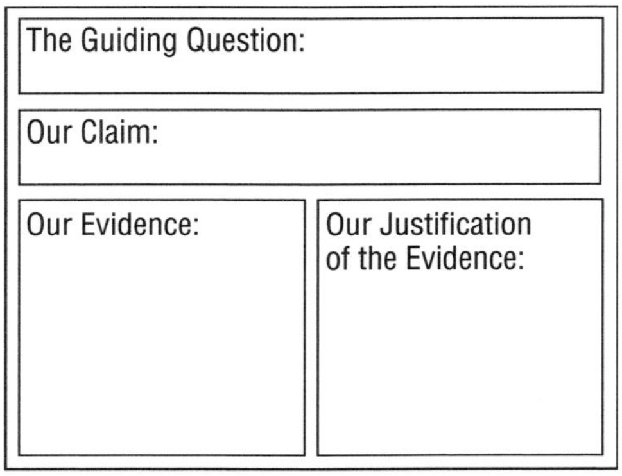

The Guiding Question:	
Our Claim:	
Our Evidence:	Our Justification of the Evidence:

Argumentation Session

The argumentation session allows all of the groups to share their arguments. One member of each group stays at the lab station to share that group's argument, while the other members of the group go to the other lab stations one at a time to listen to and critique the arguments developed by their classmates. The goal of the argumentation session is not to convince others that your argument is the best one; rather, the goal is to identify errors or instances of faulty reasoning in the initial arguments so these mistakes can be fixed. You will therefore need to evaluate the content of the claim, the quality of the evidence used to support the claim, and the strength of the justification of the evidence included in each argument that you see. To critique an argument, you might need more information than what is included on the whiteboard. You might therefore need to ask the presenter one or more follow-up questions, such as:

- How did your group collect the data? Why did you use that method?
- What did your group do to make sure the data you collected are reliable? What did you do to decrease measurement error?

National Science Teachers Association

- What did your group do to analyze the data, and why did you decide to do it that way? Did you check your calculations?

- Is that the only way to interpret the results of your group's analysis? How do you know that your interpretation of the analysis is appropriate?

- Why did your group decide to present your evidence in that manner?

- What other claims did your group discuss before deciding on that one? Why did you abandon those alternative ideas?

- How confident are you that your group's claim is valid? What could you do to increase your confidence?

Once the argumentation session is complete, you will have a chance to meet with your group and revise your original argument. Your group might need to gather more data or design a way to test one or more alternative claims as part of this process. Remember, your goal at this stage of the investigation is to develop the most valid or acceptable answer to the research question!

Report

Once you have completed your research, you will need to prepare an *investigation report* that consists of three sections that provide answers to the following questions:

1. What question were you trying to answer and why?

2. What did you do during your investigation and why did you conduct your investigation in this way?

3. What is your argument?

Your report should answer these questions in two pages or less. The report must be typed and any diagrams, figures, or tables should be embedded into the document. Be sure to write in a persuasive style; you are trying to convince others that your claim is acceptable or valid!

LAB 12

Lab 12. Magnetism and Atomic Structure: What Relationships Exist Between the Electrons in a Substance and the Strength of Magnetic Attraction?

1. Describe what you know about electron orbitals and how they are filled within an atom.

2. There are various elements that are considered metals. Many metals have shared characteristics, such as being malleable; however, one characteristic that differs among the metals is whether or not the metal is attracted to a magnet. For example, iron is very magnetic, but aluminum is not.

   Use what you know about electron orbital structure and electron spin to explain how some substances are considered paramagnetic (attracted to a strong magnetic field) and other substances are diamagnetic (not affected by a strong magnetic field).

3. Scientific knowledge is a set body of information that does not change.

   a. I agree with this statement.

   b. I disagree with this statement.

   Explain your answer, using an example from your investigation about magnetism and atomic structure.

4. Theories and laws are different kinds of scientific knowledge.

    a. I agree with this statement.

    b. I disagree with this statement.

    Explain your answer, using an example from your investigation about magnetism and atomic structure.

5. Scientists often use models to help them understand natural phenomena. Explain what a model is and why models are important, using an example from your investigation about magnetism and atomic structure.

6. In nature, the structure of an object is often related to function or the properties of that object. Explain why this is true, using an example from your investigation about magnetism and atomic structure.

LAB 13

Teacher Notes

Lab 13. Density and the Periodic Table: What Are the Densities of Germanium and Flerovium?

Purpose

The purpose of this lab is for students to *apply* what they know about physical properties and periodic trends to predict the densities of two different elements. Students will also learn about the importance of identifying patterns in science; the importance of scale, proportion, and quantity; how scientists use different methods to answer different questions; and how scientific knowledge changes over time in light of new evidence.

The Content

Periodic trends are specific patterns in the properties of elements that are present in the periodic table due to the arrangement of the elements in the table. The main periodic trends include ionization energy, electron affinity, atomic radius, and metallic character. The periodic trends that are found on the periodic table provide chemists with an invaluable tool to quickly predict the properties of any element. These trends exist because of the similar atomic structure of the elements within their respective groups or periods. Figure 13.1 provides a graphic representation of some periodic trends.

FIGURE 13.1

Examples of periodic trends

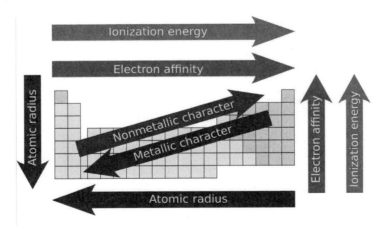

Density is a physical property of the elements and is defined as mass per unit volume. It is mathematically expressed as $d = m/v$, where d is density, m is mass, and v is volume. All elements have a specific density. The density of an element will remain the same no matter how much or how little is present. This quality makes density a useful property to chemists since it relates more to specific compounds than to specific amounts. Because density does not rely on the amount of material present, it is referred to as an *intensive property*. Intensive properties are physical properties that do not depend on size or amount. On the other hand, *extensive properties* are physical properties that are proportional to the amount of material present.

Density is often considered a quasi-periodic trend because of its differences among metals and nonmetals. In general, as atomic number increases as you move down a group, so does the density. When considering density across periods, however, the metallic nature of an element will also affect the density. Nonmetals increase in density as atomic number increases, but both their mass and volume increase in similar rates. However, most metals will have higher densities than nonmetals due to their more tightly bonded, crystalline structures. This is because their mass is increasing at a higher rate than the volume of space their crystalline structures occupy. This difference in the increases in mass and volume allows for larger densities among the metals on the periodic table.

Timeline

The instructional time needed to complete this lab investigation is 130–200 minutes. Appendix 2 (p. 501) provides options for implementing this lab investigation over several class periods. Option C (200 minutes) should be used if students are unfamiliar with scientific writing because this option provides extra instructional time for scaffolding the writing process. You can scaffold the writing process by modeling, providing examples, and providing hints as students write each section of the report. Option D (130 minutes) should be used if students are familiar with scientific writing and have the skills needed to write an investigation report on their own. In option D, students complete stage 6 (writing the investigation report) and stage 8 (revising the investigation report) of the investigation outside of class, which reduces the amount of time needed to complete the lab.

Materials and Preparation

The materials needed to implement this investigation are listed in Table 13.1 (p. 206). The consumables and equipment can be purchased from a science supply company such as Carolina, Flinn Scientific, or Ward's Science). We recommend that you use a set routine for distributing and collecting the materials during the lab investigation. The consumables and equipment for each group, for example, can be set up at each group's lab station before class begins, or one member from each group can collect them from a table or a cart when needed during class.

LAB 13

TABLE 13.1
Materials list

Item	Quantity
Consumables	
Charcoal	10 g per group
Lead shot	10 g per group
Silicon shot	10 g per group
Tin shot	10 g per group
Distilled water	As needed
Equipment and other materials	
Beakers, 250 ml	4 per group
Graduated cylinder, 25 ml	1 per group
Graduated cylinder, 50 ml	1 per group
Electronic or triple beam balance	1 per group
Periodic table	1 per student
Weighing dishes	As needed
Investigation Proposal C (optional)	1 per group
Whiteboard, 2' × 3' *	1 per group
Lab handout	1 per student
Peer-review guide and instructor scoring rubric	1 per student

*As an alternative, students can use computer and presentation software such as Microsoft PowerPoint or Apple Keynote to create their arguments.

Safety Precautions

Remind students to follow all normal lab safety rules. In addition, tell them to take the following safety precautions:

- Wear indirectly vented chemical-splash goggles when they are collecting their data.
- Handle all glassware with care.
- Wash their hands with soap and water when they are done collecting the data.

Laboratory Waste Disposal

The lead, silicon, and tin shot can be collected and reused. The charcoal can be disposed of in a landfill.

Topics for the Explicit and Reflective Discussion

Concepts That Can Be Used to Justify the Evidence

To provide an adequate justification of their evidence, students must explain why they included the evidence in their arguments and make the assumptions underlying their analysis and interpretation of the data explicit. In this investigation, students can use the following concepts to help justify their evidence:

- Physical properties
- Density
- Periodic and quasi-periodic trends

We recommend that you discuss these fundamental concepts during the explicit and reflective discussion to help students make this connection.

How to Design Better Investigations

It is important for students to reflect on the strengths and weaknesses of the investigation they designed during the explicit and reflective discussion. Students should therefore be encouraged to discuss ways to eliminate potential flaws, measurement errors, or sources of bias in their investigations. To help students be more reflective about the design of their investigation, you can ask the following questions:

- What were some of the strengths of your investigation? What made it scientific?
- What were some of the weaknesses of your investigation? What made it less scientific?
- If you were to do this investigation again, what would you do to address the weaknesses in your investigation? What could you do to make it more scientific?

Crosscutting Concepts

This investigation is well aligned with two crosscutting concepts found in *A Framework for K–12 Science Education,* and you should review these concepts during the explicit and reflective discussion.

- *Patterns:* Scientists look for patterns in nature and attempt to understand the underlying cause of these patterns. Chemists' understanding of the nature of different elements is largely due to ordering and comparison of different properties and patterns that are present in that order. The periodic table is partially organized by the increasing size of elements, and because atomic size and structure influence the chemical properties of an element, understanding those patterns helps chemists to understand their natural interactions.

- *Scale, proportion, and quantity:* It is critical for scientists to be able to recognize what is relevant at different sizes, time frames, and scales. Scientists must also be able to recognize proportional relationships between categories or quantities. The density of an element, for example, is a proportional relationship that that does not change.

The Nature of Science and the Nature of Scientific Inquiry

This investigation is well aligned with two important concepts related to the *nature of science* (NOS) *and the nature of scientific inquiry* (NOSI), and you should review these concepts during the explicit and reflective discussion.

- *Changes in scientific knowledge over time:* A person can have confidence in the validity of scientific knowledge but must also accept that scientific knowledge may be abandoned or modified in light of new evidence or because existing evidence has been reconceptualized by scientists. There are many examples in the history of science of both evolutionary changes (i.e., the slow or gradual refinement of ideas) and revolutionary changes (i.e., the rapid abandonment of a well-established idea) in scientific knowledge.

- *Methods used in scientific investigations:* Examples of methods include experiments, systematic observations of a phenomenon, literature reviews, and analysis of existing data sets; the choice of method depends on the objectives of the research. There is no universal step-by step scientific method that all scientists follow; rather, different scientific disciplines (e.g., chemistry vs. physics) and fields within a discipline (e.g., organic vs. physical chemistry) use different types of methods, use different core theories, and rely on different standards to develop scientific knowledge.

Hints for Implementing the Lab

- Allowing students to design their own procedures for collecting data gives them an opportunity to try, to fail, and to learn from their mistakes. However, you can scaffold students as they develop their procedure by having students fill out an investigation proposal. These proposals provide a way for you to offer students hints and suggestions without telling them how to do it. You can also check the proposals quickly during a class period.

- Do not tell students how to determine the volume of various elements. Encourage students to try several techniques and then select the best one. You should, however, expect students to be able to explain why they used one technique rather than another.

- Do not tell students how to predict the density of the other elements. A wide range of techniques will lead to richer discussions during the argumentation sessions.

- Do not tell students the actual values for the densities of germanium and Flerovium. Instead, ask students to think about what scientists do to increase their confidence in the validity of the claims that they make.

Topic Connections

Table 13.2 provides an overview of the scientific practices, crosscutting concepts, disciplinary core ideas, and supporting ideas at the heart of this lab investigation. In addition, it lists the NOS and NOSI concepts for the explicit and reflective discussion. Finally, it lists literacy and mathematics skills (*CCSS ELA* and *CCSS Mathematics*) that are addressed during the investigation.

TABLE 13.2 _____

Lab 13 alignment with standards

Scientific practices	• Asking questions and defining problems • Planning and carrying out investigations • Analyzing and interpreting data • Using mathematics and computational thinking • Constructing explanations and designing solutions • Engaging in argument from evidence • Obtaining, evaluating, and communicating information
Crosscutting concepts	• Patterns • Scale, proportion, and quantity
Core idea	• PS1.A: Structure and properties of matter
Supporting ideas	• Physical properties • Density • Periodic and quasi-periodic trends
NOS and NOSI concepts	• Changes in scientific knowledge over time • Methods used in scientific investigations
Literacy connections (*CCSS ELA*)	• *Reading*: Key ideas and details, craft and structure, integration of knowledge and ideas • *Writing*: Text types and purposes, production and distribution of writing, research to build and present knowledge, range of writing • *Speaking and listening*: Comprehension and collaboration, presentation of knowledge and ideas
Mathematics connection (*CCSS Mathematics*)	• Look for and express regularity in repeated reasoning

Lab Handout

Lab 13. Density and the Periodic Table: What Are the Densities of Germanium and Flerovium?

Introduction

At the time Dmitri Mendeleev proposed his periodic table for the classification of the elements in 1869, only 63 elements were known. Mendeleev arranged these 63 elements into a table of rows and columns in order of increasing atomic mass and by repeating physical properties. He also suggested that there were some missing elements that still needed to be discovered.

In Mendeleev's periodic table (see Figure L13.1), carbon and silicon were placed in the same period. Carbon appeared in the 2nd group and silicon appeared in the 3rd group. Mendeleev then proposed that there should be another element in this period. He named the missing element eka-silicium and predicted its physical properties. The German chemist Clemens Winkler discovered a new element in 1886 and called it germanium. In one of his reports of the discovery, he stated,

> It was definitely premature when I expressed such an assumption in my first notice concerning germanium; at least there was no basis for its proof. Nor would I have ventured at first to assume argyrodite to be a sulpho salt with a quadrivalent acid radical, because there were no analogies at all for such an assumption. Thus the present case shows very clearly how treacherous it can be to build upon analogies; the quadrivalency of germanium has by now become an incontrovertible fact, and there can be no longer any doubt that the new element is no other than the eka-silicium prognosticated fifteen years ago by Mendeleev. (Winkler 1886)

As this example from this history of chemistry illustrates, Mendeleev's periodic table gave chemists a powerful tool for predicting the properties of the elements. The periodic table, however, has been reorganized and rearranged over time. Carbon, silicon, and germanium are now placed in group 14 with tin and lead. Group 14 also contains another element called flerovium (atomic number 114). Flerovium is an extremely radioactive element that can only be created in the laboratory. In fact, scientists have only been able to create about 80 atoms of flerovium since it was first produced at the Flerov Laboratory of Nuclear Reactions in 1998. In this investigation, you will use the periodic table to determine the densities of two elements found in group 14.

FIGURE L13.1

Mendeleev's 1869 periodic table

ОПЫТЪ СИСТЕМЫ ЭЛЕМЕНТОВЪ.

ОСНОВАННОЙ НА ИХЪ АТОМНОМЪ ВѢСѢ И ХИМИЧЕСКОМЪ СХОДСТВѢ.

```
                    Ti = 50   Zr = 90    ? = 180.
                    V = 51    Nb = 94    Ta = 182.
                    Cr = 52   Mo = 96    W = 186.
                    Mn = 55   Rh = 104,4 Pt = 197,4.
                    Fe = 56   Rn = 104,4 Ir = 198.
                Ni = Co = 59  Pl = 106,6 O- = 199.
    H = 1                     Cu = 63,4  Ag = 108  Hg = 200.
            Be = 9,4 Mg = 24  Zn = 65,2  Cd = 112
            B = 11   Al = 27,4 ? = 68    Ur = 116  Au = 197?
            C = 12   Si = 28   ? = 70    Sn = 118
            N = 14   P = 31    As = 75   Sb = 122  Bi = 210?
            O = 16   S = 32    Se = 79,4 Te = 128?
            F = 19   Cl = 35,6 Br = 80   I = 127
    Li = 7 Na = 23   K = 39    Rb = 85,4 Cs = 133  Tl = 204.
                     Ca = 40   Sr = 87,6 Ba = 137  Pb = 207.
                     ? = 45    Ce = 92
                    ?Er = 56   La = 94
                    ?Yt = 60   Di = 95
                    ?In = 75,6 Th = 118?
```

Д. Менделѣевъ

Your Task

Determine the densities of carbon, lead, silicon, and tin by measuring the mass and volume of a sample of each of these elements. Then use this information to predict the densities of the other two elements in group 14 (germanium and flerovium).

The guiding question of this investigation is, **What are the densities of germanium and flerovium?**

Materials

You may use any of the following materials during your investigation:

Consumables	Equipment
• Charcoal	• 4 beakers (each 250 ml)
• Lead shot	• Graduated cylinders (25 ml and 50 ml)
• Silicon shot	• Electronic or triple beam balance
• Tin shot	• Periodic table
• Distilled water	• Weighing dishes

LAB 13

Safety Precautions

Follow all normal lab safety rules. Your teacher will explain relevant and important information about working with the chemicals associated with this investigation. In addition, take the following safety precautions:

- Wear chemical-splash goggles while in the laboratory.
- Handle all glassware with care.
- Wash your hands with soap and water before leaving the laboratory.

Investigation Proposal Required?   ☐ Yes   ☐ No

Getting Started

To answer the guiding question, you will need to think about what type of data you need to collect, how you will collect the data, and how you will analyze the data.

To determine *what type of data you need to collect*, think about the following questions:

- What type of information will you need to collect during your investigation to determine the densities of carbon, lead, silicon, and tin?
- How will you use the densities of carbon, lead, silicon, and tin to predict the densities of the other two elements?

To determine *how you will collect the data*, think about the following questions:

- What equipment will you need?
- How will you reduce measurement error?
- How will you keep track of the data you collect?

To determine *how you will analyze the data*, think about the following questions:

- What types of calculations will you need to make to determine the densities of carbon, lead, silicon, and tin?
- How can you use mathematics to predict the densities of the other two elements based on the densities of carbon, lead, silicon, and tin?
- What type of graph could you create to help make sense of your data?

Connections to Crosscutting Concepts, the Nature of Science, and the Nature of Scientific Inquiry

As you work through your investigation, be sure to think about

- the importance of identifying patterns;

- issues of scale, proportion, and quantity;
- how scientific knowledge can change over time in light of new evidence; and
- how scientists can use different methods to answer different types of questions.

Initial Argument

Once your group has finished collecting and analyzing your data, you will need to develop an initial argument. Your argument must include a *claim*, which is your answer to the guiding question. Your argument must also include *evidence* in support of your claim. The evidence is your analysis of the data and your interpretation of what the analysis means. Finally, you must include a *justification* of the evidence in your argument. You will therefore need to use a scientific concept or principle to explain why the evidence that you decided to use is relevant and important. You will create your initial argument on a whiteboard. Your whiteboard must include all the information shown in Figure L13.2.

FIGURE L13.2

Argument presentation on a whiteboard

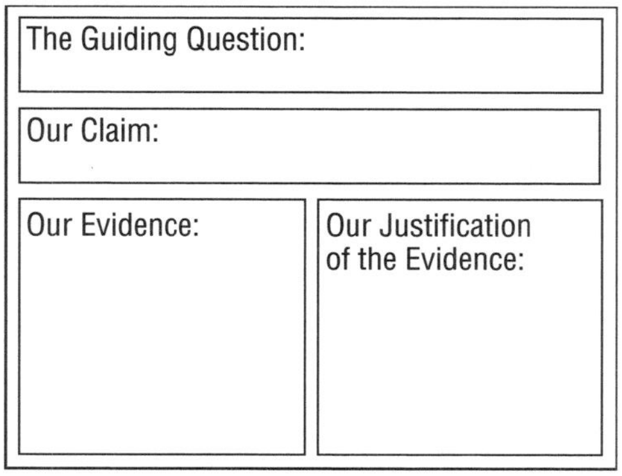

The Guiding Question:

Our Claim:

Our Evidence:	Our Justification of the Evidence:

Argumentation Session

The argumentation session allows all of the groups to share their arguments. One member of each group stays at the lab station to share that group's argument, while the other members of the group go to the other lab stations one at a time to listen to and critique the arguments developed by their classmates. The goal of the argumentation session is not to convince others that your argument is the best one; rather, the goal is to identify errors or instances of faulty reasoning in the initial arguments so these mistakes can be fixed. You will therefore need to evaluate the content of the claim, the quality of the evidence used to support the claim, and the strength of the justification of the evidence included in each argument that you see. To critique an argument, you might need more information than what is included on the whiteboard. You might therefore need to ask the presenter one or more follow-up questions, such as:

- How did your group collect the data? Why did you use that method?
- What did your group do to make sure the data you collected are reliable? What did you do to decrease measurement error?
- What did your group do to analyze the data, and why did you decide to do it that way? Did you check your calculations?
- Is that the only way to interpret the results of your group's analysis? How do you know that your interpretation of the analysis is appropriate?
- Why did your group decide to present your evidence in that manner?

LAB 13

- What other claims did your group discuss before deciding on that one? Why did you abandon those alternative ideas?
- How confident are you that your group's claim is valid? What could you do to increase your confidence?

Once the argumentation session is complete, you will have a chance to meet with your group and revise your original argument. Your group might need to gather more data or design a way to test one or more alternative claims as part of this process. Remember, your goal at this stage of the investigation is to develop the most valid or acceptable answer to the research question!

Report

Once you have completed your research, you will need to prepare an *investigation report* that consists of three sections that provide answers to the following questions:

1. What question were you trying to answer and why?

2. What did you do during your investigation and why did you conduct your investigation in this way?

3. What is your argument?

Your report should answer these questions in two pages or less. The report must be typed and any diagrams, figures, or tables should be embedded into the document. Be sure to write in a persuasive style; you are trying to convince others that your claim is acceptable or valid!

Reference

Winkler, C. 1886. About germanium. *Journal für prakische Chemie* 142 (N.F. 34): 177–229. Excerpt available online at *www.chemteam.info/Chem-History/Disc-of-Germanium.html*.

Checkout Questions

Lab 13. Density and the Periodic Table: What Are the Densities of Germanium and Flerovium?

1. What are periodic trends?

2. The following table shows the measured densities for period 2 and group 14 on the periodic table.

Element	Density	Location on the periodic table	
		Period	**Group**
Lithium	0.53 g/cm³	2	1
Beryllium	1.85 g/cm³	2	2
Boron	2.46 g/cm³	2	13
Carbon	2.26 g/cm³	2	14
Nitrogen	1.25 g/L	2	15
Oxygen	1.43 g/L	2	16
Fluorine	1.70 g/L	2	17
Neon	0.90 g/L	2	18
Silicon	2.33 g/cm³	3	14
Germanium	5.32 g/cm³	4	14
Tin	7.31 g/cm³	5	14
Lead	11.34 g/cm³	6	14

Use what you know about density and periodic trends, along with the data in the table, to explain whether or not density is a periodic trend.

3. Scientific knowledge can change over time in light of new evidence.

   a. I agree with this statement.

   b. I disagree with this statement.

   Explain your answer, using an example from your investigation about density and the periodic table.

4. An investigation must follow the scientific method to be considered scientific.

   a. I agree with this statement.

   b. I disagree with this statement.

   Explain your answer, using an example from your investigation about density and the periodic table.

5. Scientists often look for and attempt to explain patterns in nature. Explain why patterns are important, using an example from your investigation about density and the periodic table.

6. Scientists often need to look for proportional relationships. Explain why looking for a proportional relationship is often useful in science, using an example from your investigation about density and the periodic table.

LAB 14

Teacher Notes

Lab 14. Molar Relationships: What Are the Identities of the Unknown Compounds?

Purpose

The purpose of this lab is for students to *apply* what they know about the concepts of moles and molar mass to identify unknown substances. This investigation will help students understand how physical properties of chemicals (molar mass) can be used to compare a variety of substances. Through this activity, students will have an opportunity to determine the molar mass for different compounds and use that information to determine the identity of an array of unknown substances. Students will also learn about the importance of identifying patterns in science; the importance of scale, proportion, and quantity; how scientific knowledge changes over time in light of new evidence; and the difference between data and evidence.

The Content

The concept of the *mole* is central to understanding chemistry. The mole provides a measure of the number of atoms present in a sample of a compound. One mole of any element or compound contains 6.02×10^{23} atoms or molecules; this quantity is referred to as the Avogadro constant. Amedeo Avogadro was a professor of physics in Italy in the early 19th century, whose major contribution involved establishing the idea that the volume of gas at a certain temperature and pressure is proportional to the number of particles in that gas. Avogadro, however, did not actually determine the constant named for him. The value of the constant was first indicated by Johann Josef Loschmidt in 1865; subsequently, the French physicist Jean Perrin determined the constant by various methods and proposed naming it after Avogadro.

Knowing the amounts of particles allows chemists to understand how different chemicals behave during chemical reactions and predict the outcomes of reactions. Moles provide a standardized way of comparing elements. Indeed, the mole is considered one of the seven base units of measurement in the SI system. Using the Avogadro constant, chemists can use other measures, such as mass or volume, to determine the amount of particles a sample has.

To use mass to determine the number of moles of an element or molecule in a sample, you must also know the *molar mass* of that element or molecule. The molar mass refers to the total mass of an element present in one mole of that element. The unit for these masses is grams per mole (g/mol). The molar mass of an element is easily identified on most periodic tables, where it is typically listed in the box provided for a particular element. Examples of molar mass include carbon (C), 12.011 g/mol; oxygen (O), 15.994 g/mol; and gold (Au), 196.967 g/mol. To determine the molar mass for a compound made of larger molecules, you must add up the molar masses of all the atoms present in the molecular formula. For example, the molar mass of CO_2 is 43.999 g/mol, which is calculated by adding the mass of 1 mole of carbon (12.011 g/mol) and the mass of 2 moles of oxygen (15.994 g/mol + 15.994 g/mol). The subscripts in the molecular formula must also be considered when determining the molar mass of a compound.

By knowing the molar mass of a compound and the mass of a sample of that compound, you can determine the number of moles in the compound. Continuing from the example above, if you have a sample of CO_2 with a mass of 2.523 g, then you can determine the number of moles in that sample by dividing the actual mass by the molar mass (e.g., 2.523 g / 43.999 g/mol = 0.0573 moles of CO_2). Remember, moles provide a standardized unit of measure (based on the Avogadro constant) so that chemists can compare a wide variety of substances. Using the same example (2.523 g of CO_2 contains 0.0573 moles of CO_2), we can determine the number of molecules by multiplying the number of moles by the Avogadro constant (0.0573 mol × 6.02 × 10^{23} = 3.45 × 10^{22} molecules of CO_2). With CO_2 in its gaseous state, it is possible to use the number of moles to determine the volume that the sample of gas occupies (at standard temperature and pressure [STP]). The molar volume of any gas at STP is 22.4 L for every mole of compound. So there are 1.28 L of CO_2 present in the sample from the example at STP (0.0573 mol × 22.4 L/mol). Figure 14.1 summarizes the different molar conversion calculations used in the example above.

Timeline

The instructional time needed to complete this lab investigation is 130–200 minutes. Appendix 2 (p. 501) provides options for implementing this lab investigation over several class periods. Option C (200 minutes) should be used if students are unfamiliar with scientific writing because this option provides extra instructional time for scaffolding the writing process. You can scaffold the writing process by modeling, providing examples, and providing hints as students write each section of the report. Option D (130 minutes) should be used if students are familiar with scientific writing and have the skills needed to write an investigation report on

FIGURE 14.1

Molar conversions

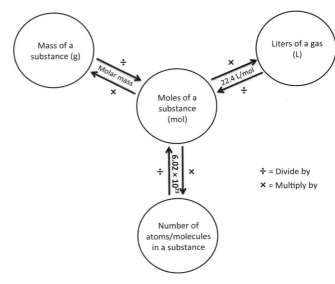

LAB 14

their own. In option D, students complete stage 6 (writing the investigation report) and stage 8 (revising the investigation report) of the investigation outside of class, which reduces the amount of time needed to complete the lab.

Materials and Preparation

The materials needed to implement this investigation are listed in Table 14.1. To make each unknown bag, use the following procedure:

- Weigh out a 5–10 g sample of a compound.
- Add the sample to a plastic bag and seal it.
- Calculate the number of moles of the sample that you added to the bag.
- Use a permanent marker to label the bag with the letter of the unknown and the number of moles of the compound (e.g., unknown A, 0.10 mol).

TABLE 14.1

Materials list

Item	Quantity
Consumables	
Sealed plastic bag of calcium acetate, $Ca(C_2H_3O_2)_2$ (unknown B)	1 per group
Sealed plastic bag of calcium oxide, CaO (unknown F)	1 per group
Sealed plastic bag of potassium sulfate, K_2SO_4 (unknown C)	1 per group
Sealed plastic bag of sodium acetate, $NaC_2H_3O_2$ (unknown E)	1 per group
Sealed plastic bag of sodium carbonate, Na_2CO_3 (unknown G)	1 per group
Sealed plastic bag of sodium chloride, NaCl (unknown A)	1 per group
Sealed plastic bag of zinc(II) oxide, ZnO (unknown D)	1 per group
Empty plastic bags (same type and size as the ones used to make the unknown samples)	5 per class
Equipment and other materials	
Electronic or triple beam balance	1 per group
Periodic table	1 per student
Whiteboard, 2' × 3' *	1 per group
Lab handout	1 per student
Peer-review guide and instructor scoring rubric	1 per student

*As an alternative, students can use computer and presentation software such as Microsoft PowerPoint or Apple Keynote to create their arguments.

Safety Precautions

Remind students to follow all normal lab safety rules. In addition, tell them to take the following safety precautions:

- Wear indirectly vented chemical-splash goggles.
- Wash their hands with soap and water when they are done collecting the data.

Laboratory Waste Disposal

There should be no waste material produced during this investigation. The sealed bags of unknown compounds can be stored for use another time.

Topics for the Explicit and Reflective Discussion

Concepts That Can Be Used to Justify the Evidence

To provide an adequate justification of their evidence, students must explain why they included the evidence in their arguments and make the assumptions underlying their analysis and interpretation of the data explicit. In this investigation, students can use the following concepts to help justify their evidence:

- Mole
- Standardization of comparisons using the Avogadro constant
- Relationship between molar values and physical properties

We recommend that you discuss these fundamental concepts during the explicit and reflective discussion to help students make this connection.

How to Design Better Investigations

It is important for students to reflect on the strengths and weaknesses of the investigation they designed during the explicit and reflective discussion. Students should therefore be encouraged to discuss ways to eliminate potential flaws, measurement errors, or sources of bias in their investigations. To help students be more reflective about the design of their investigation, you can ask the following questions:

- What were some of the strengths of your investigation? What made it scientific?
- What were some of the weaknesses of your investigation? What made it less scientific?
- If you were to do this investigation again, what would you do to address the weaknesses in your investigation? What could you do to make it more scientific?

LAB 14

Crosscutting Concepts

This investigation is well aligned with two crosscutting concepts found in *A Framework for K–12 Science Education,* and you should review these concepts during the explicit and reflective discussion.

- *Patterns:* Scientists look for patterns in nature and attempt to understand the underlying causes of these patterns. Molar values are based on the pattern of molecular amounts inherent in the Avogadro constant; that is, every substance contains 6.02×10^{23} atoms or molecules in one mole. The mole unit is a basic pattern for describing amounts of substances and allows chemists to make many predictions about how substances will react with each other.

- *Scale, proportion, and quantity:* It is critical for scientists to be able to recognize what is relevant at different sizes, time frames, and scales. Scientists must also be able to recognize proportional relationships between categories or quantities. The mole concept and relevant quantities like the Avogadro constant and molar mass are fundamental proportional relationships that help chemists understand the nature of chemical interactions.

The Nature of Science and the Nature of Scientific Inquiry

This investigation is well aligned with two important concepts related to the *nature of science* (NOS) and the *nature of scientific inquiry* (NOSI), and you should review these concepts during the explicit and reflective discussion.

- *Changes in scientific knowledge over time:* A person can have confidence in the validity of scientific knowledge but must also accept that scientific knowledge may be abandoned or modified in light of new evidence or because existing evidence has been reconceptualized by scientists. There are many examples in the history of science of both evolutionary changes (i.e., the slow or gradual refinement of ideas) and revolutionary changes (i.e., the rapid abandonment of a well-established idea) in scientific knowledge.

- *The difference between data and evidence in science:* Data are measurements, observations, and findings from other studies that are collected as part of an investigation. Evidence, in contrast, is analyzed data and an interpretation of the analysis.

Hints for Implementing the Lab

- Keep a record of the identity of compound in each sealed plastic bag (e.g., unknown A is sodium chloride).
- Students may not realize that they will need to know the mass of the plastic bag to determine the mass of the compound in each bag. Do not tell them that they will need to know the mass of the empty bags so they can subtract it from the mass

of the filled bags before the investigation begins—let them figure it out. You will, however, need to have several empty plastic bags available for students to use.

- Students' evidence should include some demonstration of their calculations for molar amounts of compounds.

Topic Connections

Table 14.2 provides an overview of the scientific practices, crosscutting concepts, disciplinary core ideas, and supporting ideas at the heart of this lab investigation. In addition, it lists the NOS and NOSI concepts for the explicit and reflective discussion. Finally, it lists literacy and mathematics skills (*CCSS ELA* and *CCSS Mathematics*) that are addressed during the investigation.

TABLE 14.2

Lab 14 alignment with standards

Scientific practices	• Asking questions and defining problems • Planning and carrying out investigations • Analyzing and interpreting data • Using mathematics and computational thinking • Constructing explanations and designing solutions • Engaging in argument from evidence • Obtaining, evaluating, and communicating information
Crosscutting concepts	• Patterns • Scale, proportion, and quantity
Core idea	• PS1.A: Structure and properties of matter
Supporting ideas	• Mole • Molar mass • Avogadro constant
NOS and NOSI concepts	• Changes in scientific knowledge over time • Difference between data and evidence
Literacy connections (CCSS ELA)	• *Reading*: Key ideas and details, craft and structure, integration of knowledge and ideas • *Writing:* Text types and purposes, production and distribution of writing, research to build and present knowledge, range of writing • *Speaking and listening*: Comprehension and collaboration, presentation of knowledge and ideas
Mathematics connections (CCSS Mathematics)	• Reason abstractly and quantitatively • Construct viable arguments and critique the reasoning of others • Use appropriate tools strategically • Attend to precision • Look for and express regularity in repeated reasoning

LAB 14

Lab 14. Molar Relationships: What Are the Identities of the Unknown Compounds?

Introduction

The concept of the mole is important for understanding chemistry. The mole provides a measure of the number of atoms present in a sample of a compound. One mole of an element or compound contains 6.02×10^{23} atoms or molecules. This quantity is referred to as the Avogadro constant. Knowing the amounts of particles allows chemists to understand how different chemicals behave during chemical reactions and predict the outcomes of reactions. Moles provide a standardized way of comparing elements. Using the Avogadro constant, chemists can use other measures, such as mass or volume, to determine the amount of particles a sample has.

To use mass to determine the number of moles of an element or molecule in a sample, you must also know the molar mass of that element or molecule. The molar mass refers to the total mass of an element present in one mole of that element. The unit for these masses is grams per mole (g/mol). The molar mass of an element is easily identified on most periodic tables, where it is typically listed in the box provided for a particular element. Examples of molar mass include carbon (C), 12.011 g/mol; oxygen (O), 15.994 g/mol; and gold (Au), 196.967 g/mol. To determine the molar mass for a compound made of larger molecules, you must add up the molar masses of all the atoms present in the molecular formula. For example, the molar mass of CO_2 is 43.999 g/mol, which is calculated by 12.011 g/mol (C) + 15.994 g/mol (O) + 15.994 g/mol (O). Remember that you have to include the total number of atoms in the molecular formula when calculating molar mass, so be mindful of the subscripts in those formulas.

By knowing the molar mass of a compound and the mass of a sample of that compound, you can determine the number of moles in the compound. Continuing from the example above, if you have a sample of CO_2 whose mass is 2.523 g, then you can determine the number of moles in that sample by dividing the actual mass by the molar mass (e.g., 2.523 g / 43.999 g/mol = 0.0573 moles of CO_2).

You will now use your understanding of the relationships between moles, molar mass, and mass of a sample to identify some unknown compounds. Remember, moles provide a standardized unit of measure (based on the Avogadro constant) so that chemists can compare a wide variety of substances, including the amount of substances needed and produced by a chemical reaction.

Your Task

You will be given seven sealed bags. Each bag will be filled with a different powder and will be labeled with the number of moles of powder that is inside the bag. Your task will be to identify the powder in each bag. The unidentified powders could be any of the following compounds:

- Calcium acetate, $Ca(C_2H_3O_2)_2$
- Calcium oxide, CaO
- Potassium sulfate, K_2SO_4
- Sodium acetate, $NaC_2H_3O_2$
- Sodium carbonate, Na_2CO_3
- Sodium chloride, NaCl
- Zinc(II) oxide, ZnO

The guiding question of this investigation is, **What are the identities of the unknown compounds?**

Materials

You may use any of the following materials during your investigation:

Consumables	Equipment
• Sealed plastic bags of unknown compounds • Empty plastic bags	• Electronic or triple beam balance • Periodic table

Safety Precautions

Follow all normal lab safety rules. Your teacher will explain relevant and important information about working with the chemicals associated with this investigation. In addition, take the following safety precautions:

- Wear indirectly vented chemical-splash goggles while in the laboratory.
- Wash your hands with soap and water before leaving the laboratory.

Investigation Proposal Required?   ☐ Yes   ☐ No

Getting Started

To answer the guiding question, you will need to design and conduct an investigation. To accomplish this task, you must first determine what type of data you need to collect, how you will collect the data, and how you will analyze the data.

To determine *what type of data you need to collect*, think about what type of measurements you will need to make during your investigation.

To determine *how you will collect the data*, think about the following questions:

- How will you make sure that your data are of high quality (i.e., how will you reduce error)?
- How will you keep track of the data you collect and how will you organize it?

To determine *how you will analyze the data*, think about the following questions:

- What type of table or graph could you create to help make sense of your data?
- What types of calculations will you need to make?

Connections to Crosscutting Concepts, the Nature of Science, and the Nature of Scientific Inquiry

As you work through your investigation, be sure to think about

- the importance of identifying patterns,
- which proportional relationships are critical to the understanding of this investigation,
- how scientific knowledge changes over time in light of new evidence, and
- the difference between data and evidence.

Initial Argument

Once your group has finished collecting and analyzing your data, you will need to develop an initial argument. Your argument must include a *claim*, which is your answer to the guiding question. Your argument must also include *evidence* in support of your claim. The evidence is your analysis of the data and your interpretation of what the analysis means. Finally, you must include a *justification* of the evidence in your argument. You will therefore need to use a scientific concept or principle to explain why the evidence that you decided to use is relevant and important. You will create your initial argument on a whiteboard. Your whiteboard must include all the information shown in Figure L14.1.

FIGURE L14.1 _____

Argument presentation on a whiteboard

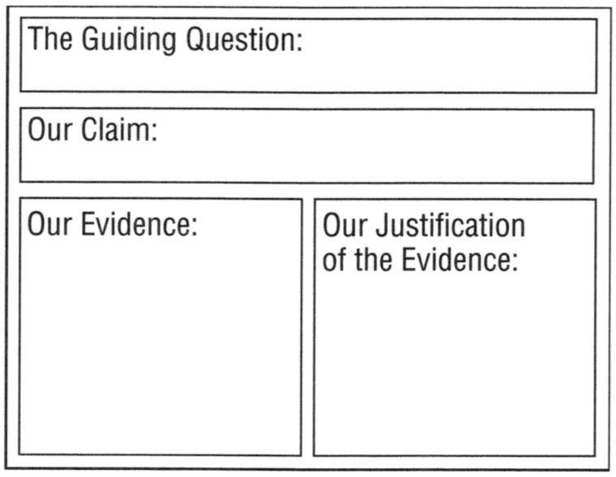

The Guiding Question:	
Our Claim:	
Our Evidence:	Our Justification of the Evidence:

Argumentation Session

The argumentation session allows all of the groups to share their arguments. One member of each group stays at the lab station to share that group's argument, while the other members of the group go to the other lab stations one at a time to listen to and critique the arguments developed by their classmates. The goal of the argumentation session is not to convince others that your argument is the best one; rather, the goal is to identify errors or instances of faulty reasoning in the initial arguments so these mistakes can be fixed. You will therefore need to evaluate the content of the claim, the quality of the evidence used to support the claim, and the strength of the justification of the evidence included in each argument that you see. To critique an argument, you might need more information than what is included on the whiteboard. You might therefore need to ask the presenter one or more follow-up questions, such as:

- How did your group collect the data? Why did you use that method?
- What did your group do to make sure the data you collected are reliable? What did you do to decrease measurement error?
- What did your group do to analyze the data? Did you check your calculations?
- Is that the only way to interpret the results of your group's analysis? How do you know that your interpretation of the analysis is appropriate?
- Why did your group decide to present your evidence in that manner?
- What other claims did your group discuss before deciding on that one? Why did you abandon those alternative ideas?
- How confident are you that your group's claim is valid? What could you do to increase your confidence?

Once the argumentation session is complete, you will have a chance to meet with your group and revise your original argument. Your group might need to gather more data or design a way to test one or more alternative claims as part of this process. Remember, your goal at this stage of the investigation is to develop the most valid or acceptable answer to the research question!

Report

Once you have completed your research, you will need to prepare an *investigation report* that consists of three sections that provide answers to the following questions:

1. What question were you trying to answer and why?
2. What did you do during your investigation and why did you conduct your investigation in this way?
3. What is your argument?

Your report should answer these questions in two pages or less. The report must be typed and any diagrams, figures, or tables should be embedded into the document. Be sure to write in a persuasive style; you are trying to convince others that your claim is acceptable or valid!

LAB 14

Lab 14. Molar Relationships: What Are the Identities of the Unknown Compounds?

1. A 1-mole sample of sugar ($C_6H_{12}O_6$) is 180 grams, but a 1-mole sample of salt (NaCl) is 58 grams. These two samples are equal when comparing the number of moles, but not equal when comparing mass. Describe why this relationship is possible.

2. The following chemical equation describes the chemical reaction of hydrogen gas and oxygen gas to create water.

$$2H_2 + O_2 \rightarrow 2H_2O$$

Use what you know about molar relationships to explain how scientists can predict the amount of water produced if they know the amounts of hydrogen and oxygen gases they have to react.

3. Scientific knowledge is a set body of information that can change over time in light of new evidence.

   a. I agree with this statement.

   b. I disagree with this statement.

Explain your answer, using an example from your investigation about molar relationships.

4. The terms *data* and *evidence* do not have the same meaning in science.

    a. I agree with this statement.

    b. I disagree with this statement.

    Explain your answer, using an example from your investigation about molar relationships.

5. Scientists often look for and attempt to explain patterns in nature. Explain why patterns are important, using an example from your investigation about molar relationships.

6. Scientists often need to look for proportional relationships. Explain why looking for proportional relationships is often useful in science, using an example from your investigation about molar relationships.

Teacher Notes

Lab 15. The Ideal Gas Law: How Can a Value of *R* for the Ideal Gas Law Be Accurately Determined Inside the Laboratory?

Purpose

The purpose of this lab is for students to *apply* their understanding of the ideal gas law to complete a design challenge. This lab gives students an opportunity to design a method that they can use to determine the ideal gas constant, *R*, and then assess how well the method works. Students will also learn about determining cause and effect, the use of models to understand complex phenomena, the difference between scientific laws and theories, and the nature and role of experiments in science.

The Content

The *ideal gas law* brings together four basic gas laws from chemistry that describe the behavior of gases in certain conditions:

- *Boyle's law:* The pressure of a fixed amount of a gas at a constant temperature is inversely proportional to the volume that gas occupies. (Pressure decreases as volume increases, and vice versa).

- *Charles' law:* The volume of a fixed amount of gas at a constant pressure is directly proportional to its absolute temperature (*K*). (As temperature increases, so does the volume, and vice versa).

- *Gay-Lussac's law:* The pressure of a fixed amount of gas at a constant volume is directly proportional to its absolute temperature (*K*). (As temperature increases, so does the pressure, and vice versa).

- *Avogadro's law:* The volume of a gas at a fixed temperature and pressure is directly proportional to the molar amount of the gas. (As the moles of gas increase, so does the volume, and vice versa).

The ideal gas law combines these four simple laws to describe the combined relationship among the pressure, volume, temperature, and number of moles of gas. Many common gases exhibit behavior very close to that of an ideal gas at ambient temperature and pressure. The ideal gas law provides chemists with a powerful predictive tool that helps them understand how gases will react in different systems. The ideal gas law is expressed mathematically as $PV = nRT$ where P is pressure in atmospheres (atm); V is volume in liters (L); n is the number of moles of gas (mol); and T is absolute temperature in Kelvin (K). The remaining component of the ideal gas law is R, which is called the ideal gas constant.

The ideal gas constant helps connect the four other components of the ideal gas law. The purpose that R serves in the equation is to help correct discrepancies in the different scales of measure inherent in the other variables (the pressure-volume scale compared with the absolute temperature scale). The standard, theoretical value of R is 0.0821 L•atm/mol•K or 8.314 L•kPa/mol•K (depending on the units that are used to measure pressure). Note that the unit designation for R relates to all four of the other equation components. R can also be expressed using units of energy ($R = 8.314$ J/mol•K). Indeed, the many different values of R are all equivalent, but their differences stem from different collections of units used to describe the value. Again, this characteristic goes back to the function of R, which is to help correct for fundamental differences among very different scales of measurement. As chemists were researching and developing these laws about the behavior of gases, they did not automatically know the value of R; it had to be determined experimentally. As a result, different values can be determined based on the variables used in the experiments.

Experiments in science are investigations that involve controlling (or keeping constant) multiple variables at the same time to determine the effect of changing one variable (the independent variable) on the outcome of another variable (the dependent variable). Controlled experiments are a critical method for the development of scientific knowledge. These types of investigations help scientists to understand how certain factors cause certain effects. By controlling as many other variables as possible (keeping them the same in each trial or setup), scientists can isolate the impact that changes to the independent variable have on the outcome value of the dependent variable. Understanding how to design a controlled experiment is critical to learning and understanding science. In this investigation, the students will need to design an experiment to determine the value of R. It is important to note that R is not the dependent variable being measured here. You will need to make sure that students understand that aspect. Two experimental methods are suggested: a volume method and a pressure method (see Figures 15.1 and 15.2, p. 232). In each approach, the independent variable is the number of moles of H_2 gas produced because students can use different amounts of magnesium metal to react with hydrochloric acid (HCl). The balanced chemical equation for this reaction is provided in the student handout. Depending on the method used by the students, either the gas's pressure or the volume it occupies will be the dependent variable that is measured. Once students have designed their setup, they should run their experiment several times. They can use the data they collect to calculate a value for R. Then they can assess the accuracy of their R value through percent error calculations. The students will then need to refine their method in order to reduce the percent error as much as possible. Once they have a method that functions as intended, they will need to conduct a formal evaluation of it. The students should then share their findings from the formal evaluation during the argumentation session. This will provide you with an opportunity to discuss the importance of iterative design in the development of new methods in science.

LAB 15

FIGURE 15.1

The volume method to determine the value of R

Procedure for volume method	Equipment setup for volume method
<ul><li>Set up materials as shown in the figure to the right.</li><li>Obtain a piece of magnesium ribbon.</li><li>Place the magnesium ribbon in the flask.</li><li>Fill a graduated cylinder to the rim with water.</li><li>Place a piece of paper over the rim of the cylinder.</li><li>Carefully invert the cylinder into a water-filled pneumatic trough (make sure no water is spilled out of the cylinder).</li><li>Insert the gas delivery tube into the cylinder.</li><li>Carefully pour 10 ml of 6 M HCl (2.19 grams of HCl) into the flask and quickly seal the flask with a rubber stopper.</li><li>Capture gas in an inverted graduated cylinder.</li></ul>	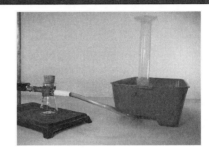

FIGURE 15.2

The pressure method to determine the value of R

Procedure for pressure method	Equipment setup for pressure method
<ul><li>Set up materials as shown in the figure to the right.</li><li>Carefully pour 10 ml of 6 M HCl (2.19 grams of HCl) into the Erlenmeyer flask.</li><li>Place the rubber stopper with the gas pressure probe in the neck of the Erlenmeyer flask.</li><li>Obtain a baseline pressure of gas inside of the flask and then remove the stopper.</li><li>Obtain a piece of magnesium ribbon.</li><li>Place the magnesium ribbon inside the flask and quickly replace the rubber stopper, and observe the change in pressure.</li></ul>	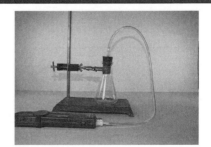

Timeline

The instructional time needed to complete this lab investigation is 200–300 minutes. Appendix 2 (p. 501) provides options for implementing this lab investigation over several class periods. Option A (250 minutes) should be used if students are unfamiliar with scientific writing because this option provides extra instructional time for scaffolding the writing process. You can scaffold the writing process by modeling, providing examples, and providing hints as students write each section of the report. Option B (200 minutes) should be used if students are familiar with scientific writing and have the skills needed to write an investigation report on their own. In option B, students complete stage 6 (writing the investigation report) and stage 8 (revising the investigation report) as homework. It is assumed that stage 7 (peer-review session) in option B will take an entire 50 minutes, rather than the 30 minutes listed in the appendix. Also, stage 2 (design a method and collect data) may take longer than two class sessions, which could require another 50 minutes, leading to a total of 300 minutes. However, this extra time will allow for students to deeply engage in experimental design and evaluation.

Materials and Preparation

The materials needed to implement this investigation are listed in Table 15.1 (p. 234). The consumables and equipment can be purchased from a science supply company such as Carolina, Flinn Scientific, or Ward's Science. We recommend that you use a set routine for distributing and collecting the materials during the lab investigation. For example, the consumables and equipment for each group can be set up at each lab station before class begins, or one member from each group can collect them from a table or a cart when needed during class.

Safety Precautions

Remind students to follow all normal lab safety rules. Hydrochloric acid is corrosive to eyes, skin, and other body tissues. You will therefore need to explain the potential hazards of working with hydrochloric acid and how to work with hazardous chemicals. In addition, tell students to take the following precautions:

- Wear indirectly vented chemical-splash goggles and chemical-resistant gloves and aprons when they are collecting their data.
- Handle all glassware with care.
- Wash their hands with soap and water when they are done collecting the data.

Laboratory Waste Disposal

We recommend following Flinn laboratory waste disposal methods 3 to dispose of the magnesium ribbon and 24b to dispose of the acid solutions. Information about laboratory

LAB 15

TABLE 15.1

Materials list

Item	Quantity
Consumables	
6 M solution of HCl	50 ml per group
Magnesium ribbon	10 g per group
Equipment and other materials	
Side-arm Erlenmeyer flask with stopper, 50 ml	1 per group
Pneumatic trough	1 per group
Plastic or rubber tubing, 50 cm long	1 per group
Electronic or triple beam balance	1 per group
Graduated cylinder, 50 ml	1 per group
Graduated cylinder, 100 ml	1 per group
Graduated cylinder, 250 ml	1 per group
Graduated cylinder, 500 ml	1 per group
Glass test tube	1 per group
Utility clamp	1 per group
Ring stand	1 per group
Gas pressure sensor	1 per group
Temperature probe	1 per group
Sensor interface	1 per group
Investigation Proposal C (optional)	1 per group
Whiteboard, 2' × 3' *	1 per group
Lab handout	1 per student
Peer-review guide and instructor scoring rubric	1 per student

*As an alternative, students can use computer and presentation software such as Microsoft PowerPoint or Apple Keynote to create their arguments.

waste disposal methods is included in the Flinn Catalog and Reference Manual; you can request a free copy at *www.flinnsci.com*.

Topics for the Explicit and Reflective Discussion

Concepts That Can Be Used to Justify the Evidence

To provide an adequate justification of their evidence, students must explain why they included the evidence in their arguments and make the assumptions underlying their analysis and interpretation of the data explicit. In this investigation, students can use the following concepts to help justify their evidence:

- Characteristics of gases (pressure, temperature, volume, and amount)
- The ideal gas law
- The importance of accuracy, precision, and repeatability in science

We recommend that you discuss these fundamental concepts during the explicit and reflective discussion to help students make this connection.

How to Design Better Investigations

It is important for students to reflect on the strengths and weaknesses of the investigation they designed during the explicit and reflective discussion. Students should therefore be encouraged to discuss ways to eliminate potential flaws, measurement errors, or sources of bias in their investigations. To help students be more reflective about the design of their investigations, you can ask the following questions:

- What were some of the strengths of your investigation? What made it scientific?
- What were some of the weaknesses of your investigation? What made it less scientific?
- If you were to do this investigation again, what would you do to address the weaknesses in your investigation? What could you do to make it more scientific?

Crosscutting Concepts

This investigation is well aligned with two crosscutting concepts found in *A Framework for K–12 Science Education,* and you should review these concepts during the explicit and reflective discussion.

- *Cause and effect: Mechanism and explanation:* One of the main objectives of science is to identify and establish relationships between a cause and an effect.
- *Systems and system models:* Scientists often need to use models to understand complex phenomena. In this investigation, the students are directed to develop

LAB 15

a model to help explain what is happening during a chemical reaction at the submicroscopic level.

The Nature of Science and the Nature of Scientific Inquiry

This investigation is well aligned with two important concepts related to the *nature of science* (NOS) and the *nature of scientific inquiry* (NOSI), and you should review these concepts during the explicit and reflective discussion.

- *The difference between laws and theories in science:* A scientific law describes the behavior of a natural phenomenon or a generalized relationship under certain conditions; a scientific theory is a well-substantiated explanation of some aspect of the natural world. Theories do not become laws even with additional evidence; they explain laws. However, not all scientific laws have an accompanying explanatory theory. It is also important for students to understand that scientists do not discover laws or theories; the scientific community develops them over time.

- *The nature and role of experiments.* Scientists use experiments to test the validity of a hypothesis (i.e., a tentative explanation) for an observed phenomenon. Experiments include a test and the formulation of predictions (expected results) if the test is conducted and the hypothesis is valid. The experiment is then carried out and the predictions are compared with the observed results of the experiment. If the predictions match the observed results, then the hypothesis is supported. If the observed results do not match the prediction, then the hypothesis is not supported. A signature feature of an experiment is the control of variables to help eliminate alternative explanations for observed results.

Hints for Implementing the Lab

- Allowing students to design their own procedures for collecting data gives students an opportunity to try, to fail, and to learn from their mistakes. However, you can scaffold students as they develop their procedure by having them fill out an investigation proposal. These proposals provide a way for you to offer students hints and suggestions without telling them how to do it. You can also check the proposals quickly during a class period.

- Allow students to play around with the size of the graduated cylinder that they will need to collect gas. If you do not have large enough graduated cylinders to capture all the gas that is produced, encourage students to think of ways to decrease the amount of gas produced without changing the mole ratios of the reactants.

- Reliability and repeatability are important goals for scientific experiments, so emphasize to your students the need for multiple trials. They will also want to use the data from those trials to calculate the R value for each trial.

Topic Connections

Table 15.2 provides an overview of the scientific practices, crosscutting concepts, disciplinary core ideas, and supporting ideas at the heart of this lab investigation. In addition, it lists the NOS and NOSI concepts for the explicit and reflective discussion. Finally, it lists literacy and mathematics skills (*CCSS ELA* and *CCSS Mathematics*) that are addressed during the investigation.

TABLE 15.2

Lab 15 alignment with standards

Scientific practices	• Asking questions and defining problems • Developing and using models • Planning and carrying out investigations • Analyzing and interpreting data • Using mathematics and computational thinking • Constructing explanations and designing solutions • Engaging in argument from evidence • Obtaining, evaluating, and communicating information
Crosscutting concepts	• Cause and effect: Mechanism and explanation • Systems and system models
Core idea	• PS1.A: Structure and properties of matter
Supporting ideas	• Characteristics of gases • Ideal gas law • Accuracy, precision, and repeatability
NOS and NOSI concepts	• Scientific laws and theories • Nature and role of experiments
Literacy connections (*CCSS ELA*)	• *Reading:* Key ideas and details, craft and structure, integration of knowledge and ideas • *Writing:* Text types and purposes, production and distribution of writing, research to build and present knowledge, range of writing • *Speaking and listening:* Comprehension and collaboration, presentation of knowledge and ideas
Mathematics connections (*CCSS Mathematics*)	• Make sense of problems and persevere in solving them • Reason abstractly and quantitatively • Model with mathematics • Attend to precision

Lab Handout

Lab 15. The Ideal Gas Law: How Can a Value of *R* for the Ideal Gas Law Be Accurately Determined Inside the Laboratory?

Introduction

A *gas* is the state of matter that is characterized by having neither a fixed shape nor a fixed volume. Gases exert pressure, are compressible, have low densities, and diffuse rapidly when mixed with other gases. On a submicroscopic level, the molecules in a gas are separated by large distances and are in constant, random motion. A gas can be described using four measurable properties: pressure (P), defined as the force exerted by a gas per unit area; volume (V), defined as the quantity of space a gas occupies; temperature (T), defined as the average kinetic energy of the molecules that make up a gas; and the number of moles of gas (n). The relationships among these properties are summarized by the gas laws, as shown in Table L15.1.

TABLE L15.1 _____

The gas laws

Gas law	Relationship	Equation
Boyle's law	$V \propto 1/P$ (*T* and *n* are held constant). As gas pressure increases, gas volume decreases.	$P_1V_1 = P_2V_2$
Charles' law	$V \propto T$ (*P* and *n* are held constant). As gas temperature increases, gas volume increases.	$V_1/T_1 = V_2/T_2$
Gay-Lussac's law	$P \propto T$ (*V* and *n* are held constant). As gas pressure increases, gas temperature increases.	$P_1/T_1 = P_2/T_2$
Avogadro's law	$V \propto n$ (*P* and *T* are held constant). As the number of moles of gas increase, gas volume increases.	$V_1/n_1 = V_2/n_2$
Combined law	$V \propto T/P$ (*n* is held constant); obtained by combining Boyle's law, Charles' law, and Gay-Lussac's law.	$(P_1V_1)/T_1 = (P_2V_2)/T_2$

The ideal gas law combines Boyle's law, Charles' law, Gay-Lussac's law, and Avogadro's law to describe the relationship among the pressure, volume, temperature, and number of moles of gas. Émile Clapeyron is often given the credit for developing this law. The ideal gas law provides chemists with a powerful predictive tool that helps them understand how gases will react in different systems. The ideal gas law is expressed mathematically as $PV = nRT$. P is pressure in atmospheres (atm); V is volume in liters (L); n is the number of moles of gas (mol); and T is absolute temperature in Kelvin (K). The remaining component of the ideal gas law is R, which is called the ideal gas constant. The theoretical value of R that is often reported in textbooks and handbooks is 0.0821 L•atm/mol•K or 8.314 L•kPa/mol•K.

As chemists worked to determine an exact value for R in the mid-1800s, they were faced with numerous challenges. First, they had to develop a method that they could use to generate the experimental data they needed to calculate a value for R from the ideal gas law. Second, they needed to improve the precision of their measurements. The French chemist Henri Victor Regnault was able to overcome many of these challenges and generate some of the most precise experimental data about the properties of gases at that time. Rudolf Clausius, a German physicist, then used Regnault's data to calculate the earliest published value for R. This value for R, however, was not very precise by current standards. Fortunately, there have been numerous advancements in the methods and tools that chemists use to measure the properties of gases, and the value of R has become increasingly precise over time. There is, however, always room for improvement. In this investigation, you will have an opportunity to follow in the footsteps of Clapeyron, Regnault, and Clausius by designing, conducting trials of, refining, and then evaluating a method that can be used to calculate a precise value for the ideal gas constant.

Your Task

Design a method that can be used to calculate an accurate value of R inside the lab by generating a specific number of moles of gas at room temperature and then measuring the pressure or volume of the gas. As part of this process you will need to test, evaluate, and then refine your method. Your method should allow you to produce a consistent and accurate value for R.

The guiding question of this investigation is, **How can a value of R for the ideal gas law be accurately determined inside the laboratory?**

Materials

You may use any of the following materials during your investigation:

Consumables	Equipment
• 6 M hydrochloric acid (HCl) solution • Magnesium (Mg) ribbon	• Side-arm Erlenmeyer flask with stopper (50 ml) • Pneumatic trough • Plastic or rubber tubing (50 cm long) • Electronic or triple beam balance • Graduated cylinders (one each 50 ml, 100 ml, 250 ml, and 500 ml) • Glass test tube • Utility clamp • Ring stand • Gas pressure sensor • Temperature probe • Sensor interface

Safety Precautions

Follow all normal lab safety rules. Hydrochloric acid is corrosive to eyes, skin, and other body tissues. Your teacher will explain relevant and important information about working with the chemicals associated with this investigation. In addition, take the following safety precautions:

- Wear indirectly vented chemical-splash goggles and chemical-resistant gloves and apron while in the laboratory.
- Handle all glassware with care.
- Wash your hands with soap and water before leaving the laboratory.

Investigation Proposal Required?   ☐ Yes   ☐ No

Getting Started

In this lab, you will react magnesium metal with hydrochloric acid to produce a sample of hydrogen gas. The hydrogen gas produced by this reaction behaves mostly like an ideal gas. The equation for this chemical reaction is

$$Mg + 2HCl \rightarrow MgCl_2 + H_2$$

The first step in your investigation is to design a method that will allow you to obtain the pressure, volume, temperature, and number of moles of a sample of hydrogen so you can use these data to calculate the gas constant (R). There are several approaches that you can use. One approach is to produce a specific amount of gas (in moles) and then measure the pressure of that gas while holding the volume and temperature constant. A second approach is to produce a specific amount of gas (in moles) and then measure its volume while holding pressure and temperature constant. It is important for you to consider how you might be able to measure the various properties of a sample of hydrogen gas using the equipment available. As you design your method, you should also think about the following questions:

- What type of measurements or observations will you need to record during your investigation?
- How often will you collect data and when will you do it?
- How will you make sure that your data are of high quality (i.e., how will you reduce error)?
- How will you keep track of the data you collect and how will you organize it?
- How will you determine if there is a difference between the two methods?
- What type of calculations will you need to make?

The second step in your investigation is to test and refine your method. To do this, use your method to obtain information about a sample of hydrogen gas and then use this information to calculate a value for *R*. The value that you calculate will likely be rather imprecise due to flaws in your method or poor measurements. Use what you have learned from your initial test to refine your method. Once you have refined your method, you will need to test it again. You should continue this process of testing and refining until it functions as intended. Your method will therefore go through numerous iterations.

The last step in this investigation will be to conduct a formal evaluation of your method. As part of the evaluation of your method, you will need to determine if you can use it to produce a consistent and accurate value for *R*. You will therefore need to determine what data you need to collect and how you will analyze it as part of the evaluation to show that your method works.

Connections to Crosscutting Concepts, the Nature of Science, and the Nature of Scientific Inquiry

As you work through your investigation, be sure to think about

- the importance of identifying causal relationships in science,
- how scientists develop and use system models to understand complex phenomena,
- the difference between laws and theories, and
- the nature and role of experiments in science.

Initial Argument

Once your group has finished collecting and analyzing your data, you will need to develop an initial argument. Your argument must include a *claim*, which is your answer to the guiding question. Your argument must also include *evidence* in support of your claim. The

LAB 15

Argument presentation on a whiteboard

The Guiding Question:	
Our Claim:	
Our Evidence:	Our Justification of the Evidence:

evidence is your analysis of the data and your interpretation of what the analysis means. Finally, you must include a *justification* of the evidence in your argument. You will therefore need to use a scientific concept or principle to explain why the evidence that you decided to use is relevant and important. You will create your initial argument on a whiteboard. Your whiteboard must include all the information shown in Figure L15.1.

Argumentation Session

The argumentation session allows all of the groups to share their arguments. One member of each group stays at the lab station to share that group's argument, while the other members of the group go to the other lab stations one at a time to listen to and critique the arguments developed by their classmates. The goal of the argumentation session is not to convince others that your argument is the best one; rather, the goal is to identify errors or instances of faulty reasoning in the initial arguments so these mistakes can be fixed. You will therefore need to evaluate the content of the claim, the quality of the evidence used to support the claim, and the strength of the justification of the evidence included in each argument that you see. To critique an argument, you might need more information than what is included on the whiteboard. You might therefore need to ask the presenter one or more follow-up questions, such as:

- How did your group collect the data? Why did you use that method?
- What did your group do to make sure the data you collected are reliable? What did you do to decrease measurement error?
- What did your group do to analyze the data, and why did you decide to do it that way? Did you check your calculations?
- Is that the only way to interpret the results of your group's analysis? How do you know that your interpretation of the analysis is appropriate?
- Why did your group decide to present your evidence in that manner?
- What other claims did your group discuss before deciding on that one? Why did you abandon those alternative ideas?
- How confident are you that your group's claim is valid? What could you do to increase your confidence?

Once the argumentation session is complete, you will have a chance to meet with your group and revise your original argument. Your group might need to gather more data or design a way to test one or more alternative claims as part of this process. Remember, your goal at this stage of the investigation is to develop the most valid or acceptable answer to the research question!

Report

Once you have completed your research, you will need to prepare an *investigation report* that consists of three sections. Each section should provide an answer for the following questions:

1. What question were you trying to answer and why?

2. What did you do during your investigation and why did you conduct your investigation in this way?

3. What is your argument?

Your report should answer these questions in two pages or less. The report must be typed and any diagrams, figures, or tables should be embedded into the document. Be sure to write in a persuasive style; you are trying to convince others that your claim is acceptable or valid!

LAB 15

Lab 15. The Ideal Gas Law: How Can a Value of *R* for the Ideal Gas Law Be Accurately Determined Inside the Laboratory?

1. Describe the relationship between the pressure, volume, temperature, and number of moles of a gas.

2. One day when Jessica was at the grocery store she noticed two types of birthday balloons she could buy for her friend's upcoming party. One type of balloon was made out of latex, which is a type of rubber, and the other was made out of a substance called Mylar, which is like a very thin aluminum foil. Jessica bought one balloon of each type and had the store clerk fill them both with the same amount of helium. When she released the balloons in her house they both floated all the way to the ceiling, but the next morning only the Mylar balloon was still at the ceiling while the latex balloon was midway between the ceiling and the floor.

   Use what you know about the ideal gas law to explain what conditions must have been present for the balloons to behave in this manner.

3. Measuring the changes in volume in response to changes in temperature to test if they are directly related is an example of an experiment.

    a. I agree with this statement.

    b. I disagree with this statement.

    Explain your answer, using an example from your investigation about ideal gas law.

4. Laws are theories that have been developed and refined over time.

    a. I agree with this statement.

    b. I disagree with this statement.

    Explain your answer, using an example from your investigation about the ideal gas law.

5. An important goal in science is to develop causal explanations for observations. Explain what a casual explanation is and why these explanations are important, using an example from your investigation about the ideal gas law.

LAB 15

6. Scientists often use models to help them understand natural phenomena. Explain what a model is and why models are important, using an example from your investigation about the ideal gas law.

7. Theories can become laws over time.

   a. I agree with this statement.
   b. I disagree with this statement.

   Explain your answer, using an example from your investigation about bond character and molecular polarity.

8. Scientists often use models to help them understand natural phenomena. Explain what a model is and why models are important, using an example from your investigation about bond character and molecular polarity.

9. Scientists often look for and attempt to explain patterns in nature. Explain why patterns are important, using an example from your investigation about bond character and molecular polarity.

10. In nature, the structure of an object is often related to the function or properties of that object. Explain why this is true, using an example from your investigation about bond character and molecular polarity.

SECTION 3
Physical Sciences
Core Idea 1.B

Chemical Reactions

Introduction Labs

LAB 16

Teacher Notes

Lab 16. Development of a Reaction Matrix: What Are the Identities of the Unknown Chemicals?

Purpose

The purpose of this lab is to *introduce* students to the idea of chemical reactions and how they can be used to identify unknown substances. Completing a series of reactions and carefully observing the results will allow students to identify a set of unknown solutions based on their interactions with other substances within the group. Students will also learn about the nature and role of experiments and the role that observations and inferences play in science.

The Content

Physical and chemical properties of matter are often used as a means to identify unknown substances. Chemists can observe and compare *physical properties* of known and unknown substances, such as boiling point or density; likewise, it is possible to compare *chemical properties* such as how an unknown reacts with a known substance. Comparing the physical properties of pure substances can be useful, but such an approach proves problematic for nonpure substances like solutions. Two different solutions may have the same density or even the same boiling point; in such a case those properties would not be helpful for differentiating between the two substances. In this situation, it may be more helpful to use observations of chemical reactions involving known substances and the unknown substance to determine what is in the solution.

There are two key principles in chemistry that allow this approach to be useful: the law of conservation of mass, which states that matter is neither created nor destroyed during chemical processes; and the *law of definite proportions*, which states that samples of a compound will always contain the same proportion of elements by mass. When these two laws are applied together, in the context of chemical reactions, it is possible to predict the outcome of reactions as well as be confident that a particular reaction will consistently result in the same products.

For example, consider a double replacement reaction between a solution of sodium chloride (NaCl) and a solution of silver nitrate ($AgNO_3$). The balanced chemical equation for this reaction is shown below. When the two aqueous solutions are mixed together, the Ag^+ ions and the Cl^- ions combine to form the insoluble ionic compound silver chloride (AgCl). The remaining Na^+ ions and NO_3^- ions form an aqueous solution of sodium nitrate ($NaNO_3$).

$$NaCl(aq) + AgNO_3(aq) \rightarrow AgCl(s) + NaNO_3(aq)$$

This reaction is predictable based on the notions that no mass can be lost, we must have the same numbers of total atoms before and after the reaction, and the ions in the compounds will arrange in consistent proportions (e.g., 1 Ag^+:1 Cl^-). Additionally, if we consider the role of ion charges in this example, we know that the ions will not rearrange to form a sodium-silver compound because both ions have a positive charge. In the case of ionic compounds it is necessary for the sum of the positive and negative changes to equal zero, resulting in a neutral substance; likewise, there is no chlorine-nitrate compound formed because both ions carry a negative charge.

It is important to note, however, that it is not always possible to determine the identity of a set of reactants by observing the products of a chemical reaction without a reference point. If you mix two clear solutions and they form a white precipitate, you cannot assume that the solutions were sodium chloride and silver nitrate as in the earlier example; there are many combinations of clear solutions that produce a white precipitate when mixed. However, you can be certain that whenever these two solutions are mixed, the same white precipitate will be produced. Therefore, you can use a *reaction matrix* to identify the reactants. A reaction matrix is simply a table of possible reactions and their results. If you know how specific known chemicals react with each other, then it is possible to identify unknown chemicals by comparison.

Timeline

The instructional time needed to complete this lab investigation is 130–180 minutes. Appendix 2 (p. 501) provides options for implementing this lab investigation over several class periods. Option H (180 minutes) incorporates additional time on investigation day 2 to complete data collection during stage 2. The data collection process is not extensive for this investigation; however, organization is essential, as are quality observations, so additional time during stage 2 may be necessary. Option D (130 minutes) provides a timeline that is appropriate for students who are able to work quickly and maintain organization. Each of these options assumes that the students have experience with the scientific writing process and have the skills needed to write an investigation report on their own. In options D and H, students complete stage 6 (writing the investigation report) and stage 8 (revising the investigation report) as homework.

Materials and Preparation

The materials needed to implement this investigation are listed in Table 16.1 (p. 255). Preparation for this investigation largely involves preparation of the solutions students will use to generate their reaction matrix.

Table 16.1 provides a list of several common salts that can be used in this investigation; you can choose the six that you prefer or perhaps already have available. If you are unsure if a reaction will occur and form a precipitate, consult the general solubility rules in

Table 16.2 (p. 256). It would also be appropriate to include combinations that result in no precipitate formation due to both potential products being soluble in water.

It is important that the six solutions you choose for the known solutions be the same as the six unknown solutions that the students will identify using the data they collect when generating their reaction matrix. The known set of solutions should be labeled with their actual contents, but the unknown solutions should simply be labeled with the letters A–F. The concentration of each solution is not critical; concentrations between 0.1 M and 1.0 M are sufficient. It is important, however, that the concentration of the known solutions and the unknown solutions be the same.

We recommend putting the solutions in small dropper bottles so that each lab group may have their own set of knowns and unknowns. Using dropper bottles cuts down on waste from disposable pipettes and reduces the likelihood of contamination due to using pipettes and stock solutions for the class.

Safety Precautions

Remind students to follow all normal lab safety rules. Some of the chemicals may stain skin and clothing. You will therefore need to explain the potential hazards of working with these chemicals and how to work with hazardous chemicals. In addition, tell students to take the following safety precautions:

- Wear indirectly vented chemical-splash goggles and chemical-resistant gloves and aprons when they are collecting their data.
- Handle all glassware with care.
- Wash their hands with soap and water when they are done collecting the data.

Laboratory Waste Disposal

Waste disposal will vary depending on which chemicals you choose to use during this investigation. Some of the substances listed in Table 16.1 and potential products generated during the reactions should not be disposed of down the drain. Given the small amounts of solution the students will use, it is best to simply dump the waste from the well plates onto paper towels and discard the towels as solid waste. Clean the well plates with damp paper towels or cotton swabs and dispose of the towels or swabs as solid waste. For additional disposal methods and information, consult the Flinn Catalog and Reference Manual (available at *www.flinnsci.com*). Retain the stock solutions for use when the investigation is repeated.

TABLE 16.1

Materials list

Item	Quantity
Consumables (potential solutions, choose six)	
Aluminum nitrate, $Al(NO_3)_3$	2–3 ml of solution per group
Barium nitrate, $Ba(NO_3)_2$	2–3 ml of solution per group
Calcium chloride, $CaCl_2$	2–3 ml of solution per group
Calcium nitrate, $Ca(NO_3)_2$	2–3 ml of solution per group
Copper(II) nitrate, $Cu(NO_3)_2$	2–3 ml of solution per group
Iron(III) nitrate, $Fe(NO_3)_3$	2–3 ml of solution per group
Nickel(II) chloride, $NiCl_2$	2–3 ml of solution per group
Silver nitrate, $AgNO_3$	2–3 ml of solution per group
Sodium chloride, $NaCl$	2–3 ml of solution per group
Sodium chromate, Na_2CrO_4	2–3 ml of solution per group
Sodium hydroxide, $NaOH$	2–3 ml of solution per group
Sodium nitrate, $NaNO_3$	2–3 ml of solution per group
Sodium sulfate, Na_2SO_4	2–3 ml of solution per group
Trisodium phosphate, Na_3PO_4	2–3 ml of solution per group
Zinc nitrate, $Zn(NO_3)_2$	2–3 ml of solution per group
Equipment and other materials	
Well plates*	1 per group
Toothpicks	As needed
Dropper bottles for known solutions[†]	6 per group
Dropper bottles for unknown solutions (labeled A–F)[†]	6 per group
Whiteboard, 2' x 3' [‡]	1 per group
Lab handout	1 per student
Peer-review guide and instructor scoring rubric	1 per student

* It is recommended that you use well plates, but test tubes can be used as a substitute for well plates; if using test tubes, you will need 6 per group.

[†] It is recommended that you use dropper bottles for known and unknown solutions, but pipettes can be used as a substitute for dropper bottles; if using pipettes, you will need 12 per group.

[‡] As an alternative, students can use computer and presentation software such as Microsoft PowerPoint or Apple Keynote to create their arguments.

LAB 16

TABLE 16.2

Solubility rules for ionic compounds in water

Ion	Soluble?	Exceptions
NO_3^-	Yes	None
ClO_4^-	Yes	None
Cl^-	Yes	Ag^+, Hg_2^{2+}, Pb^{2+}
I^-	Yes	Ag^+, Hg_2^{2+}, Pb^{2+}
SO_4^{2-}	Yes	Ca^{2+}, Ba^{2+}, Sr^{2+}, Ag^+, Hg^{2+}, Pb^{2+}
CO_3^{2-}	No	Group IA and NH_4^+
PO_4^{3-}	No	Group IA and NH_4^+
OH^-	No	Group IA, Ca^{2+} (slightly soluble), Ba^{2+}, Sr^{2+}
S^{2-}	No	Groups IA and IIA and NH_4^+
Na^+	Yes	None
NH_4^+	Yes	None
K^+	Yes	None

Topics for the Explicit and Reflective Discussion

Concepts That Can Be Used to Justify the Evidence

To provide an adequate justification of their evidence, students must explain why they included the evidence in their arguments and make the assumptions underlying their analysis and interpretation of the data explicit. In this investigation, students can use the following concepts to help justify their evidence:

- Solutions, solvents, solutes
- Chemical reactions
- Ions and ionic compounds

We recommend that you discuss these fundamental concepts during the explicit and reflective discussion to help students make this connection.

How to Design Better Investigations

It is important for students to reflect on the strengths and weaknesses of the investigation they designed during the explicit and reflective discussion. Students should therefore be encouraged to discuss ways to eliminate potential flaws, measurement errors, or sources of bias in their investigations. To help students be more reflective about the design of their investigation, you can ask the following questions:

- What were some of the strengths of your investigation? What made it scientific?

- What were some of the weaknesses of your investigation? What made it less scientific?

- If you were to do this investigation again, what would you do to address the weaknesses in your investigation? What could you do to make it more scientific?

Crosscutting Concepts

This investigation is well aligned with one crosscutting concept found in *A Framework for K–12 Science Education,* and you should review this concept during the explicit and reflective discussion.

Patterns: Observed patterns in nature, such as how chemicals interact with other chemicals, guide the way scientists organize and classify chemical properties and reactions. Scientists also explore the relationships between and the underlying causes of the patterns they observe in nature.

The Nature of Science and the Nature of Scientific Inquiry

This investigation is well aligned with two important concepts related to the *nature of science* (NOS) and the *nature of scientific inquiry* (NOSI), and you should review these concepts during the explicit and reflective discussion.

- *The difference between observations and inferences*: An observation is a descriptive statement about a natural phenomenon, whereas an inference is an interpretation of an observation. Students should also understand that current scientific knowledge and the perspectives of individual scientists guide both observations and inferences. Thus, different scientists can have different but equally valid interpretations of the same observations due to differences in their perspectives and background knowledge.

- *The nature and role of experiments:* Scientists use experiments to test the validity of a hypothesis (i.e., a tentative explanation) for an observed phenomenon. Experiments include a test and the formulation of predictions (expected results) if the test is conducted and the hypothesis is valid. The experiment is then carried out and the predictions are compared with the observed results of the experiment. If the predictions match the observed results, then the hypothesis is supported. If the observed results do not match the prediction, then the hypothesis is not supported. A signature feature of an experiment is the control of variables to help eliminate alternative explanations for observed results.

LAB 16

Hints for Implementing the Lab

- Be sure the students have an understanding of their procedures before they begin; being organized and keeping clear observation data are critical for this investigation. However, "sloppy" observation records may lead to diverse arguments among the student groups, which is positive.

- Stress that the students will not have access to the known solutions once they complete their initial data collection and turn them in. If they know they can't get them back they will likely take more care with their initial observations, which they must use to identify the unknown solutions.

- Encourage the students to use as little of the solutions as possible to conserve materials; using well plates will help limit the physical amount of each solution that can be mixed.

- An alternative approach to identifying unknown solutions is to provide the students with only one unknown or a smaller portion of the original set if time is a limiting factor; this approach, however, is less rigorous.

Topic Connections

Table 16.3 provides an overview of the scientific practices, crosscutting concepts, disciplinary core ideas, and supporting ideas at the heart of this lab investigation. In addition, it lists NOS and NOSI concepts for the explicit and reflective discussion. Finally, it lists literacy and mathematics skills (*CCSS ELA* and *CCSS Mathematics*) that are addressed during the investigation.

TABLE 16.3

Lab 16 alignment with standards

Scientific practices	• Asking questions and defining problems • Planning and carrying out investigations • Analyzing and interpreting data • Constructing explanations and designing solutions • Engaging in argument from evidence • Obtaining, evaluating, and communicating information
Crosscutting concept	• Patterns
Core idea	• PS1.B: Chemical reactions
Supporting ideas	• Solutions, solvents, solutes • Chemical reactions • Ions and ionic compounds
NOS and NOSI concepts	• Observations and inferences • Nature and role of experiments
Literacy connections (*CCSS ELA*)	• *Reading:* Key ideas and details, craft and structure, integration of knowledge and ideas • *Writing:* Text types and purposes, production and distribution of writing, research to build and present knowledge, range of writing • *Speaking and listening:* Comprehension and collaboration, presentation of knowledge and ideas
Mathematics connections (*CCSS Mathematics*)	• Reason abstractly and quantitatively • Look for and express regularity in repeated reasoning

Lab Handout

Lab 16. Development of a Reaction Matrix: What Are the Identities of the Unknown Chemicals?

Introduction

In chemistry it is often necessary to identify unknown substances. There are many procedures and pieces of equipment that can help a chemist to identify unknown substances, such as mass spectrometry or gas chromatography. While some of these methods and equipment are very complex and others are simpler, a common feature of each of these approaches is the comparison of properties of known substances with those of the unknown substances. Chemists can observe and compare *physical properties* of known and unknown substances, such as boiling point or density; likewise, it is possible to compare *chemical properties* such as how an unknown substance reacts with a known substance. One problem with testing and comparing physical properties of a substance involves solutions. It is possible that two different solutions will have the same density or even the same boiling point; in such a case those properties would not be helpful for differentiating between the two substances. In this situation, it may be more helpful to use observations of chemical reactions involving the unknown substance to determine what is in the solution.

An *aqueous solution* is created when a substance, the *solute*, is dissolved into water, the *solvent*. When different aqueous solutions are mixed together, it is possible that the substances dissolved into the water will undergo a chemical reaction, which may produce a *precipitate*. A precipitate is an insoluble ionic compound. Adding a few drops of an unknown solution to a powder may also cause a reaction, which might result in the formation of a gas. Other observable outcomes of a chemical reaction might be a change in color or a change in temperature. Since the chemical properties of a given substance are consistent when chemicals react with each other, the resulting products are also consistent and predictable; therefore, combining solutions together can be used as a way to identify them. This systematic process can be used to develop a table of reactions and their results, or a *reaction matrix*. If you know how specific known chemicals react with each other, it is possible to identify unknown chemicals by comparison.

Your Task

You will be given six labeled bottles to test and record data in order to generate a reaction matrix. Once your observations are complete and your reaction matrix has been developed, you will turn in your original set of chemicals. Then you will be given a second set of the same chemicals that are missing the correct labels. These unknowns will only be labeled as A–F. Your task will be to match the known solutions to the unknown solutions by comparing reactions with your written observations and the results in your reaction matrix.

The guiding question of this investigation is, **What are the identities of the unknown chemicals?**

Materials

You may use any of the following materials during your investigation:

Consumables	Equipment
• 6 known solutions • 6 unknown solutions (labeled A–F)	• Well plates or test tubes • Toothpicks • Dropper bottles or disposable pipettes

Safety Precautions

Follow all normal lab safety rules. Your teacher will explain relevant and important information about working with the chemicals associated with this investigation. Some of these chemicals may stain your skin and clothing. Take the following safety precautions in addition to any precautions specified by your teacher:

- Wear indirectly vented chemical-splash goggles and chemical-resistant gloves and apron while in the laboratory.

- Handle all glassware with care.

- Wash your hands with soap and water before leaving the laboratory.

Investigation Proposal Required?　　☐ Yes　　　☐ No

Getting Started

To answer the guiding question, you will need to generate a reaction matrix. A reaction matrix is a specific style of data table that allows you to systematically document what happens when you react a series of chemicals with each other (see Table L16.1, p. 262). The reaction matrix accounts for all of the different possible reactions within the set of substances you have available.

LAB 16

Example of a reaction matrix

Solution	Solution				
	A	**B**	**C**	**D**	**E**
A	Observation 1	Observation 2	Observation 3	Observation 4	Observation 5
B	Observation 6	Observation 7	Observation 8	Observation 9	Observation 10
C	Observation 11	Observation 12	Observation 13	Observation 14	Observation 15

The solutions being tested are listed as entries in the first column and as headers in the remaining columns. Observations made during each test can be recorded in the boxes.

To generate a reaction matrix, you must also determine what type of data you will need to collect, how you will collect the data, and how you will analyze the data.

To determine *what type of data you need to collect*, think about what types of observations or measurements will be most useful to include in your reaction matrix.

To determine *how you will collect the data*, think about the following questions:

- How much of each chemical will you need to mix together?
- How often will you collect data and when will you do it?
- How will you make sure that your data are of high quality (i.e., how will you reduce error)?
- How will you keep track of the data you collect and how will you organize it?

To determine *how you will analyze the data*, think about how you will determine if a known substance and an unknown substance match.

Connections to Crosscutting Concepts, the Nature of Science, and the Nature of Scientific Inquiry

As you work through your investigation, be sure to think about

- the importance of identifying patterns in science,
- the difference between observations and inferences in science, and
- the nature and role of experiments in science.

Initial Argument

Once your group has finished collecting and analyzing your data, you will need to develop an initial argument. Your argument must include a *claim*, which is your answer to the guiding question. Your argument must also include *evidence* in support of your claim. The evidence is your analysis of the data and your interpretation of what the analysis means. Finally, you must include a *justification* of the evidence in your argument. You will therefore need to use a scientific concept or principle to explain why the evidence that you decided to use is relevant and important. You will create your initial argument on a whiteboard. Your whiteboard must include all the information shown in Figure L16.1.

FIGURE L16.1 _____

Argument presentation on a whiteboard

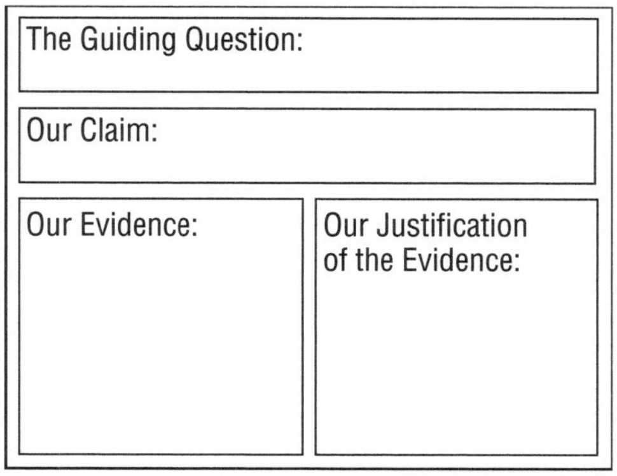

The Guiding Question:	
Our Claim:	
Our Evidence:	Our Justification of the Evidence:

Argumentation Session

The argumentation session allows all of the groups to share their arguments. One member of each group stays at the lab station to share that group's argument, while the other members of the group go to the other lab stations one at a time to listen to and critique the arguments developed by their classmates. The goal of the argumentation session is not to convince others that your argument is the best one; rather, the goal is to identify errors or instances of faulty reasoning in the initial arguments so these mistakes can be fixed. You will therefore need to evaluate the content of the claim, the quality of the evidence used to support the claim, and the strength of the justification of the evidence included in each argument that you see. To critique an argument, you might need more information than what is included on the whiteboard. You might, therefore, need to ask the presenter one or more follow-up questions, such as:

- What did your group do to analyze the data, and why did you decide to do it that way?

- Is that the only way to interpret the results of your group's analysis? How do you know that your interpretation of the analysis is appropriate?

- Why did your group decide to present your evidence in that manner?

- What other claims did your group discuss before deciding on that one? Why did you abandon those alternative ideas?

- How confident are you that your group's claim is valid? What could you do to increase your confidence?

Once the argumentation session is complete, you will have a chance to meet with your group and revise your original argument. Your group might need to gather more data or design a way to test one or more alternative claims as part of this process. Remember, your goal at this stage of the investigation is to develop the most valid or acceptable answer to the research question!

LAB 16

Report

Once you have completed your research, you will need to prepare an *investigation report* that consists of three sections that provide answers to the following questions:

1. What question were you trying to answer and why?

2. What did you do during your investigation and why did you conduct your investigation in this way?

3. What is your argument?

Your report should answer these questions in two pages or less. The report must be typed and any diagrams, figures, or tables should be embedded into the document. Be sure to write in a persuasive style; you are trying to convince others that your claim is acceptable or valid!

Checkout Questions

Lab 16. Development of a Reaction Matrix: What Are the Identities of the Unknown Chemicals?

1. Describe why chemical properties are useful for identifying matter.

2. Sometimes in chemistry two clear aqueous solutions, A and B, can be mixed and a reaction will occur. The result of the reaction is a new solid compound, C, that settles to the bottom of the container. Even though the initial solutions were clear, the new solid compound can have a very different color such as yellow, green, blue, or white.

   Use what you know about chemical reactions to explain where the matter for compound C comes from.

LAB 16

3. In science, observations are facts, whereas inferences are just guesses.

    a. I agree with this statement.
    b. I disagree with this statement.

    Explain your answer, using an example from your investigation about the development of a reaction matrix.

4. All investigations in chemistry are experiments.

    a. I agree with this statement.
    b. I disagree with this statement.

    Explain your answer, using an example from your investigation about the development of a reaction matrix.

National Science Teachers Association

5. Scientists conduct many investigations in hopes of identifying a pattern within nature. Using an example from your investigation about the development of a reaction matrix, describe how patterns can be useful for predicting the outcome of a new investigation.

LAB 17

Lab 17. Limiting Reactants: Why Does Mixing Reactants in Different Mole Ratios Affect the Amount of the Product and the Amount of Each Reactant That Is Left Over?

Purpose

The purpose of this lab is to *introduce* students to the concept of mole ratios, limiting reactants, and excess reactants. This lab gives students an opportunity to use atomic theory to develop a model that can help them explain why mixing reactants in different mole ratios affects the amount of product that is observed at the end of a chemical reaction. Students will also learn about the important role that laws, theories, and models play in science and why scientists need to be creative and have a good imagination to excel in science.

The Content

Chemical equations give the ideal stoichiometric relationship among reactants and products. In real life, however, the reactants for a reaction are not necessarily mixed together in an exact stoichiometric ratio. In a chemical reaction, reactants that are not used up when the reaction is finished are called excess reactants. The reactant that is completely used up is called the limiting reactant because its quantity limits the amount of product formed. To illustrate, consider the reaction between sodium (Na), which is a metal, and chlorine (Cl), which is a diatomic gas. The reaction can be represented by the following equation:

$$2\,Na + Cl_2 \rightarrow 2\,NaCl$$

This balanced chemical equation indicates that two Na atoms react with two Cl atoms or one Cl_2 molecule. Therefore, if six Na atoms are available, three Cl_2 molecules will be needed to create six molecules of NaCl. If there are more than three Cl_2 molecules available, then they will remain unreacted. We can also state that 6 moles of sodium metal will require 3 moles of Cl_2 gas. If there are more than 3 moles of Cl_2 gas, some will remain as an excess reactant. The sodium metal, in contrast, will be a limiting reactant because it limits the amount of the product that will be formed when the two substances are mixed.

Timeline

The instructional time needed to complete this lab investigation is 180–250 minutes. Appendix 2 (p. 501) provides options for implementing this lab investigation over several class periods. Option E (250 minutes) should be used if students are unfamiliar with

Limiting Reactants

Why Does Mixing Reactants in Different Mole Ratios Affect the Amount of the Product and the Amount of Each Reactant That Is Left Over?

scientific writing because this option provides extra instructional time for scaffolding the writing process. You can scaffold the writing process by modeling, providing examples, and providing hints as students write each section of the report. Option F (180 minutes) should be used if students are familiar with scientific writing and have the skills needed to write an investigation report on their own. In option F, students complete stage 6 (writing the investigation report) and stage 8 (revising the investigation report) as homework.

Materials and Preparation

The materials needed to implement this investigation are listed in Table 17.1. The consumables and equipment can be purchased from a science supply company such as Carolina, Flinn Scientific, or Ward's Science.

TABLE 17.1

Materials list

Item	Quantity
Consumables	
1 M solution of acetic acid, CH_3COOH	20 ml per group
Sodium bicarbonate, $NaHCO_3$	10 g per group
pH paper	10 pieces per group
Equipment and other materials	
Side-arm Erlenmeyer flask with stopper, 50 ml	1 per group
Pneumatic trough	1 per group
Plastic or rubber tubing, 50 cm long	1 per group
Electronic or triple beam balance	1 per group
Graduated cylinder, 500 ml	1 per group
Graduated cylinder, 250 ml	1 per group
Spatula or chemical scoop	As needed
Weighing paper or dishes	As needed
Investigation Proposal C (optional but recommended)	3 per group
Whiteboard, 2' × 3' *	1 per group
Lab handout	1 per student
Peer-review guide and instructor scoring rubric	1 per student

*As an alternative, students can use computer and presentation software such as Microsoft PowerPoint or Apple Keynote to create their arguments.

LAB 17

You will need to show students how to use a pneumatic trough to collect gas as part of the tool talk. The basic steps are as follows: (1) fill a graduated cylinder with water, (2) cover the opening of the graduated cylinder with an index card, (3) invert the cylinder and place into the pneumatic trough already containing water, (4) remove the index card, and (5) insert the outlet tube from the gas-generating apparatus into the opening of the graduated cylinder so gas can bubble into the cylinder and displace the water.

We recommend that you use a set routine for distributing and collecting the materials during the lab investigation. For example, the consumables and equipment for each group can be set up at each group's lab station before class begins, or one member from each group can collect them from a table or a cart when needed during class.

Safety Precautions

Remind students to follow all normal lab safety rules. Acetic acid is a skin and eye irritant, so you will need to explain the potential hazards of working with acetic acid and how to work with hazardous chemicals. In addition, tell students to take the following safety precautions:

- Wear indirectly vented chemical-splash goggles and chemical-resistant gloves and aprons when they are collecting their data.
- Handle all glassware with care.
- Wash their hands with soap and water when they are done collecting the data.

Laboratory Waste Disposal

We recommend following Flinn laboratory waste disposal method 26b to dispose of waste solutions. Information about laboratory waste disposal methods is included in the Flinn Catalog and Reference Manual; you can request a free copy at *www.flinnsci.com*.

Topics for the Explicit and Reflective Discussion

Concepts That Can Be Used to Justify the Evidence

To provide an adequate justification of their evidence, students must explain why they included the evidence in their arguments and make the assumptions underlying their analysis and interpretation of the data explicit. In this investigation, students can use the following concepts to help justify their evidence:

- Atomic theory
- How chemical equations are used to represent chemical reactions
- Moles and mole ratios

Limiting Reactants

Why Does Mixing Reactants in Different Mole Ratios Affect the Amount of the Product and the Amount of Each Reactant That Is Left Over?

We recommend that you discuss these fundamental concepts during the explicit and reflective discussion to help students make this connection.

How to Design Better Investigations

It is important for students to reflect on the strengths and weaknesses of the investigation they designed during the explicit and reflective discussion. Students should therefore be encouraged to discuss ways to eliminate potential flaws, measurement errors, or sources of bias in their investigations. To help students be more reflective about the design of their investigations, you can ask the following questions:

- What were some of the strengths of your investigation? What made it scientific?

- What were some of the weaknesses of your investigation? What made it less scientific?

- If you were to do this investigation again, what would you do to address the weaknesses in your investigation? What could you do to make it more scientific?

Crosscutting Concepts

This investigation is well aligned with two crosscutting concepts found in *A Framework for K–12 Science Education,* and you should review these concepts during the explicit and reflective discussion.

- *Scale, proportion, and quantity:* It is critical for scientists to be able to recognize what is relevant at different sizes, time frames, and scales. Scientists must also be able to recognize proportional relationships between categories or quantities. In this investigation, for example, students must determine the quantity of sodium bicarbonate needed to mix with acetic acid based on a specified proportional relationship between the reactants.

- *Systems and system models:* Scientists often need to use models to understand complex phenomena. In this investigation, students are directed to develop a model to help explain what is happening during a chemical reaction at the submicroscopic level.

The Nature of Science and the Nature of Scientific Inquiry

This investigation is well aligned with two important concepts related to the *nature of science* (NOS) and the *nature of scientific inquiry* (NOSI), and you should review these concepts during the explicit and reflective discussion.

- *The difference between laws and theories in science:* A scientific law describes the behavior of a natural phenomenon or a generalized relationship under certain conditions; a scientific theory is a well-substantiated explanation of some aspect of the natural world. Theories do not become laws even with additional evidence; they explain laws. However, not all scientific laws have an accompanying explanatory theory. It is also important for students to understand that scientists do not discover laws or theories; the scientific community develops them over time.

- *The importance of imagination and creativity in science:* Students should learn that developing explanations for or models of natural phenomena and then figuring out how they can be put to the test of reality is as creative as writing poetry, composing music, or designing skyscrapers. Scientists must also use their imagination and creativity to figure out new ways to test ideas and collect or analyze data.

Hints for Implementing the Lab

- Allowing students to design their own procedures for collecting data gives students an opportunity to try, to fail, and to learn from their mistakes. However, you can scaffold students as they develop their procedure by having them fill out an investigation proposal. These proposals provide a way for you to offer students hints and suggestions without telling them how to do it. You can also check the proposals quickly during a class period.

- Allow students to play around with the size of the graduated cylinder that they will need to collect gas. If you do not have large enough graduated cylinders to capture all the gas that is produced, encourage students to think of ways to decrease the amount of gas produced without changing the mole ratios of the reactants.

- Remind students that they can use pH paper to determine if any acid remains in the flask after the reaction is complete.

Topic Connections

Table 17.2 provides an overview of the scientific practices, crosscutting concepts, disciplinary core ideas, and supporting ideas at the heart of this lab investigation. In addition, it lists the NOS and NOSI concepts for the explicit and reflective discussion. Finally, it lists literacy and mathematics skills (*CCSS ELA* and *CCSS Mathematics*) that are addressed during the investigation.

TABLE 17.2

Lab 17 alignment with standards

Scientific practices	• Asking questions and defining problems • Developing and using models • Planning and carrying out investigations • Analyzing and interpreting data • Using mathematics and computational thinking • Constructing explanations and designing solutions • Engaging in argument from evidence • Obtaining, evaluating, and communicating information
Crosscutting concepts	• Scale, proportion, and quantity • Systems and system models
Core idea	• PS1.B: Chemical reactions
Supporting ideas	• Atomic theory • How chemical equations are used to represent chemical reactions • Moles and mole ratios • Limiting and excess reactants
NOS and NOSI concepts	• Scientific laws and theories • Imagination and creativity in science
Literacy connections (CCSS ELA)	• *Reading:* Key ideas and details, craft and structure, integration of knowledge and ideas • *Writing:* Text types and purposes, production and distribution of writing, research to build and present knowledge, range of writing • *Speaking and listening:* Comprehension and collaboration, presentation of knowledge and ideas
Mathematics connections (CCSS Mathematics)	• Reason abstractly and quantitatively • Model with mathematics

Lab Handout

Lab 17. Limiting Reactants: Why Does Mixing Reactants in Different Mole Ratios Affect the Amount of the Product and the Amount of Each Reactant That Is Left Over?

Introduction

Atomic theory is a model that has been developed over time to explain the properties and behavior of matter. This model consists of five important principles, as listed below (see also Figure L17.1):

1. All matter is composed of submicroscopic particles called atoms.

2. All atoms of a given element are identical.

3. All atoms of one element have the same mass, and atoms from different elements have different masses.

4. Atoms can be combined with other atoms to form molecules, and molecules can be split apart into individual atoms.

5. Atoms are not created or destroyed during a chemical reaction that results in the production of a new substance.

A chemical reaction, according to atomic theory, is simply the rearrangement of atoms. The substances (elements and/or compounds) that are changed into other substances during a chemical reaction are called *reactants*. The substances that are produced as a result of a chemical reaction are called *products*. Chemical equations show the reactants and products of a chemical reaction. A chemical equation includes the chemical formulas of the reactants and the products. The products and reactants are separated by an arrow symbol ($\rightarrow$), and each individual substance's chemical formula is separated by a plus sign (+).

Atomic theory, as noted earlier, indicates that atoms are not created or destroyed during a chemical reaction. Thus, each side of the chemical equation must include the same number of each type of atom. When there is an equal number of each type of atom on each side of the equation, the equation is described as balanced. Chemists balance chemical equations by changing the number of each type of substance involved in the reaction; they do not change the number of atoms within each substance. The number of atoms found within a substance cannot be changed because it would change the nature of the substances involved in the reaction. For example, suppose a chemist needs to balance the following equation for the reaction of nitrogen and hydrogen gas:

$$N_2(g) + H_2(g) \rightarrow NH_3(g)$$

In this case, he or she cannot simply add another atom of nitrogen and take an atom of hydrogen away from the chemical formula for ammonia (NH_3) because this would change ammonia (NH_3) to diazene (N_2H_2). Chemists therefore use stoichiometric coefficients to indicate how much of each substance is involved in the reaction without changing the nature of those substances. The balanced chemical equation for the reaction of nitrogen and hydrogen gas, as a result, is denoted as follows:

$$N_2(g) + 3\,H_2(g) \rightarrow 2\,NH_3(g)$$

FIGURE L17.1

The basic principles of atomic theory

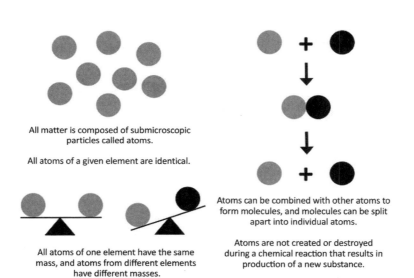

All matter is composed of submicroscopic particles called atoms.

All atoms of a given element are identical.

All atoms of one element have the same mass, and atoms from different elements have different masses.

Atoms can be combined with other atoms to form molecules, and molecules can be split apart into individual atoms.

Atoms are not created or destroyed during a chemical reaction that results in production of a new substance.

The stoichiometric coefficients indicate the relative amount of each substance involved in the chemical reaction in terms of molecules or moles. This equation, as a result, can be read as, 1 molecule of nitrogen gas reacts with 3 molecules of hydrogen gas to yield 2 molecules of ammonia gas. The equation can also be read as, 1 mole of nitrogen gas reacts with 3 moles of hydrogen gas to yield 2 moles of ammonia gas.

The stoichiometric coefficients are also used to determine the *mole ratio*, which is the relationship between the amounts of any two compounds, in moles, that are involved in a chemical reaction. For example, the mole ratio of the reactants (N_2 and H_2) in the reaction listed above is 1:3 (1 mole of N_2 reacts with 3 moles of H_2) and the mole ratio between hydrogen gas and ammonia gas is 3:2 (3 moles of H_2 yields 2 moles of NH_3). Chemists use mole ratios as a conversion factor in many chemistry problems. Mole ratios are also useful when a chemist needs to create a specific amount of a product but must also minimize the amount of reactants that need to be purchased and limit the amount of waste that is left over.

To be able to use mole ratios to create a specific amount of a product, minimize costs, and limit waste, it is important to understand how varying the mole ratio of the reactants affects the amount of the product that is formed and the amount of the reactants that are left over at the end of a chemical reaction. You will therefore explore how mixing acetic acid and sodium bicarbonate in different mole ratios affects the amount of carbon dioxide (CO_2) gas that is produced and determine which reactant, if any, is left over at the end of the chemical reaction. You will then develop a conceptual model that you can use to explain your observations and predict the amount of CO_2 gas that will be produced in other conditions.

Your Task

Determine how varying the mole ratio of the reactants affects the amount of the product that is produced and the amount of the reactants that remain at the end of a chemical reaction. You will then develop a conceptual model that can be used to explain why mixing reactants in different mole ratios will affect the amount of product that is produced and the amount of each reactant left over. Once you have developed your conceptual model, you will need to test it to determine if it allows you to predict the dissolution rate of another solute under various conditions.

The guiding question for this investigation is, **Why does mixing reactants in different mole ratios affect the amount of the product and the amount of each reactant that is left over?**

Materials

You may use any of the following materials during your investigation:

Consumables	Equipment
• 1 M solution of acetic acid, CH_3COOH • Sodium bicarbonate, $NaHCO_3$ • pH paper	• Side-arm Erlenmeyer flask with stopper (50 ml) • Pneumatic trough • Tubing (50 cm long) • Electronic or triple beam balance • Graduated cylinder (500 ml) • Graduated cylinder (250 ml) • Spatula or chemical scoop • Weighing paper or dishes

Safety Precautions

Follow all normal lab safety rules. Your teacher will explain relevant and important information about working with the chemicals associated with this investigation. In addition, take the following safety precautions:

Limiting Reactants

Why Does Mixing Reactants in Different Mole Ratios Affect the Amount of the Product and the Amount of Each Reactant That Is Left Over?

- Wear indirectly vented chemical-splash goggles and chemical-resistant gloves and apron while in the laboratory.

- Handle all glassware with care.

- Wash your hands with soap and water before leaving the laboratory.

Investigation Proposal Required?   ☐ Yes   ☐ No

Getting Started

The first step in developing your model is to design and carry an experiment to determine how varying the mole ratio of the reactants affects the amount of the product that is formed and the amount of the reactants that remain at the end of a chemical reaction. To conduct this experiment, you will focus on the reaction of acetic acid and sodium bicarbonate. Acetic acid (CH_3COOH) reacts with sodium bicarbonate ($NaHCO_3$) according to the following equation:

$$CH_3COOH(aq) + NaHCO_3(s) \rightarrow NaCH_3CO_2(aq) + CO_2(g) + H_2O(l)$$

You will need to react sodium bicarbonate and acetic acid in different molar ratios while keeping everything else the same during your experiment. To do this, you will first need to decide which mole ratios (e.g., 2:1, 1:1, 1:3) and how many different mole ratios to test. You should, however, test at least five different mole ratios to have enough useful comparisons. Next, you will need to determine the amount of each reactant you need to use in each reaction. You should use the same amount of acetic acid (5 ml or 0.005 moles) in each reaction and vary the amount of sodium bicarbonate. You will therefore need to first determine the number of moles of sodium bicarbonate that you will need to use in each test. You can then calculate the mass of sodium bicarbonate that you will need to react with the 5 ml of acetic acid.

Once you have determined the amount of sodium bicarbonate you will need to use for each of the reactions, you will need to devise a way to determine how much product is produced during each reaction and if there are one or more reactants left over after the reaction is complete. You will therefore need to think about *what type of data you need to collect* during your experiment. To accomplish this task, think about the following questions:

- How will you know if there is any sodium bicarbonate left over at the end of the reaction?

- How will you know if there is any acetic acid left over at the end of the reaction?

- How will you measure the amount of product that is produced?

The easiest way to determine how much product is produced is to measure the amount of CO_2 that is formed after you combine the acetic acid and sodium bicarbonate. To accom-

LAB 17

Gas collection using water displacement

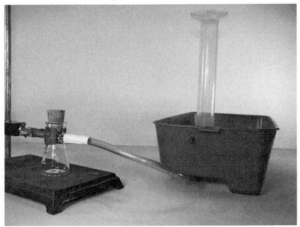

plish this task, you will need to collect the CO_2 gas by water displacement. Figure L17.2 shows how to collect gas by water displacement.

Once you have collected your data, you will need to think about *how you will analyze the data*. The following questions may be helpful:

- What types of comparisons will you make?
- What type of calculations will you need to make?
- What type of graph could you create to help you make sense of your data?

Once you have carried out your experiments, your group will need to develop a conceptual model that can be used to explain why mixing reactants in different mole ratios will affect the amount of product that is produced and the amount of each reactant left over. The model also needs to be able to explain what is happening during the reaction at the submicroscopic level.

The last step in this investigation is to test your model. To accomplish this goal, you can use a different mole ratio to determine if your model leads to accurate predictions about the amount of product produced and which reactant or reactants will be left over under different conditions. If you are able to use your model to make accurate predictions under different conditions, then you will be able to generate the evidence you need to convince others that the conceptual model you developed is valid.

Connections to Crosscutting Concepts, the Nature of Science, and the Nature of Scientific Inquiry

As you work through your investigation, be sure to think about

- why it is important to look for proportional relationships,
- how models are used to help understand natural phenomena,
- the difference between laws and theories in science, and
- the role of imagination and creativity in science.

Initial Argument

Once your group has finished collecting and analyzing your data, you will need to develop an initial argument. Your argument must include a *claim*, which is your answer to the guiding question. Your argument must also include *evidence* in support of your claim. The evidence is your analysis of the data and your interpretation of what the analysis means.

Finally, you must include a *justification* of the evidence in your argument. You will therefore need to use a scientific concept or principle to explain why the evidence that you decided to use is relevant and important. You will create your initial argument on a whiteboard. Your whiteboard must include all the information shown in Figure L17.3.

Argumentation Session

The argumentation session allows all of the groups to share their arguments. One member of each group stays at the lab station to share that group's argument, while the other members of the group go to the other lab stations one at a time to listen to and critique the arguments developed by their classmates. The goal of the argumentation session is not to convince others that your argument is the best one; rather, the goal is to identify errors or instances of faulty reasoning in the initial arguments so these mistakes can be fixed. You will therefore need to evaluate the content of the claim, the quality of the evidence used to support the claim, and the strength of the justification of the evidence included in each argument that you see. To critique an argument, you might need more information than what is included on the whiteboard. You might, therefore, need to ask the presenter one or more follow-up questions, such as:

FIGURE L17.3

Argument presentation on a whiteboard

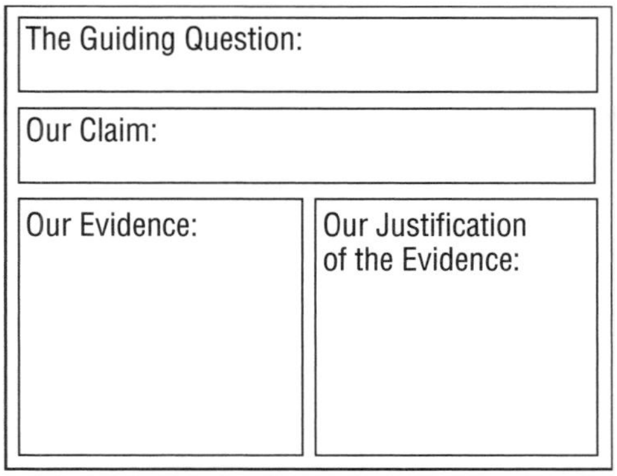

The Guiding Question:	
Our Claim:	
Our Evidence:	Our Justification of the Evidence:

- How did your group collect the data? Why did you use that method?

- What did your group do to make sure the data you collected are reliable? What did you do to decrease measurement error?

- What did your group do to analyze the data, and why did you decide to do it that way? Did you check your calculations?

- Is that the only way to interpret the results of your group's analysis? How do you know that your interpretation of the analysis is appropriate?

- Why did your group decide to present your evidence in that manner?

- What other claims did your group discuss before deciding on that one? Why did you abandon those alternative ideas?

- How confident are you that your group's claim is valid? What could you do to increase your confidence?

Once the argumentation session is complete, you will have a chance to meet with your group and revise your original argument. Your group might need to gather more data or design a way to test one or more alternative claims as part of this process. Remember, your goal at this stage of the investigation is to develop the most valid or acceptable answer to the research question!

LAB 17

Report

Once you have completed your research, you will need to prepare an *investigation report* that consists of three sections that provide answers to the following questions:

1. What question were you trying to answer and why?

2. What did you do during your investigation and why did you conduct your investigation in this way?

3. What is your argument?

Your report should answer these questions in two pages or less. The report must be typed and any diagrams, figures, or tables should be embedded into the document. Be sure to write in a persuasive style; you are trying to convince others that your claim is acceptable or valid!

Limiting Reactants

Why Does Mixing Reactants in Different Mole Ratios Affect the Amount of the Product and the Amount of Each Reactant That Is Left Over?

Checkout Questions

Lab 17. Limiting Reactants: Why Does Mixing Reactants in Different Mole Ratios Affect the Amount of the Product and the Amount of Each Reactant That Is Left Over?

Use the following information to answer questions 1 and 2. Iron(III) chloride reacts with sodium hydroxide as follows:

$$FeCl_3(aq) + 3\ NaOH(aq) \rightarrow Fe(OH)_3(s) + 3\ NaCl(aq)$$

A student starts with 50 g of $FeCl_3$ and adds NaOH in 1-gram increments. He or she measures the mass of $Fe(OH)_3$ produced in each reaction and then plots the mass of $Fe(OH)_3$ as a function of the mass of NaOH added (see graph below).

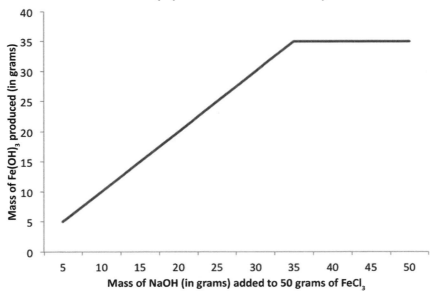

Reaction of iron(III) chloride with sodium hydroxide

y-axis: Mass of $Fe(OH)_3$ produced (in grams)

x-axis: Mass of NaOH (in grams) added to 50 grams of $FeCl_3$

LAB 17

1. Describe the concept of a limiting reactant.

2. Use what you know about limiting reactants to explain why the mass of $Fe(OH)_3$ no longer changes when more than 37 g of NaOH is added to 50 g of $FeCl_3$.

3. Imagination and creativity play an important role in science.

    a. I agree with this statement.
    b. I disagree with this statement.

    Explain your answer, using an example from your investigation about limiting reactants.

National Science Teachers Association

4. Scientists use laws to describe and theories to explain natural phenomena.

   a. I agree with this statement
   b. I disagree with this statement

   Explain your answer, using an example from your investigation about limiting reactants.

5. Scientists often need to look for proportional relationships. Explain why looking for a proportional relationship is useful in science, using an example from your investigation about limiting reactants.

6. Models are important in science. Explain what a model is and why models are important, using an example from your investigation about limiting reactants.

LAB 18

Teacher Notes

Lab 18. Characteristics of Acids and Bases: How Can the Chemical Properties of an Aqueous Solution Be Used to Identify It as an Acid or a Base?

Purpose

The purpose of this lab is to *introduce* students to the physical and chemical properties of acids and bases. You can use this lab at the beginning of a unit on acid-base chemistry to set the stage for future lessons and investigations. This lab gives students an opportunity to devise, test, and refine a method that can be used to classify an aqueous solution as being an acid or a base using the physical or chemical properties of the solution. Students will also learn about the difference between observations and inferences in science and the different methods used in scientific investigations.

The Content

Acids and bases have several unique physical and chemical properties. These properties, which stem from the atomic structure of the compounds, can be used to classify an aqueous solution as an acid or a base. Properties of acids include

- conducting electricity;
- reacting with active metals to form hydrogen gas and solutions of metal ions;
- reacting with carbonates to form a salt, water, and carbon dioxide gas; and
- reacting with a base to form a salt and water.

Properties of bases include

- conducting electricity,
- reacting with oils and greases, and
- reacting with an acid to form a salt and water.

In addition to these properties, acids and bases also interact with acid-base indicators. Acid-base indicators are organic dyes that change color in acidic or basic solutions. The color of an acid-base indicator depends on the concentration of H_3O^+ ions in a solution. The concentration of H_3O^+ ions in a solution is often described using the pH scale. The mathematical relationship between pH and H_3O^+ ion concentration is $pH = -\log[H_3O^+]$.

The typical H_3O^+ concentration in water ranges from 1 M in a 1 M solution of a strong acid (such as hydrochloric acid [HCl]) to 10^{-14} M in a 1 M solution of a strong base (such as sodium hydroxide, NaOH). In pure water, the H_3O^+ concentration is equal to 10^{-7} M. The logarithm of the concentration is the "power of 10" exponent in these concentration terms. Thus, the negative logarithms of typical H_3O^+ concentrations are positive numbers from 0 to 14. The pH scale, as a result, ranges from 0 to 14. Acids have pH values less than 7, whereas bases have pH values greater than 7.

Within the pH range of acid solutions, a strong acid or a more concentrated acid solution will have a lower pH value than a weak acid or a less concentrated acid solution. Thus, the pH value of 0.1 M HCl (a strong acid) is 1 and the pH value of 0.01 M HCl is 2, whereas the pH value of 0.1 M acetic acid (a weak acid) is about 3. On the basic side of the pH scale, a strong base or a more concentrated base solution will have a higher pH value than a weaker or a less concentrated base solution. The pH value of a 0.1 M NaOH (which is a strong base) is 13 and the pH value of 0.01 M NaOH is 12, whereas the pH of 0.1 M ammonia (a weak base) is about 11. It is important to remember that the pH scale is logarithmic, so a solution of pH 3 is 10 times more acidic than a solution of pH 4 and 100 times more acidic than a solution of pH 5.

Although acid-base indicators are useful for broadly classifying substances as acids or bases, they are not able to distinguish among different strengths of acids and bases. By using combinations of different indicators, however, it is possible to obtain a spectrum of color changes over a wide range of acidity levels. Table 18.1 lists the colors associated with the five different indicators that the students will use in this lab investigation.

TABLE 18.1

Indicators used in this lab investigation

Indicator	Color associated with each level of pH													
	1	2	3	4	5	6	7	8	9	10	11	12	13	14
Thymol blue	R	P	Y	Y	Y	Y	Y	Y	B	B	B	B	B	B
Bromphenol blue	Y	Y	Y	G	B	B	B	B	B	B	B	B	B	B
Bromthymol blue	Y	Y	Y	Y	Y	Y	G	B	B	B	B	B	B	B
Methyl red	R	R	R	P	P	Y	Y	Y	Y	Y	Y	Y	Y	Y
Phenol red	Y	Y	Y	Y	Y	P	P	R	R	R	R	R	R	R

B = blue; G = green; P = pink; R = red; V = violet; Y = yellow

LAB 18

Timeline

The instructional time needed to complete this lab investigation is 130–200 minutes. Appendix 2 (p. 501) provides options for implementing this lab investigation over several class periods. Option C (200 minutes) should be used if students are unfamiliar with scientific writing because this option provides extra instructional time for scaffolding the writing process. You can scaffold the writing process by modeling, providing examples, and providing hints as students write each section of the report. Option D (130 minutes) should be used if students are familiar with scientific writing and have the skills needed to write an investigation report on their own. In option D, students complete stage 6 (writing the investigation report) and stage 8 (revising the investigation report) as homework.

Materials and Preparation

The materials needed to implement this investigation are listed in Table 18.2. The consumables and equipment can be purchased from a science supply company such as Carolina, Flinn Scientific, or Ward's Science. We recommend using buffer capsules to prepare standard acid-base solutions of known pH. Buffer capsules contain preweighed amounts of stable, dry powders that dissolve in distilled or deionized water to give solutions of known, constant pH. They can be purchased separately or as a set. The buffer solutions, once prepared, can then be divided evenly between several 30 ml dropper bottles for students to use. The acid-base solutions for this lab are listed in Table 18.2.

We also recommend that you use a set routine for distributing and collecting the materials during the lab investigation. For example, the consumables and equipment for each group can be set up at each group's lab station before class begins, or one member from each group can collect them from a central table or a cart when needed during class.

Safety Precautions

Remind students to follow all normal lab safety rules. The acids are corrosive and toxic by ingestion, and the bases are body tissue irritants. You will therefore need to explain the potential hazards of working with acids and bases and how to work with hazardous chemicals. In addition, tell students to take the following safety precautions:

- Wear indirectly vented chemical-splash goggles and chemical-resistant gloves and aprons when they are collecting their data.
- Handle all glassware with care.
- Wash their hands with soap and water when they are done collecting the data.

Laboratory Waste Disposal

The solutions may be flushed down the drain with a large quantity of water according to Flinn laboratory waste disposal method 26b. Use Flinn laboratory waste disposal method 3 to dispose of any extra zinc that has been mixed with an acid. Information about laboratory

TABLE 18.2

Materials list

Item	Quantity
Consumables	
Zinc	10 ml per group (1 dropper bottle)
1 M solution of sodium bicarbonate, $NaHCO_3$	10 ml per group (1 dropper bottle)
1 M solution of HCl	10 ml per group (1 dropper bottle)
Acid solution 1 (buffer solution, pH 2)	10 ml per group (1 dropper bottle)
Acid solution 2 (buffer solution, pH 3)	10 ml per group (1 dropper bottle)
Acid solution 3 (buffer solution, pH 4)	10 ml per group (1 dropper bottle)
Acid solution 4 (buffer solution, pH 5)	10 ml per group (1 dropper bottle)
Base solution 1 (buffer solution, pH 9)	10 ml per group (1 dropper bottle)
Base solution 2 (buffer solution, pH 10)	10 ml per group (1 dropper bottle)
Base solution 3 (buffer solution, pH 11)	10 ml per group (1 dropper bottle)
Base solution 4 (buffer solution, pH 12)	10 ml per group (1 dropper bottle)
Acid test solution A (buffer solution, pH 2)	10 ml per group (1 dropper bottle)
Acid test solution B (buffer solution, pH 6)	10 ml per group (1 dropper bottle)
Base test solution A (buffer solution, pH 8)	10 ml per group (1 dropper bottle)
Base test solution B (buffer solution, pH 11)	10 ml per group (1 dropper bottle)
Thymol blue	10 ml per group (1 dropper bottle)
Bromphenol blue	10 ml per group (1 dropper bottle)
Bromthymol blue	10 ml per group (1 dropper bottle)
Methyl red	10 ml per group (1 dropper bottle)
Phenol red	10 ml per group (1 dropper bottle)
Equipment and other materials	
Conductivity tester or probe	1 per group
Reaction plate	1 or 2 per group
Small beaker(s), 50 ml	1 or 2 per group
Whiteboard, 2' x 3' *	1 per group
Lab handout	1 per student
Peer-review guide and instructor scoring rubric	1 per student

*As an alternative, students can use computer and presentation software such as Microsoft PowerPoint or Apple Keynote to create their arguments.

waste disposal methods is included in the Flinn Catalog and Reference Manual; you can request a free copy at *www.flinnsci.com.*

Topics for the Explicit and Reflective Discussion

Concepts That Can Be Used to Justify the Evidence

To provide an adequate justification of their evidence, students must explain why they included the evidence in their arguments and make the assumptions underlying their analysis and interpretation of the data explicit. In this investigation, students can use the following concepts to help justify their evidence:

- The nature of chemical properties
- How atomic structure determines chemical properties

We recommend that you discuss these fundamental concepts during the explicit and reflective discussion to help students make this connection.

How to Design Better Investigations

It is important for students to reflect on the strengths and weaknesses of the investigation they designed during the explicit and reflective discussion. Students should therefore be encouraged to discuss ways to eliminate potential flaws, measurement errors, or sources of bias in their investigations. To help students be more reflective about the design of their investigation, you can ask the following questions:

- What were some of the strengths of your investigation? What made it scientific?
- What were some of the weaknesses of your investigation? What made it less scientific?
- If you were to do this investigation again, what would you do to address the weaknesses in your investigation? What could you do to make it more scientific?

Crosscutting Concepts

This investigation is well aligned with two crosscutting concepts found in *A Framework for K–12 Science Education,* and you should review these concepts during the explicit and reflective discussion.

- *Patterns:* A major objective in chemistry is to identify patterns. Once the patterns are identified, they are often used to guide classification systems and prompt questions about the underlying cause of the observed patterns. In this investigation, for example, students need to identify patterns in the physical and chemical properties of acids and bases and then use these patterns to classify them.

- *Structure and function:* The way an object is shaped or structured determines many of its properties and functions. The observable physical and chemical properties of acids and bases, for example, are determined by the atomic structure of these molecules.

The Nature of Science and the Nature of Scientific Inquiry

This investigation is well aligned with two important concepts related to the *nature of science* (NOS) and the *nature of scientific inquiry* (NOSI), and you should review these concepts during the explicit and reflective discussion.

- *The difference between observations and inferences*: An observation is a descriptive statement about a natural phenomenon, whereas an inference is an interpretation of an observation. Students should also understand that current scientific knowledge and the perspectives of individual scientists guide both observations and inferences. Thus, different scientists can have different but equally valid interpretations of the same observations due to differences in their perspectives and background knowledge.

- *Methods used in scientific investigations*: Examples of methods include experiments, systematic observations of a phenomenon, literature reviews, and analysis of existing data sets; the choice of method depends on the objectives of the research. There is no universal step-by-step scientific method that all scientists follow; rather, different scientific disciplines (e.g., chemistry vs. physics) and fields within a discipline (e.g., organic vs. physical chemistry) use different types of methods, use different core theories, and rely on different standards to develop scientific knowledge

Hints for Implementing the Lab

- Allowing students to design their own methods for identifying acids and bases gives students an opportunity to try, to fail, and to learn from their mistakes. Encourage students to try things out and refine their method when they uncover a flaw or weakness with it.

- Use reaction plates rather than test tubes for the tests. It is safer and limits the amount of consumables that are used during the investigation.

- Use small dropper bottles for acid solutions, base solutions, and indicators. They prevent students from using too much and decrease the chance of spills.

Topic Connections

Table 18.3 (p. 290) provides an overview of the scientific practices, crosscutting concepts, disciplinary core ideas, and supporting ideas at the heart of this lab investigation. In addition,

LAB 18

it lists NOS and NOSI concepts for the explicit and reflective discussion. Finally, it lists literacy skills (*CCSS ELA*) that are addressed during the investigation.

TABLE 18.3

Lab 18 alignment with standards

Scientific practices	• Asking questions and defining problems • Planning and carrying out investigations • Analyzing and interpreting data • Constructing explanations and designing solutions • Engaging in argument from evidence • Obtaining, evaluating, and communicating information
Crosscutting concepts	• Patterns • Structure and function
Core ideas	• PS1.A: Structure and properties of matter • PS1.B: Chemical reactions
Supporting idea	• Properties of acids and bases
NOS and NOSI concepts	• Observations and inferences • Methods used in scientific investigations
Literacy connections (*CCSS ELA*)	• *Reading:* Key ideas and details, craft and structure, integration of knowledge and ideas • *Writing:* Text types and purposes, production and distribution of writing, research to build and present knowledge, range of writing • *Speaking and listening:* Comprehension and collaboration, presentation of knowledge and ideas

Lab Handout

Lab 18. Characteristics of Acids and Bases: How Can the Chemical Properties of an Aqueous Solution Be Used to Identify It as an Acid or a Base?

Introduction

Acids and bases represent two important classes of chemical compounds. These compounds play a significant role in many atmospheric and geological processes. In addition, acid-base reactions affect many of the physiological processes that take place within the human body. Acids and bases are important in atmospheric, geological, and physiological processes because they have unique chemical properties. Acids and bases have unique chemical properties because of the atomic composition of these compounds and how these compounds interact with other atoms and molecules.

Some of the unique chemical properties of acids and bases include how they interact with metals, carbonates, and a class of compounds called acid-base indicators. A *metal* is a solid material that is hard, shiny, malleable, and ductile. Metals are also good electrical and thermal conductors. *Carbonates* are compounds that contain a carbonate ion (CO_3^{2-}), such as calcium carbonate ($CaCO_3$), potassium carbonate (K_2CO_3), and sodium bicarbonate ($NaHCO_3$). An acid-base indicator is a dye or a pigment that changes color when it is mixed with an acid or a base. People have used indicators to identify acids and bases for hundreds of years. For example, in the 17th century Sir Robert Boyle described how different indicators could be used to identify acids and bases (Boyle 1664).

In this investigation you will explore some of the unique chemical properties of acids and bases. You will then develop a method that can be used to identify acidic or basic aqueous solutions. This is important because, like all chemists, you will need to be able to determine if an aqueous solution is acidic or basic as part of your future investigations. This is an important aspect of doing acid-base chemistry.

Your Task

Devise, test, and then, if needed, refine a method that can be used to determine if an aqueous solution is acidic or a basic. For this method to be useful, it should provide consistent and accurate results but should also be simple and quick to perform inside the lab.

The guiding question for this investigation is, **How can the chemical properties of an aqueous solution be used to identify it as an acid or a base?**

LAB 18

Materials

You may use any of the following materials during this investigation:

Consumables	Indicators	Equipment	Aqueous solutions for developing a method	Aqueous solutions for testing a method
• Zinc • 1 M solution of NaHCO₃ • 1 M solution of hydrochloric acid, HCl	• Thymol blue • Bromphenol blue • Bromthymol blue • Methyl red • Phenol red	• Conductivity tester or probe • Reaction plate • Small beakers	• Acid solution 1 • Acid solution 2 • Acid solution 3 • Acid solution 4 • Base solution 1 • Base solution 2 • Base solution 3 • Base solution 4	• Acid test solution A • Acid test solution B • Base test solution A • Base test solution B

Safety Precautions

Follow all normal lab safety rules. All of the acids you will use are corrosive to eyes, skin, and other body tissues. They are also toxic when ingested. Your teacher will explain relevant and important information about working with the chemicals associated with this investigation. In addition, take the following safety precautions:

- Wear indirectly vented chemical-splash goggles and chemical-resistant gloves and apron while in the laboratory.
- Handle all glassware with care.
- Wash your hands with soap and water before leaving the laboratory.

Investigation Proposal Required?   ☐ Yes   ☐ No

Getting Started

To answer the guiding question, you will first need to learn more about the unique chemical properties of acids and bases. You will therefore need to explore how aqueous solutions that are classified as acids or as bases react with metal, a solution of sodium bicarbonate, and a solution of hydrochloric acid. You will then determine if these same solutions are able to conduct electricity. Finally, and perhaps most important, you will examine how different acidic and basic solutions interact with different indicators. You goal is to learn more about the chemical properties of aqueous solutions that are classified as being acids or bases so you can use these unique properties to classify other aqueous solutions. To accomplish this task, you will need to design and conduct a series of systematic observations.

Be sure to think about *how you will collect your data and how you will analyze the data you collect* before you begin your investigation. One way to collect data is to add a small amount (about 5 to 10 drops) of each acid or base solution to the wells in a reaction plate.

You can then add a small piece of metal or other solution to each well and observe what happens. You can also create a reaction matrix to help stay organized. A reaction matrix is a chart that allows you to record your observations (see Table L18.1 for an example). Only use the solutions found under the heading "Aqueous Solutions for Developing a Method" in the "Materials" section during this stage of your investigation.

TABLE L18.1

Example of a reaction matrix

Compound	Test				
	Zinc	**Conductivity**	**HCl**	**Bromthymol blue**	**Methyl red**
Acid solution 1	Observation 1	Observation 2	Observation 3	Observation 4	Observation 5
Acid solution 2	Observation 6	Observation 7	Observation 8	Observation 9	Observation 10
Base solution 1	Observation 11	Observation 12	Observation 13	Observation 14	Observation 15

Notice that the compounds being tested are included in the first column and each test is labeled as a header in the remaining columns. Observations made during each test can be recorded in the boxes.

Once you have made your observations about the chemical properties of acids and bases, you will need to use what you have learned to devise a method for classifying an unknown as either an acid or a base. You can then test your method using the solutions found under the heading "Aqueous Solutions for Testing a Method" in the "Materials" section. If you are able to use your method to accurately classify all four of these solutions, then you will be able to provide evidence that the method you devised will provide accurate results. If you cannot accurately classify all four of the test solutions, you will need to refine your method and test it again. Keep in mind that your method needs to be a simple and quick way to classify an unknown aqueous solution based on its chemical properties.

Connections to Crosscutting Concepts, the Nature of Science, and the Nature of Scientific Inquiry

As you work through your investigation, be sure to think about

- the importance of looking for, using, and explaining patterns in science;
- the relationship between structure and function in nature;
- the difference between observations and inferences in science; and
- the different methods used in scientific investigations.

LAB 18

Initial Argument

Once your group has finished collecting and analyzing your data, you will need to develop an initial argument. Your argument must include a *claim*, which is your answer to the guiding question. Your argument must also include *evidence* in support of your claim. The evidence is your analysis of the data and your interpretation of what the analysis means. Finally, you must include a *justification* of the evidence in your argument. You will therefore need to use a scientific concept or principle to explain why the evidence that you decided to use is relevant and important. You will create your initial argument on a whiteboard. Your whiteboard must include all the information shown in Figure L18.1.

FIGURE L18.1 _____

Argument presentation on a whiteboard

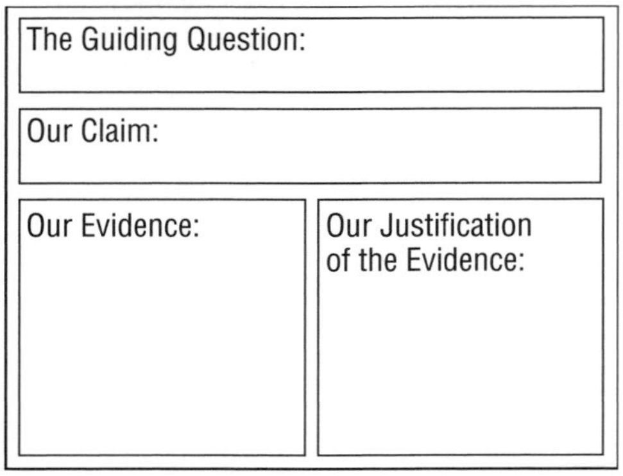

The Guiding Question:	
Our Claim:	
Our Evidence:	Our Justification of the Evidence:

Argumentation Session

The argumentation session allows all of the groups to share their arguments. One member of each group stays at the lab station to share that group's argument, while the other members of the group go to the other lab stations one at a time to listen to and critique the arguments developed by their classmates. The goal of the argumentation session is not to convince others that your argument is the best one; rather, the goal is to identify errors or instances of faulty reasoning in the initial arguments so these mistakes can be fixed. You will therefore need to evaluate the content of the claim, the quality of the evidence used to support the claim, and the strength of the justification of the evidence included in each argument that you see. To critique an argument, you might need more information than what is included on the whiteboard. You might, therefore, need to ask the presenter one or more follow-up questions, such as:

- How did your group collect the data? Why did you use that method?
- What did your group do to make sure the data you collected are reliable?
- What did your group do to analyze the data, and why did you decide to do it that way?
- Is that the only way to interpret the results of your group's analysis?
- Why did your group decide to present your evidence in that manner?
- What other claims did your group discuss before deciding on that one? Why did you abandon those alternative ideas?
- How confident are you that your group's claim is valid? What could you do to increase your confidence?

Once the argumentation session is complete, you will have a chance to meet with your group and revise your original argument. Your group might need to gather more data or design a way to test one or more alternative claims as part of this process. Remember, your goal at this stage of the investigation is to develop the most valid or acceptable answer to the research question!

Report

Once you have completed your research, you will need to prepare an *investigation report* that consists of three sections that provide answers to the following questions:

1. What question were you trying to answer and why?

2. What did you do during your investigation and why did you conduct your investigation in this way?

3. What is your argument?

Your report should answer these questions in two pages or less. The report must be typed and any diagrams, figures, or tables should be embedded into the document. Be sure to write in a persuasive style; you are trying to convince others that your claim is acceptable or valid!

Reference

Boyle, R. 1664. *Experiments and considerations touching colours first occasionally written, among some other essays to a friend, and now suffer'd to come abroad as the beginning of an experimental history of colours.* London: Henry Herringman.

LAB 18

Lab 18. Characteristics of Acids and Bases: How Can the Chemical Properties of an Aqueous Solution Be Used to Identify It as an Acid or a Base?

1. Describe three characteristics of acids and three characteristics of bases.

2. An unknown solution conducts electricity but the indicators thymol blue and bromphenol blue do not change color when they are added to it. Should this solution be classified as an acid or a base?

   a. Acid
   b. Base
   c. Not enough information to determine

   Explain your answer.

3. "The solution is an acid" is an example of an observation.

   a. I agree with this statement.
   b. I disagree with this statement.

   Explain your answer, using an example from your investigation about the characteristics of acids and bases.

4. An investigation must follow the scientific method to be considered scientific.

   a. I agree with this statement.
   b. I disagree with this statement.

   Explain your answer, using an example from your investigation about the characteristics of acids and bases.

5. Scientists often look for and attempt to explain patterns in nature. Explain why patterns are important, using an example from your investigation about the characteristics of acids and bases.

6. In nature, the structure of an object is often related to function or the properties of that object. Explain why this is true, using an example from your investigation about the characteristics of acids and bases.

LAB 19

Teacher Notes

Lab 19. Strong and Weak Acids: Why Do Strong and Weak Acids Behave in Different Manners Even Though They Have the Same Chemical Properties?

Purpose

The purpose of this lab is to *introduce* students to the Brønsted-Lowry acid-base theory and the difference between strong and weak acids. This lab gives students an opportunity to develop a model that can help them explain why the behavior of strong and weak acids differs based on the interactions that take place between acid and base molecules on the submicroscopic level. Students will also learn about what counts as an experiment in science and why scientists need to be creative and have a good imagination to excel in science.

The Content

The strength of an acid is based on its tendency to lose a proton. A strong acid will completely ionize (or dissociate) in a solution. One mole of a strong acid (HA) will therefore yield one mole of H^+ and one mole of the conjugate base (A^-) when it is dissolved in water. In this situation, none of the acid (HA) will remain in the solution. Examples of strong acids include hydrochloric acid (HCl), hydrobromic acid (HBr), nitric acid (HNO_3), and sulfuric acid (H_2SO_4). A weak acid, in contrast, only partially dissociates in a solution. Examples of weak acids include acetic acid (CH_3COOH) and carbonic acid (H_2CO_3). Weak acids tend to be less conductive and react with reactive metals at a slower rate than strong acids because they do not completely dissociate in solution like strong acids. The two key factors that influence how easy it is to remove a proton from an acid is the polarity of the bond between H^+ and A^- and the size of A^-.

Timeline

The instructional time needed to complete this lab investigation is 180–250 minutes. Appendix 2 (p. 501) provides options for implementing this lab investigation over several class periods. Option E (250 minutes) should be used if students are unfamiliar with scientific writing because this option provides extra instructional time for scaffolding the writing process. You can scaffold the writing process by modeling, providing examples, and providing hints as students write each section of the report. Option F (180 minutes) should be used if students are familiar with scientific writing and have the skills needed to write an investigation report on their own. In option F, students complete stage 6 (writing the investigation report) and stage 8 (revising the investigation report) as homework.

Materials and Preparation

The materials needed to implement this investigation are listed in Table 19.1. The consumables and equipment can be purchased from a science supply company such as Carolina, Flinn Scientific, or Ward's Science. We recommend that you use a set routine for distributing and collecting the materials during the lab investigation. For example, the consumables and equipment for each group can be set up at each group's lab station before class begins, or one member from each group can collect them from a table or a cart when needed during class.

TABLE 19.1

Materials list

Item	Quantity
Consumables	
1 M solution of CH_3COOH	30 ml per group
1 M solution of HCl	30 ml per group
1 M solution of H_2SO_4	30 ml per group
Magnesium ribbon, 1 g piece	9 per group
Equipment and other materials	
Conductivity tester or probe	1 per group
Pneumatic trough and tubing	1 per group
Graduated cylinder, 250 ml	1 per group
Graduated cylinder, 10 ml	1 per group
Filtering flask, 50 ml, with stopper*	1 per group
Stopwatch	1 per group
Electronic or triple beam balance	1 per group
Investigation Proposal A (optional but recommended)	3 per group
Whiteboard, 2' x 3' †	1 per group
Lab handout	1 per student
Peer-review guide and instructor scoring rubric	1 per student

* As an alternative, students can use a 50 ml Erlenmeyer flask and a one-hole rubber stopper with a glass tube inserted into it.

† As an alternative, students can use computer and presentation software such as Microsoft PowerPoint or Apple Keynote to create their arguments.

LAB 19

Safety Precautions

Remind students to follow all normal lab safety rules. All of the acids used during this investigation are corrosive to eyes, skin, and other body tissues, and they are also toxic by ingestion. Magnesium metal is a flammable solid and burns with an intense flame. You will therefore need to explain the potential hazards of working with acids and magnesium metal and how to work with hazardous chemicals. In addition, tell students to take the following safety precautions:

- Wear indirectly vented chemical-splash goggles and chemical-resistant gloves and aprons when they are collecting their data.
- Handle all glassware with care.
- Wash their hands with soap and water when they are done collecting the data.

Laboratory Waste Disposal

We recommend following Flinn laboratory waste disposal methods 3 to dispose of any extra magnesium that has been mixed with an acid and 24b to dispose of the acids. Information about laboratory waste disposal methods is included in the Flinn Catalog and Reference Manual; you can request a free copy at *www.flinnsci.com*.

Topics for the Explicit and Reflective Discussion

Concepts That Can Be Used to Justify the Evidence

To provide an adequate justification of their evidence, students must explain why they included the evidence in their arguments and make the assumptions underlying their analysis and interpretation of the data explicit. In this investigation, students can use the following concepts to help justify their evidence:

- Brønsted-Lowry acid-base theory
- The nature of ions and ionic compounds
- Bond and bond energy

We recommend that you discuss these fundamental concepts during the explicit and reflective discussion to help students make this connection.

How to Design Better Investigations

It is important for students to reflect on the strengths and weaknesses of the investigation they designed during the explicit and reflective discussion. Students should therefore be encouraged to discuss ways to eliminate potential flaws, measurement errors, or sources of bias in their investigations. To help students be more reflective about the design of their investigation, you can ask the following questions:

- What were some of the strengths of your investigation? What made it scientific?

- What were some of the weaknesses of your investigation? What made it less scientific?

- If you were to do this investigation again, what would you do to address the weaknesses in your investigation? What could you do to make it more scientific?

Crosscutting Concepts

This investigation is well aligned with two crosscutting concepts found in *A Framework for K–12 Science Education,* and you should review these concepts during the explicit and reflective discussion.

- *Cause and effect: Mechanism and explanation*: A major goal of science is to determine or describe the underlying cause of natural phenomena. Some causes are simple and some are multifaceted, so it is important for scientists to develop and test potential explanations for what is observed. In this investigation, for example, students need to determine how acid strength influences electrical conductivity and reactivity with metal and then develop an explanation for their observations.

- *Systems and system models*: Scientists often need to use models to understand complex phenomena. In this investigation, the students are directed to develop a model to help explain the difference between strong and weak acids at the submicroscopic level.

The Nature of Science and the Nature of Scientific Inquiry

This investigation is well aligned with two important concepts related to the *nature of science* (NOS) and the *nature of scientific inquiry* (NOSI), and you should review these concepts during the explicit and reflective discussion.

- *The importance of imagination and creativity in science:* Students should learn that developing explanations for or models of natural phenomena and then figuring out how they can be put to the test of reality is as creative as writing poetry, composing music, or designing skyscrapers. Scientists must also use their imagination and creativity to figure out new ways to test ideas and collect or analyze data.

- *The nature and role of experiments:* Scientists use experiments to test the validity of a hypothesis (i.e., a tentative explanation) for an observed phenomenon. Experiments include a test and the formulation of predictions (expected results) if the test is conducted and the hypothesis is valid. The experiment is then carried out and the predictions are compared with the observed results of the experiment. If the predictions match the observed results, then the hypothesis is supported. If the observed results do not match the prediction, then the hypothesis is not supported. A signature feature of an experiment is the control of variables to help eliminate alternative explanations for observed results.

Hints for Implementing the Lab

- Allowing students to design their own procedures for collecting data gives students an opportunity to try, to fail, and to learn from their mistakes. However, you can scaffold students as they develop their procedure by having them fill out an investigation proposal. These proposals provide a way for you to offer students hints and suggestions without telling them how to do it. You can also check the proposals quickly during a class period.

- Students should decide how to measure the rate of reaction when the acids come in contact with the magnesium. The rate reaction can be measured in several ways. Students can measure the amount of time required for a set amount of hydrogen gas to be produced or how much gas is produced in a set amount of time. They can also measure the amount of time it takes for the reaction to finish. The students should be able to explain why they decided to use one technique rather than another.

- When checking the students' proposals, make sure that they use the same volume of acid and the same mass of magnesium across all test conditions.

Topic Connections

Table 19.2 provides an overview of the scientific practices, crosscutting concepts, disciplinary core ideas, and supporting ideas at the heart of this lab investigation. In addition, it lists NOS and NOSI concepts for the explicit and reflective discussion. Finally, it lists literacy and mathematics skills (*CCSS ELA* and *CCSS Mathematics*) that are addressed during the investigation.

TABLE 19.2

Lab 19 alignment with standards

Scientific practices	• Asking questions and defining problems • Developing and using models • Planning and carrying out investigations • Analyzing and interpreting data • Using mathematics and computational thinking • Constructing explanations and designing solutions • Engaging in argument from evidence • Obtaining, evaluating, and communicating information
Crosscutting concepts	• Cause and effect: Mechanism and explanation • Systems and system models
Core ideas	• PS1.A: Structure and properties of matter • PS1.B: Chemical reactions
Supporting ideas	• Brønsted-Lowry acid-base theory • The nature of ions and ionic compounds • Bond and bond energy
NOS and NOSI concepts	• Imagination and creativity in science • Nature and role of experiments
Literacy connections (*CCSS ELA*)	• *Reading:* Key ideas and details, craft and structure, integration of knowledge and ideas • *Writing:* Text types and purposes, production and distribution of writing, research to build and present knowledge, range of writing • *Speaking and listening:* Comprehension and collaboration, presentation of knowledge and ideas
Mathematics connections (*CCSS Mathematics*)	• Reason abstractly and quantitatively • Use appropriate tools strategically • Attend to precision • Look for and express regularity in repeated reasoning

Lab Handout

Lab 19. Strong and Weak Acids: Why Do Strong and Weak Acids Behave in Different Manners Even Though They Have the Same Chemical Properties?

Introduction

Johannes Nicolaus Brønsted and Thomas Martin Lowry published nearly identical explanations for the nature of acids and bases in 1923. These two explanations were later combined into a single explanation, which is now known as the Brønsted-Lowry acid-base theory. This theory defines acids and bases in terms of how molecules interact with *hydrogen ions*. A hydrogen ion is just a proton (see Figure L19.1). An *acid*, according to the Brønsted-Lowry definition, is any substance from which a proton can be removed, and a base is any substance that can remove a proton from an acid molecule. In an acid, the hydrogen ion is bonded to the rest of the molecule. It therefore takes energy to break that bond. So an acid molecule does not "give up" or "donate" a proton, it has it taken away. When a base molecule interacts with an acid molecule, it will (if it is strong enough) rip the proton off the acid molecule.

FIGURE L19.1 _____

A hydrogen atom and a hydrogen ion

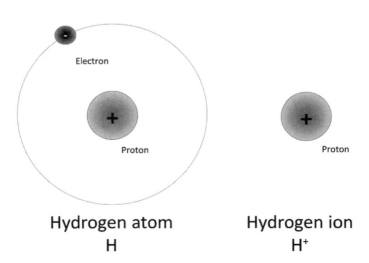

Hydrogen atom
H

Hydrogen ion
H⁺

To illustrate how this works, consider what happens when hydrogen chloride (HCl) is mixed with water. In this situation, the water is able to remove a proton from the hydrogen chloride. The hydrogen chloride is therefore an acid and the water is therefore a base. In Brønsted's original explanation for the nature of an acid-base interaction, he used H^+ to describe how the proton is removed from an acid by a base. This interaction between the molecules of hydrogen chloride and water can therefore be represented as

$$HCl + H_2O \leftarrow \rightarrow H^+ + Cl^- + H_2O$$

A hydrogen ion (H^+), however, does not exist for very long in water because the proton affinity of H_2O is approximately 799 kJ/mol. As a result, a hydrogen ion quickly combines with a water molecule once it is removed from an acid molecule. This can be represented as

$$H_2O + H^+ \rightarrow H_3O^+$$

Lowry therefore used H_3O^+ rather than H^+ to describe the transfer of protons in his explanation of what happens on a submicroscopic level during an acid-base reaction. In his explanation, Lowry explained that when an acid is added to water, a proton from the acid is split off and taken up by water (the base) to produce hydronium ions (H_3O^+). For the hydrogen chloride and water example, this can be represented as

$$HCl + H_2O \leftarrow \rightarrow Cl^- + H_3O^+$$

The Brønsted-Lowry acid-base theory was a groundbreaking idea when it was first introduced because it was able to explain a wide range of macroscopic observations about the behavior of acids and bases by providing a model for how acid and base molecules interact with each other on the submicroscopic level. In this investigation, you will use the Brønsted-Lowry definition for an acid and a base as a starting point to develop a model that can be used to explain the behavior of strong and weak acids.

Your Task

Develop a model that can be used to explain why strong and weak acids behave in a different manner even though they have the same chemical properties. The two chemical properties of acids that you will focus on during this investigation are electrical conductivity and reactivity with metal. To develop your model, you will first need to determine how acid strength affects electrical conductivity and reaction rate. You will then need to determine how to explain your observations on the macroscopic level by describing the nature or behavior of strong and weak acids on the submicroscopic level. The Brønsted-Lowry definition of acids and bases will serve as the theoretical foundation for your model.

The guiding question for this investigation is, **Why do strong and weak acids behave in different manners even though they have the same chemical properties?**

LAB 19

Materials

You may use any of the following materials during this investigation:

Consumables	Equipment
• 1 M solution of acetic acid, CH_3COOH (weak acid) • 1 M solution of hydrochloric acid, HCl (strong acid) • 1 M solution of sulfuric acid, H_2SO_4 (strong acid) • Magnesium ribbon	• Conductivity tester or probe • Pneumatic trough and tubing • Graduated cylinder (250 ml) • Graduated cylinder (10 ml) • Filtering flask (50 ml) and rubber stopper • Stopwatch • Electronic or triple beam balance

Safety Precautions

Follow all normal lab safety rules. All of the acids you will use are corrosive to eyes, skin, and other body tissues. They are also toxic by ingestion. Magnesium metal is a flammable solid and burns with an intense flame. Keep away from flames. Your teacher will explain relevant and important information about working with the chemicals associated with this investigation. In addition, take the following safety precautions:

- Wear indirectly vented chemical-splash goggles and chemical-resistant gloves and apron while in the laboratory.

- Handle all glassware with care.

- Wash your hands with soap and water before leaving the laboratory.

Investigation Proposal Required? ☐ Yes ☐ No

Getting Started

To answer the guiding question, you will need to first determine how the behavior of strong and weak acids differ in terms of electrical conductivity and reactivity with metals. A conductivity tester or probe can be used to measure the conductivity of the three different acid solutions. You can design and carry out an experiment to determine how acid strength affects reactivity with metal. All of the acids that you will be using react with magnesium to produce hydrogen gas. Your goal is to determine the relationship between acid strength and the rate of this reaction. You will therefore need to determine what type of data to collect, how you will collect the data, and how you will analyze the data to accomplish your goal.

To determine *what type of data you need to collect*, think about the following questions:

- What type of measurements will you need to make during your investigation? You could, for example, measure the amount of H_2 gas that is produced, the time it takes to produce a set amount of gas, or the time it takes for the reaction to go to completion.

- When will you need to take your measurements?

To determine *how you will collect the data*, think about the following questions:

- What equipment can you use to capture and measure the volume of a gas? You could, for example, capture and measure the volume of a gas using water displacement (see Figure L19.2).
- What types of test conditions will you need to set up and how will you do it?
- How will you eliminate confounding variables?
- How will you make sure that your data are of high quality (i.e., how will you reduce error)?
- How often will you collect data and when will you do it?
- How will you keep track of the data you collect and how will you organize it?

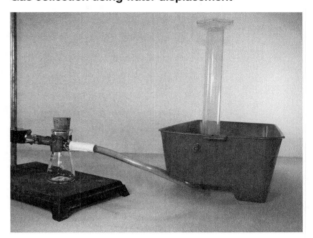

FIGURE L19.2
Gas collection using water displacement

To determine how you will analyze the data, think about the following questions:

- How will you determine if there is a difference between the test conditions?
- What type of calculations will you need to make?
- What type of graph could you create to help make sense of your data?

Once you have carried out your experiment, your group will need to develop your conceptual model. The model should be able to explain why strong and weak acids differ in terms of electrical conductivity and reactivity with metal. The model should also include a description of the interactions that take place between molecules and should be based on the Brønsted-Lowry definition of acids and bases.

Connections to Crosscutting Concepts, the Nature of Science, and the Nature of Scientific Inquiry

As you work through your investigation, be sure to think about

- the importance of developing causal explanations for observations,
- how models are used to help understand natural phenomena,
- the importance of imagination and creativity in science, and
- the nature and role of experiments in science.

LAB 19

Initial Argument

Once your group has finished collecting and analyzing your data, you will need to develop an initial argument. Your argument must include a *claim*, which is your answer to the guiding question. Your argument must also include *evidence* in support of your claim. The evidence is your analysis of the data and your interpretation of what the analysis means. Finally, you must include a *justification* of the evidence in your argument. You will therefore need to use a scientific concept or principle to explain why the evidence that you decided to use is relevant and important. You will create your initial argument on a whiteboard. Your whiteboard must include all the information shown in Figure L19.3.

FIGURE L19.3 _____

Argument presentation on a whiteboard

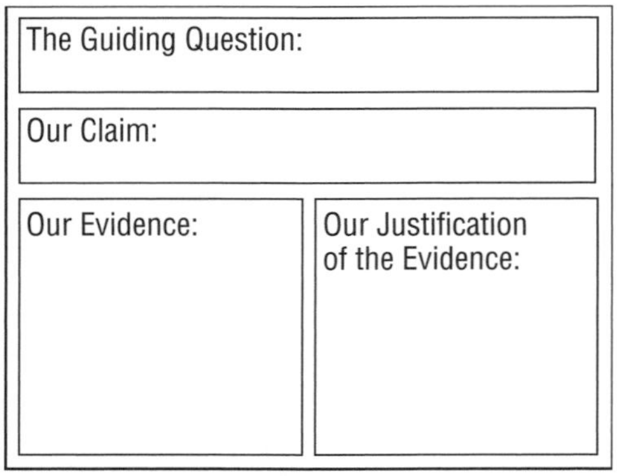

The Guiding Question:	
Our Claim:	
Our Evidence:	Our Justification of the Evidence:

Argumentation Session

The argumentation session allows all of the groups to share their arguments. One member of each group stays at the lab station to share that group's argument, while the other members of the group go to the other lab stations one at a time to listen to and critique the arguments developed by their classmates. The goal of the argumentation session is not to convince others that your argument is the best one; rather, the goal is to identify errors or instances of faulty reasoning in the initial arguments so these mistakes can be fixed. You will therefore need to evaluate the content of the claim, the quality of the evidence used to support the claim, and the strength of the justification of the evidence included in each argument that you see. To critique an argument, you might need more information than what is included on the whiteboard. You might, therefore, need to ask the presenter one or more follow-up questions, such as:

- How did your group collect the data? Why did you use that method?
- What did your group do to make sure the data you collected are reliable? What did you do to decrease measurement error?
- What did your group do to analyze the data, and why did you decide to do it that way? Did you check your calculations?
- Is that the only way to interpret the results of your group's analysis? How do you know that your interpretation of the analysis is appropriate?
- Why did your group decide to present your evidence in that manner?
- What other claims did your group discuss before deciding on that one? Why did you abandon those alternative ideas?
- How confident are you that your group's claim is valid? What could you do to increase your confidence?

Once the argumentation session is complete, you will have a chance to meet with your group and revise your original argument. Your group might need to gather more data or design a way to test one or more alternative claims as part of this process. Remember, your goal at this stage of the investigation is to develop the most valid or acceptable answer to the research question!

Report

Once you have completed your research, you will need to prepare an *investigation report* that consists of three sections that provide answers to the following questions:

1. What question were you trying to answer and why?

2. What did you do during your investigation and why did you conduct your investigation in this way?

3. What is your argument?

Your report should answer these questions in two pages or less. The report must be typed and any diagrams, figures, or tables should be embedded into the document. Be sure to write in a persuasive style; you are trying to convince others that your claim is acceptable or valid!

LAB 19

Lab 19. Strong and Weak Acids: Why Do Strong and Weak Acids Behave in Different Manners Even Though They Have the Same Chemical Properties?

1. Acids and bases are useful reactants in the chemistry laboratory and play an important role in biology and nature. According to the Brønsted-Lowry acid-base theory, what is an acid and what is a base?

2. Strong acids react with reactive metals at a faster rate and conduct electricity better than weak acids. Use what you know about the characteristics of strong and weak acids to explain why the strong acids have a faster reaction rate and are better at conducting electricity than weak acids.

3. Measuring the electrical conductivity of a solution is an example of an experiment.

   a. I agree with this statement.
   b. I disagree with this statement.

   Explain your answer, using an example from your investigation about strong and weak acids.

4. Scientists do not need to be creative or have a good imagination to excel in science.

   a. I agree with this statement.
   b. I disagree with this statement.

   Explain your answer, using an example from your investigation about strong and weak acids.

5. An important goal in science is to develop causal explanations for observations. Explain what a casual explanation is and why these explanations are important, using an example from your investigation about strong and weak acids.

6. Scientists often use models to help them understand natural phenomena. Explain what a model is and why models are important, using an example from your investigation about strong and weak acids.

LAB 20

Teacher Notes

Lab 20. Enthalpy Change of Solution: How Can Chemists Use the Properties of a Solute to Predict If an Enthalpy Change of Solution Will Be Exothermic or Endothermic?

Purpose

The purpose of this lab is to *introduce* students to the concepts of enthalpy, enthalpy change, and the difference between exothermic and endothermic processes. This lab gives students an opportunity to develop a rule for predicting if an enthalpy change of solution will be endothermic or exothermic. Students will also learn about the differences between observations and inferences and between laws and theories in science.

The Content

Enthalpy is a measure of the heat content of a system. The transfer of heat into or out of a system results in a change in enthalpy. The enthalpy change of solution refers to the overall amount of heat energy that is released or absorbed during the dissolution of a solute (at constant pressure). The *enthalpy change of solution* can be either negative (an *exothermic* change) or positive (an *endothermic* change). The enthalpy change of solution is referred to as $\Delta H_{solution}$ and is equal in magnitude to the heat lost from or gained by the surroundings.

When an ionic compound is dissolved in water, the enthalpy change is the net result of two processes. First, an energy input breaks the attractive forces between the ions (ΔH_1) and disrupts the intermolecular forces that exist between the molecules of water (ΔH_2). Second, energy is released as ion-dipole attractive forces form between the dissociated ions and the water molecules (ΔH_3). The enthalpy change of solution can be written as an equation:

$$\Delta H_{solution} = \Delta H_1 + \Delta H_2 + \Delta H_3$$

The dissolution process is exothermic ($\Delta H_{solution} < 0$) when the amount of energy released during the formation of the hydrated ions (ΔH_3) is greater than the amount of energy required to separate the solute particles and solvent particles ($\Delta H_1 + \Delta H_2$). The dissolution process is endothermic ($\Delta H_{solution} > 0$) when the amount of energy released in the formation of the hydrated ions (ΔH_3) is less than the amount of energy required to separate the solute particles and solvent particles ($\Delta H_1 + \Delta H_2$). Figures 20.1 and 20.2 help explain this concept. Figure 20.1 is for an endothermic reaction, where $\Delta H_{solution} > 0$. Figure 20.2 is an exothermic reaction, where $\Delta H_{solution} < 0$. The chemical and physical properties of the solute and the

Enthalpy Change of Solution

How Can Chemists Use the Properties of a Solute to Predict If an Enthalpy Change of Solution Will Be Exothermic or Endothermic?

solvent can therefore be used to predict if an enthalpy change of solution will be endothermic or exothermic.

FIGURE 20.1

Endothermic $\Delta H_{solution}$

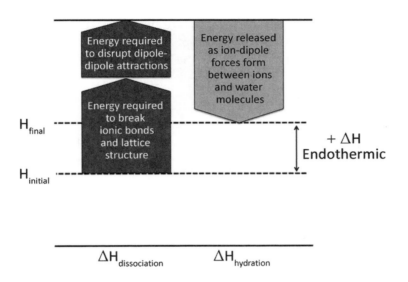

FIGURE 20.2

Exothermic $\Delta H_{solution}$

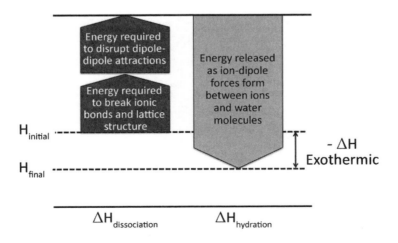

LAB 20

Timeline

The instructional time needed to complete this lab investigation is 180–250 minutes. Appendix 2 (p. 501) provides options for implementing this lab investigation over several class periods. Option E (250 minutes) should be used if students are unfamiliar with scientific writing because this option provides extra instructional time for scaffolding the writing process. You can scaffold the writing process by modeling, providing examples, and providing hints as students write each section of the report. Option F (180 minutes) should be used if students are familiar with scientific writing and have the skills needed to write an investigation report on their own. In option F, students complete stage 6 (writing the investigation report) and stage 8 (revising the investigation report) as homework.

Materials and Preparation

The materials needed to implement this investigation are listed in Table 20.1. The consumables and equipment can be purchased from a science supply company such as Carolina, Flinn Scientific, or Ward's Science. We recommend that you use a set routine for distributing and collecting the materials during the lab investigation. For example, the consumables and equipment for each group can be set up at each group's lab station before class begins, or one member from each group can collect them from a table or a cart when needed during class.

Safety Precautions

Remind students to follow all normal lab safety rules. Calcium chloride, cesium chloride, lithium chloride, sodium chlorate, and sodium iodide are moderately toxic by ingestion and tissue irritants. You will therefore need to explain the potential hazards of working with these chemicals and how to work with hazardous chemicals. In addition, tell students to take the following safety precautions:

- Wear indirectly vented chemical-splash goggles and chemical-resistant gloves and aprons when they are collecting their data.
- Handle all glassware (including thermometers) with care.
- Wash their hands with soap and water when they are done collecting the data.

Laboratory Waste Disposal

The aqueous salt solutions can be disposed of down a drain if the drain is connected to a municipal sewage system. We recommend following Flinn laboratory waste disposal method 26b for these liquids. Information about laboratory waste disposal methods is included in the Flinn Catalog and Reference Manual; you can request a free copy at *www.flinnsci.com*.

Enthalpy Change of Solution

How Can Chemists Use the Properties of a Solute to Predict If an Enthalpy Change of Solution Will Be Exothermic or Endothermic?

TABLE 20.1

Materials list

Item	Quantity
Consumables	
Calcium chloride, $CaCl_2$	5 g per group
Cesium chloride, CsCl	5 g per group
Lithium chloride, LiCl	5 g per group
Potassium chloride, KCl	5 g per group
Sodium chlorate, $NaClO_4$	5 g per group
Sodium chloride, NaCl	5 g per group
Sodium iodide, NaI	5 g per group
Distilled water	200 ml per group
Equipment and other materials	
Polystyrene cups*	2 per group
Thermometer†	1 per group
Graduated cylinder, 25 ml	1 per group
Beaker, 250 ml	3 per group
Stirring rod	1 per group
Electronic or triple beam balance	1 per group
Timer or stopwatch	1 per group
Support stand	1 per group
Ring clamp	1 per group
Chemical scoop	1 per group
Weighing paper or dishes	1 per group
Investigation Proposal C (optional but recommended)	3 per group
Whiteboard, 2' x 3' ‡	1 per group
Lab handout	1 per student
Peer-review guide and instructor scoring rubric	1 per student

* Polystyrene cups are recommended, but a calorimeter can be used instead.

† A temperature probe and sensor interface can be used instead of a thermometer.

‡ As an alternative, students can use computer and presentation software such as Microsoft PowerPoint or Apple Keynote to create their arguments.

LAB 20

Topics for the Explicit and Reflective Discussion

Concepts That Can Be Used to Justify the Evidence

To provide an adequate justification of their evidence, students must explain why they included the evidence in their arguments and make the assumptions underlying their analysis and interpretation of the data explicit. In this investigation, students can use the following concepts to help justify their evidence:

- Characteristics of solutions and the process of dissolution
- Properties of polar molecules (such as water) and ionic compounds
- The nature of intra- and intermolecular attractive forces
- The difference between heat and temperature
- Enthalpy and enthalpy change

We recommend that you discuss these fundamental concepts during the explicit and reflective discussion to help students make this connection.

How to Design Better Investigations

It is important for students to reflect on the strengths and weaknesses of the investigation they designed during the explicit and reflective discussion. Students should therefore be encouraged to discuss ways to eliminate potential flaws, measurement errors, or sources of bias in their investigations. To help students be more reflective about the design of their investigation, you can ask the following questions:

- What were some of the strengths of your investigation? What made it scientific?
- What were some of the weaknesses of your investigation? What made it less scientific?
- If you were to do this investigation again, what would you do to address the weaknesses in your investigation? What could you do to make it more scientific?

Crosscutting Concepts

This investigation is well aligned with two crosscutting concepts found in *A Framework for K–12 Science Education,* and you should review these concepts during the explicit and reflective discussion.

- *Systems and system models:* Scientists often need to use models to understand complex phenomena. In this investigation, for example, students need to be able to understand the difference between the system and the surroundings, and they rely heavily on a conceptual model that explains what is happening when a solute dissolves in a solvent at the submicroscopic level in order to answer the guiding question.

- *Energy and matter: Flows, cycles and conservation.* In science it is important to track how energy and matter flow into, out of, and within a system. In this investigation, for example, students need to be able to track how heat energy flows into and out of a system to develop a rule that can be used to predict if the enthalpy change of solution will be endothermic or exothermic.

The Nature of Science and the Nature of Scientific Inquiry

This investigation is well aligned with two important concepts related to the *nature of science* (NOS) and the *nature of scientific inquiry* (NOSI), and you should review these concepts during the explicit and reflective discussion.

- *The difference between observations and inferences*: An observation is a descriptive statement about a natural phenomenon, whereas an inference is an interpretation of an observation. Students should also understand that current scientific knowledge and the perspectives of individual scientists guide both observations and inferences. Thus, different scientists can have different but equally valid interpretations of the same observations due to differences in their perspectives and background knowledge.

- *The difference between laws and theories in science:* A scientific law describes the behavior of a natural phenomenon or a generalized relationship under certain conditions; a scientific theory is a well-substantiated explanation of some aspect of the natural world. Theories do not become laws even with additional evidence; they explain laws. However, not all scientific laws have an accompanying explanatory theory. It is also important for students to understand that scientists do not discover laws or theories; the scientific community develops them over time.

Hints for Implementing the Lab

- The $\Delta H_{hydration}$ value for the ionic compounds is equal to the sum of the $\Delta H_{hydration}$ values of the individual ions.

- The $\Delta H_{dissociation}$ values represent the combined ionic bond strength of the ionic compound and the hydrogen bonds between the water molecules. The values were calculated from published $\Delta H_{solution}$ values for the ionic compounds and $\Delta H_{hydration}$ values for individual ions. These values can be used because the students only use water as the solvent throughout the investigation.

- Allowing students to design their own procedures for collecting data gives students an opportunity to try, to fail, and to learn from their mistakes. However, you can scaffold students as they develop their procedure by having them fill out an investigation proposal. The proposal provides a way for you to offer students hints and suggestions without telling them how to do it. You can also check the proposals quickly during a class period.

- Allow groups to use different procedures to collect data. For example, some groups might decide to use the same mass of the ionic compounds, while others may decide to use the same number of moles of each ionic compound across their tests. This often results in rich discussion during the argumentation session.

- Encourage students to use small amounts of each ionic compound and water in the calorimeters. One gram of ionic compound and 10 ml of water work well.

- Students should be encouraged to present their rule as a mathematical equation.

Topic Connections

Table 20.2 provides an overview of the scientific practices, crosscutting concepts, disciplinary core ideas, and supporting ideas at the heart of this lab investigation. In addition, it lists NOS and NOSI concepts for the explicit and reflective discussion. Finally, it lists literacy and mathematics skills (*CCSS ELA* and *CCSS Mathematics*) that are addressed during the investigation.

TABLE 20.2

Lab 20 alignment with standards

Scientific practices	• Asking questions and defining problems • Developing and using models • Planning and carrying out investigations • Analyzing and interpreting data • Using mathematics and computational thinking • Constructing explanations and designing solutions • Engaging in argument from evidence • Obtaining, evaluating, and communicating information
Crosscutting concepts	• Systems and system models • Energy and matter: Flows, cycles, and conservation
Core ideas	• PS1.A: Structure and properties of matter • PS1.B: Chemical reactions
Supporting ideas	• Solutes • Solvents • Solutions • Dissolution • Endothermic • Exothermic • Attractive forces • Ions and ionic compounds • Molecular-kinetic theory of matter
NOS and NOSI concepts	• Observations and inferences • Scientific laws and theories
Literacy connections (*CCSS ELA*)	• *Reading:* Key ideas and details, craft and structure, integration of knowledge and ideas • *Writing:* Text types and purposes, production and distribution of writing, research to build and present knowledge, range of writing • *Speaking and listening:* Comprehension and collaboration, presentation of knowledge and ideas
Mathematics connections (*CCSS Mathematics*)	• Reason abstractly and quantitatively • Model with mathematics • Attend to precision

LAB 20

Lab Handout

Lab 20. Enthalpy Change of Solution: How Can Chemists Use the Properties of a Solute to Predict If an Enthalpy Change of Solution Will Be Exothermic or Endothermic?

Introduction

Thermodynamics is the study of energy changes in a system. Thermodynamics is an important field of study in chemistry because energy changes occur during chemical reactions, when solutes are dissolved in solvents, and when matter goes through a change of state. Chemists often describe the energy changes that take place in these situations in terms of heat content. *Enthalpy* is a measure of the heat content of a system. The transfer of heat into or out of a system results in a change in enthalpy. This change in the heat content of a system is symbolized as ΔH (delta H). The unit of measurement for an enthalpy change is kilojoules per mole (kJ/mol).

The enthalpy change that occurs when a solute is dissolved in water is called the heat of solution or the *enthalpy change of solution* ($\Delta H_{solution}$). The enthalpy change of solution is equal in magnitude to the heat energy lost from or gained by the surroundings. When heat energy is lost from the system and gained by the surroundings, the enthalpy change of solution is described as *exothermic*. An *endothermic* enthalpy change of solution, in contrast, occurs when the system gains heat energy from the surroundings.

The overall energy change that occurs when a solute is dissolved in water (i.e., the $\Delta H_{solution}$) is the result of two key processes. First, an input of energy breaks the attractive forces holding the particles in the solute together and disrupts the hydrogen bonds holding the water molecules together. The system *gains* energy and the surroundings *lose* energy during this process. For the purposes of this investigation, this change in energy will be called the *enthalpy change of dissociation* ($\Delta H_{dissociation}$). Second, energy is released as attractive forces form between the particles of the solute and the molecules of water. The system *loses* energy and the surroundings *gain* energy during this process. The energy change that occurs during this process is called the *enthalpy change of hydration* ($\Delta H_{hydration}$). An illustration of the energy inputs and outputs associated with dissolution of an ionic compound is provided in Figure L20.1.

As described earlier, the dissolution of a solute involves both a gain and a loss of energy, and the $\Delta H_{solution}$ of a solute can be either endothermic or exothermic depending on the net amount of energy that is lost from or gained by the system. The net energy change of the system will depend, in part, on the unique properties of the solute. Solutes can be composed

FIGURE L20.1

The dissociation process and hydration process that take place when an ionic compound dissolves in water

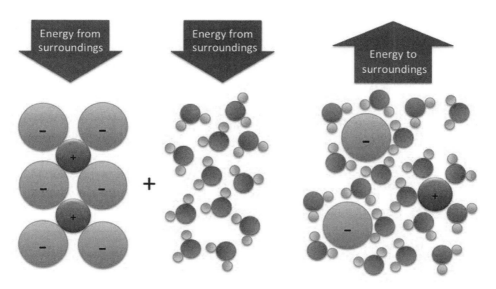

Enthalpy change of dissociation
Energy is absorbed by the system
Ions separate and water molecules separate

Enthalpy change of hydration
Energy is released by the system
Ions and water molecules combine

of different types of particles, and attractive forces hold these particles together. The nature and strength of these attractive forces will influence the amount of energy that is required to break apart the solute. In addition, the particles that make up a solute will differ in terms of the strength of their electrical charge. Some solute particles, as a result, will attract water molecules better than others. The strength of attraction that exists between solute particles and water molecules will influence the amount of energy that is released when the solute particles and the water molecules combine. The physical and chemical properties of the solute, as a result, will affect the $\Delta H_{solution}$.

In this investigation, you will be given five different ionic compounds. You will then determine if the $\Delta H_{solution}$ for each ionic compound is endothermic or exothermic by mixing the solute with water and measuring the resulting temperature change. Next, you will use a table of $\Delta H_{dissociation}$ and $\Delta H_{hydration}$ values to develop a rule that you can use to predict if the $\Delta H_{solution}$ of other ionic compounds will be endothermic or exothermic. The $\Delta H_{dissociation}$ values reflect the amount of energy needed to separate the ions in the solute and the amount of energy needed to disrupt the attractive forces between the water molecules. The $\Delta H_{hydration}$ values reflect the energy that is released when ion-dipole

LAB 20

forces form between the individual ions and the water molecules. You will then be given an opportunity to test your rule with two other ionic compounds to determine if you can use it to make accurate predictions.

Your Task

Develop a rule that chemists can use to determine if the enthalpy change of solution ($\Delta H_{solution}$) for a given ionic compound will be endothermic or exothermic based on the properties of the solute.

The guiding question for this investigation is, **How can chemists use the properties of a solute to predict if an enthalpy change of solution will be exothermic or endothermic?**

Materials

You may use any of the following materials during this investigation:

Consumables	Equipment
• Calcium chloride, $CaCl_2$, 5 grams • Cesium chloride, CsCl, 5 grams • Lithium chloride, LiCl, 5 grams • Potassium chloride, KCl, 5 grams • Sodium chlorate, $NaClO_4$, 5 grams • Sodium chloride, NaCl, 5 grams • Sodium iodide, NaI, 5 grams • Distilled water	• 2 polystyrene cups (or a calorimeter) • Thermometer (or temperature probe and sensor interface) • Graduated cylinder (25 ml) • 3 beakers (each 250 ml) • Stirring rod • Electronic or triple beam balance • Timer or stopwatch • Support stand and ring clamp • Chemical scoop • Weighing paper or dishes

Safety Precautions

Follow all normal lab safety rules. Lithium chloride, calcium chloride, cesium chloride, sodium chlorate, and sodium iodide are all moderately toxic by ingestion and are tissue irritants. Your teacher will explain relevant and important information about working with the chemicals associated with this investigation. In addition, take the following safety precautions:

- Wear indirectly vented chemical-splash goggles and chemical-resistant gloves and apron while in the laboratory.

- Handle all glassware with care.

- Wash your hands with soap and water before leaving the laboratory.

Enthalpy Change of Solution

How Can Chemists Use the Properties of a Solute to Predict If an Enthalpy Change of Solution Will Be Exothermic or Endothermic?

Investigation Proposal Required? ☐ Yes ☐ No

Getting Started

The first step in developing your rule is to determine if the $\Delta H_{solution}$ for LiCl, CaCl$_2$, KCl, NaCl, and CsCl is endothermic or exothermic. To accomplish this step, you will need to dissolve each ionic compound in water and measure the resulting temperature change using a calorimeter. A calorimeter is an insulated container that is designed to prevent heat loss to the atmosphere. A simple calorimeter can be made from two polystyrene cups, a support stand, and a ring clamp (see Figure L20.2). Once you have set up a simple calorimeter, you must determine what type of data you need to collect, how you will collect the data, and how you will analyze the data.

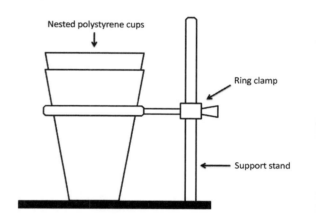

FIGURE L20.2

A simple calorimeter

To determine *what type of data you need to collect,* think about the following questions:

- What type of measurements or observations will you need to record during each test?
- How often will you need to make these measurements or observations?

To determine *how you will collect the data,* think about the following questions:

- How much water will you use in the calorimeter?
- Will the same amount of water be used for each test?
- How much of each ionic compound will you need to use?
- Will you need to use the same amount of each ionic compound for each test? If so, does it need to be the same amount in terms of mass or in terms of moles?
- What will you do to reduce measurement error?
- How will you keep track of the data you collect and how will you organize it?

To determine *how you will analyze the data,* think about the following questions:

- What type of calculations will you need to make?
- What type of graph could you create to help make sense of your data?

Once you have carried out your series of tests, your group will need to develop your rule. Table L20.1 (p. 324) provides the $\Delta H_{dissociation}$ and $\Delta H_{hydration}$ values for each ionic compound. Positive values indicate that the energy is being absorbed by the system, and

LAB 20

negative values indicate that energy is being released by the system. As you develop your rule, think about how you could present it as a mathematical equation.

TABLE L20.1

$\Delta H_{dissociation}$ and $\Delta H_{hydration}$ values for the ionic compounds used in this investigation

Ionic compound	$\Delta H_{dissociation}$ (kJ/mol)*	$\Delta H_{hydration}$ (kJ/mol)
LiCl	853	−883
$CaCl_2$	2,258	−2378
KCl	715	−685
NaCl	788	−784
CsCl	657	−639
$NaClO_4$	658	−644
NaI	693	−701

* This value reflects the energy required to break apart the ions in the solute and the energy required to disrupt the hydrogen bonds between the water molecules.

The last step is to test your rule. To accomplish this goal, you will need to determine if you can use your rule to make accurate predictions. You should, in other words, be able to use your rule to predict if the $\Delta H_{solution}$ of $NaClO_4$ and NaI will be exothermic or endothermic. If you are able to make accurate predictions about the $\Delta H_{solution}$ for these two ionic compounds, then you will be able to generate the evidence you need to convince others that the rule that you developed is valid.

Connections to Crosscutting Concepts, the Nature of Science, and the Nature of Scientific Inquiry

As you work through your investigation, be sure to think about

- the importance of defining the system under study and then using a model to make sense of it,
- the importance of tracking how matter and energy move into and within a system,
- the difference between observations and inferences in science, and
- the difference between laws and theories in science.

Initial Argument

Once your group has finished collecting and analyzing your data, you will need to develop an initial argument. Your argument must include a *claim*, which is your answer to the guiding question. Your argument must also include *evidence* in support of your claim. The evidence is your analysis of the data and your interpretation of what the analysis means. Finally, you must include a *justification* of the evidence in your argument. You will therefore need to use a scientific concept or principle to explain why the evidence that you decided to use is relevant and important. You will create your initial argument on a whiteboard. Your whiteboard must include all the information shown in Figure L20.3.

Argumentation Session

The argumentation session allows all of the groups to share their arguments. One member of each group stays at the lab station to share that group's argument, while the other members of the group go to the other lab stations one at a time to listen to and critique the arguments developed by their classmates. The goal of the argumentation session is not to convince others that your argument is the best one; rather, the goal is to identify errors or instances of faulty reasoning in the initial arguments so these mistakes can be fixed. You will therefore need to evaluate the content of the claim, the quality of the evidence used to support the claim, and the strength of the justification of the evidence included in each argument that you see. To critique an argument, you might need more information than what is included on the whiteboard. You might, therefore, need to ask the presenter one or more follow-up questions, such as:

FIGURE L20.3

Argument presentation on a whiteboard

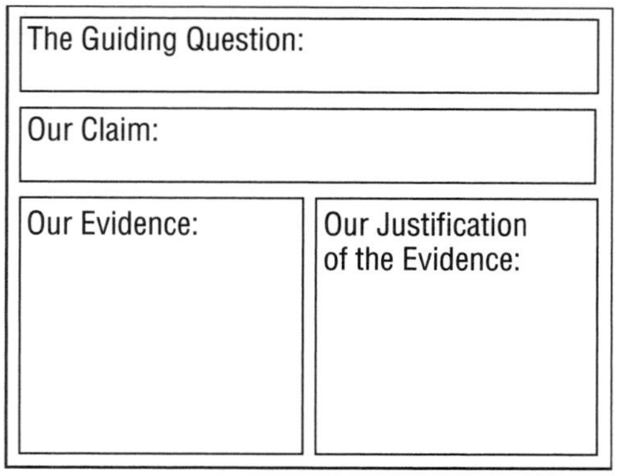

The Guiding Question:	
Our Claim:	
Our Evidence:	Our Justification of the Evidence:

- What did your group do to make sure the data you collected are reliable? What did you do to decrease measurement error?
- What did your group do to analyze the data, and why did you decide to do it that way? Did you check your calculations?
- Is that the only way to interpret the results of your group's analysis? How do you know that your interpretation of the analysis is appropriate?
- Why did your group decide to present your evidence in that manner?
- What other claims did your group discuss before deciding on that one? Why did you abandon those alternative ideas?
- How confident are you that your group's claim is valid? What could you do to increase your confidence?

Once the argumentation session is complete, you will have a chance to meet with your group and revise your original argument. Your group might need to gather more data or design a way to test one or more alternative claims as part of this process. Remember, your goal at this stage of the investigation is to develop the most valid or acceptable answer to the research question!

LAB 20

Report

Once you have completed your research, you will need to prepare an *investigation report* that consists of three sections that provide answers to the following questions:

1. What question were you trying to answer and why?

2. What did you do during your investigation and why did you conduct your investigation in this way?

3. What is your argument?

Your report should answer these questions in two pages or less. The report must be typed and any diagrams, figures, or tables should be embedded into the document. Be sure to write in a persuasive style; you are trying to convince others that your claim is acceptable or valid!

National Science Teachers Association

Checkout Questions

Lab 20. Enthalpy Change of Solution: How Can Chemists Use the Properties of a Solute to Predict If an Enthalpy Change of Solution Will Be Exothermic or Endothermic?

Use the following information to answer questions 1–3. When chromium(II) chloride, $CrCl_2$, is dissolved in water, the temperature of the water decreases.

1. Is this an endothermic or exothermic process?

   a. Endothermic

   b. Exothermic

   Explain why.

2. What is the $\Delta H_{solution}$?

   a. > 0

   b. < 0

   Explain why.

LAB 20

3. Which is stronger, the attractive force between the water molecules and the chromium and chloride ions or the combined ionic bond strength of the $CrCl_2$ and the intermolecular forces between the water molecules?

   a. The attractive forces between the water molecules and chromium and chloride ions

   b. The combined ionic bond strength of the $CrCl_2$ and the intermolecular forces between the water molecules

   Use what you know about the process of dissolving to explain your answer.

4. "Heat flowed into the system" is an observation.

   a. I agree with this statement.

   b. I disagree with this statement.

   Explain your answer, using an example from your investigation about enthalpy change of solution.

5. Theories can turn into laws.

   a. I agree with this statement.
   b. I disagree with this statement.

   Explain your answer, using an example from your investigation about enthalpy
   change of solution.

6. Scientists often need to track how matter and energy move into, out of, and within
   a system. Explain why this is important, using an example from your investigation
   about enthalpy change of solution.

7. Scientists often need to define a system under study in order to study it. Explain
   why it is important to define a system under study, using an example from your
   investigation about enthalpy change of solution.

LAB 21

Lab 21. Reaction Rates: Why Do Changes in Temperature and Reactant Concentration Affect the Rate of a Reaction?

Purpose

The purpose of this lab is to *introduce* students to the concept of a reaction rate and some of the factors that affect it. This lab gives students an opportunity to use the molecular-kinetic theory of matter to develop a model that can help them explain how temperature and reactant concentration affects a reaction rate. Students will also learn about the role experiments play in science and why scientists need to be creative and have a good imagination.

The Content

The collision theory of reaction rates suggests that reactant molecules must collide with each other for a reaction to happen. Not all collisions between reactant molecules, however, will result in the formation of new molecules. For colliding molecules to break apart and then form product molecules, two conditions must be satisfied: (1) the reactant molecules must collide in an appropriate orientation, and (2) the collision energy must exceed a specific energy level, which is called the *activation energy* of the reaction. Reactions that occur quickly tend to have a low activation energy level. Almost all the collisions that take place between molecules during these reactions, as a result, have enough energy to overcome the energy barrier for the reaction. It therefore takes less time for reactants to transform into products when a reaction has a low activation energy level. Reactions that occur slowly, in contrast, tend to have a high activation energy level. When the activation energy of a reaction is high, only a small fraction of the colliding molecules will have sufficient energy to overcome the energy barrier. It therefore takes a greater amount of time for the reactants to transform into products under these conditions.

The collision theory of reaction rates also explains why a change in temperature or a change in reactant concentration will result in a change in the rate of a reaction. A change in temperature influences the rate of a reaction because it changes the average kinetic energy of the reactant molecules. The fraction of molecules with sufficient kinetic energy to overcome the activation energy barrier of a reaction, as a result, goes up when the temperature increases and goes down when temperature decreases. A change in the concentration of the reactant, in contrast, changes the rate of collisions that occur between the reactant molecules. An increase in the rate of collisions results in an increase in the total number of effective collisions that occur during a reaction. An increase in the rate of effective collisions, however, does not change the fraction of collisions that are effective (which depends on the orientation and the amount of kinetic energy). Therefore, as the reactant concentration goes up, the rate of reaction tends to increase as well. Chemists, as a result, can use these two factors to control the rate of a reaction.

Timeline

The instructional time needed to complete this lab investigation is 180–250 minutes. Appendix 2 (p. 501) provides options for implementing this lab investigation over several class periods. Option E (250 minutes) should be used if students are unfamiliar with scientific writing because this option provides extra instructional time for scaffolding the writing process. You can scaffold the writing process by modeling, providing examples, and providing hints as students write each section of the report. Option F (180 minutes) should be used if students are familiar with scientific writing and have the skills needed to write an investigation report on their own. In option F, students complete stage 6 (writing the investigation report) and stage 8 (revising the investigation report) as homework.

Materials and Preparation

The materials needed to implement this investigation are listed in Table 21.1 (p. 332). The consumables and equipment can be purchased from a science supply company such as Carolina, Flinn Scientific, or Ward's Science. We recommend that you use a set routine for distributing and collecting the materials during the lab investigation. For example, the consumables and equipment for each group can be set up at each group's lab station before class begins, or one member from each group can collect them from a table or a cart as needed during class.

Safety Precautions

Remind students to follow all normal lab safety rules. Hydrochloric acid is a corrosive liquid, and magnesium metal is a flammable solid and burns with an intense flame. You will therefore need to explain the potential hazards of working with hydrochloric acid and magnesium and how to work with hazardous chemicals. In addition, tell students to take the following safety precautions:

- Wear indirectly vented chemical-splash goggles and chemical-resistant gloves and aprons when they are collecting their data.
- Heat the hydrochloric acid using a hot bath and keep the temperature of the bath between 20°C and 60°C.
- Use caution when working with hot plates, and keep them away from water and other liquids.
- Handle all glassware (including thermometers) with care.
- Wash their hands with soap and water when they are done collecting the data.

LAB 21

TABLE 21.1
Materials list

Item	Quantity
Consumables	
18-gauge copper wire, 20 cm long	2 pieces per group
Magnesium ribbon	30 cm per group
3 M hydrochloric acid (HCl)	50 ml per group
2 M HCl	50 ml per group
1 M HCl	50 ml per group
0.5 M HCl	50 ml per group
0.1 M HCl	50 ml per group
Ice (for ice baths)	As needed
Equipment and other materials	
Pyrex test tubes	2 per group
Test tube rack	1 per group
Graduated cylinder, 25 ml	1 per group
Beakers, 250 ml	2 per group
Thermometer (or temperature probe)	1 per group
Hot plate	1 per group
Electronic or triple beam balance	1 per group
pH paper	As needed
Investigation Proposal A (optional but recommended)	2 per group
Whiteboard, 2' x 3' *	1 per group
Lab handout	1 per student
Peer-review guide and instructor scoring rubric	1 per student

*As an alternative, students can use computer and presentation software such as Microsoft PowerPoint or Apple Keynote to create their arguments.

Laboratory Waste Disposal

We recommend following Flinn laboratory waste disposal methods 3 to dispose of the magnesium ribbon and 24b to dispose of the waste solutions. Information about laboratory waste disposal methods is included in the Flinn Catalog and Reference Manual; you can request a free copy at *www.flinnsci.com*.

Topics for the Explicit and Reflective Discussion

Concepts That Can Be Used to Justify the Evidence

To provide an adequate justification of their evidence, students must explain why they included the evidence in their arguments and make the assumptions underlying their analysis and interpretation of the data explicit. In this investigation, students can use the following concepts to help justify their evidence:

- Molecular-kinetic theory of matter
- Collision theory of reactions
- How chemical equations are used to represent chemical reactions

We recommend that you discuss these fundamental concepts during the explicit and reflective discussion to help students make this connection.

How to Design Better Investigations

It is important for students to reflect on the strengths and weaknesses of the investigation they designed during the explicit and reflective discussion. Students should therefore be encouraged to discuss ways to eliminate potential flaws, measurement errors, or sources of bias in their investigations. To help students be more reflective about the design of their investigations, you can ask the following questions:

- What were some of the strengths of your investigation? What made it scientific?
- What were some of the weaknesses of your investigation? What made it less scientific?
- If you were to do this investigation again, what would you do to address the weaknesses in your investigation? What could you do to make it more scientific?

Crosscutting Concepts

This investigation is well aligned with two crosscutting concepts found in *A Framework for K–12 Science Education,* and you should review these concepts during the explicit and reflective discussion.

LAB 21

- *Cause and effect: Mechanism and explanation*: Natural phenomena have causes, and uncovering causal relationships (e.g., why changes in temperature and reactant concentration affect the rate of a chemical reaction) is a major activity of science.
- *Systems and system models:* Scientists often need to use models to understand complex phenomena. In this investigation, students are directed to develop a model to help explain what is happening during a chemical reaction at the submicroscopic level.

The Nature of Science and the Nature of Scientific Inquiry

This investigation is well aligned with two important concepts related to the *nature of science* (NOS) and the *nature of scientific inquiry* (NOSI), and you should review these concepts during the explicit and reflective discussion.

- *The importance of imagination and creativity in science:* Students should learn that developing explanations for or models of natural phenomena and then figuring out how they can be put to the test of reality is as creative as writing poetry, composing music, or designing skyscrapers. Scientists must also use their imagination and creativity to figure out new ways to test ideas and collect or analyze data.
- *Nature and role of experiments:* Scientists use experiments to test the validity of a hypothesis (i.e., a tentative explanation) for an observed phenomenon. Experiments include a test and the formulation of predictions (expected results) if the test is conducted and the hypothesis is valid. The experiment is then carried out and the predictions are compared with the observed results of the experiment. If the predictions match the observed results, then the hypothesis is supported. If the observed results do not match the prediction, then the hypothesis is not supported. A signature feature of an experiment is the control of variables to help eliminate alternative explanations for observed results.

Hints for Implementing the Lab

- Allowing students to design their own procedures for collecting data gives students an opportunity to try, to fail, and to learn from their mistakes. However, you can scaffold students as they develop their procedure by having them fill out an investigation proposal. These proposals provide a way for you to offer students hints and suggestions without telling them how to do it. You can also check the proposals quickly during a class period. We recommend that students fill out one investigation proposal for each of their experiments.
- We recommend that students design and conduct the temperature experiment first and then design and conduct the concentration experiment. The students will

be able to design a better concentration experiment after doing the temperature experiment because they will be learn from their mistakes.

- Students will need to test at least three different temperatures in their first experiment and three different concentrations of reactants in their second experiment.

- Remind students to include multiple trials in each experiment and to average their results to help reduce measurement error.

- Be sure that students are calculating rates in terms of moles per second. Students will therefore need to determine the moles of magnesium and hydrochloric acid that they are using during the experiments. Make sure they are including a way to determine the moles of reactants in their proposals.

- We recommend that you show students how to make a cage out the copper wire as part of the tool talk. The students need to use the cage to keep the magnesium ribbon submerged in the hydrochloric acid.

- Show students how to make a hot-water bath and an ice bath as part of the tool talk. Make sure students do not heat the hydrochloric acid directly on the hot plate. The temperature of the hot plate should not exceed 60°C.

- Students only need to use about 15 ml of hydrochloric acid and 1 cm of magnesium per test.

- Remind students that they will need to test their models. They will therefore need to save some of the magnesium ribbon and at least one molarity of hydrochloric acid for these tests.

- Allow the groups to measure reaction rate in different ways. Students can measure the reaction rate by timing how long it takes for the magnesium to dissolve or by timing how long bubbles of gas evolve after the magnesium and the hydrochloric acid are mixed. Students will find that one of the two approaches is more prone to measurement error, which will stimulate interesting discussions during the argumentation sessions.

- Remind students that they can use pH paper to determine if any acid remains in the flask after the reaction is complete.

Topic Connections

Table 21.2 (p. 336) provides an overview of the scientific practices, crosscutting concepts, disciplinary core ideas, and supporting ideas at the heart of this lab investigation. In addition, it lists NOS and NOSI concepts for the explicit and reflective discussion. Finally, it lists literacy and mathematics skills (*CCSS ELA* and *CCSS Mathematics*) that are addressed during the investigation.

LAB 21

TABLE 21.2 _____

Lab 21 alignment with standards

Scientific practices	• Asking questions and defining problems • Developing and using models • Planning and carrying out investigations • Analyzing and interpreting data • Using mathematics and computational thinking • Constructing explanations and designing solutions • Engaging in argument from evidence • Obtaining, evaluating, and communicating information
Crosscutting concepts	• Cause and effect: Mechanism and explanation • Systems and system models
Core idea	• PS1.B: Chemical reactions
Supporting ideas	• Reaction rates • Molecular-kinetic theory of matter • Collision theory of reaction • Moles and molar mass
NOS and NOSI concepts	• Imagination and creativity in science • Nature and role of experiments
Literacy connections (*CCSS ELA*)	• *Reading:* Key ideas and details, craft and structure, integration of knowledge and ideas • *Writing:* Text types and purposes, production and distribution of writing, research to build and present knowledge, range of writing • *Speaking and listening:* Comprehension and collaboration, presentation of knowledge and ideas
Mathematics connections (*CCSS Mathematics*)	• Reason abstractly and quantitatively • Model with mathematics

Lab Handout

Lab 21. Reaction Rates: Why Do Changes in Temperature and Reactant Concentration Affect the Rate of a Reaction?

Introduction

The molecular-kinetic theory of matter suggests that all matter is made up of submicroscopic particles called atoms that are constantly in motion. These atoms can be joined together to form molecules. Atoms have kinetic energy because they move and vibrate. The more kinetic energy an atom has, the faster it moves or vibrates. *Temperature* is a measurement of the average kinetic energy of all the atoms in a substance. The average kinetic energy of the particles within a substance increases and decreases as it changes temperature. *Heat*, in contrast, is the total kinetic energy of all the particles in a substance.

A chemical reaction, as you have learned, is simply the rearrangement of atoms. The substances (elements and/or compounds) that are changed into other substances during a chemical reaction are called *reactants*. The substances that are produced as a result of a chemical reaction are called *products*. Chemical equations show the reactants and products of a chemical reaction. A chemical equation includes the chemical formulas of the reactants and the products. The products and reactants are separated by an arrow symbol ($\rightarrow$), and the chemical formula for each individual substance is separated by a plus sign (+).

A balanced chemical equation tells us the nature of the products and the amount of product that is formed from a given amount of reactants. A balanced chemical equation, however, tells us little about how long it takes for the reaction to happen. Some chemical reactions, such as the rusting of iron, happen slowly over time, while others, such as the burning of gasoline, are almost instantaneous. The speed of any reaction is indicated by its *reaction rate*, which is a measure of how quickly the reactants transform into products. As shown in Figure L21.1 (p. 338), a reaction begins with only reactant molecules. Over time, the reactant molecules interact with each other and form product molecules. The concentration of reactant molecules and the product molecules, as a result, will change during the process of a reaction. The rate of a reaction can therefore be calculated by measuring how the concentration of the reactants decreases or the concentration of the products increases as a function of time. The rate of a reaction can also be measured by timing how long it takes for a product to appear or for a reactant to disappear once the reaction begins.

LAB 21

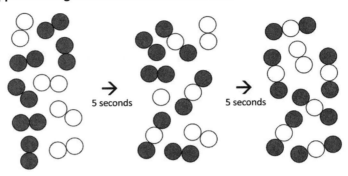

FIGURE L21.1

A model of what happens during a chemical reaction over time

5 seconds

5 seconds

It is important for chemists to understand how and why different factors affect the rate of a chemical reaction so they can make a wide range of products in a safe and economical manner. You will therefore explore two factors that affect the rate of a specific reaction and then develop a conceptual model that you can use to explain your observations and predict the rate of this reaction under different conditions.

Your Task

Determine how temperature and changes in the concentration of a reactant affect the rate of the reaction between magnesium (Mg) and hydrochloric acid (HCl). Then develop a conceptual model that can be used to explain *why* these factors influence reaction rate. Once you have developed your model, you will need to determine if it is consistent with the rates of reaction that you observe under other conditions.

The guiding question of this investigation is, **Why do changes in temperature and reactant concentration affect the rate of a reaction?**

Materials

You may use any of the following materials during your investigation:

Consumables	Equipment
• 2 pieces of 18-gauge copper wire (20 cm) • Magnesium ribbon (30 cm) • 3 M HCl • 2 M HCl • 1 M HCl • 0.5 M HCl • 0.1 M HCl • Ice	• 2 Pyrex test tubes • Test tube rack • Graduated cylinder (25 ml) • 2 beakers (each 250 ml) • Thermometer (or temperature probe) • Hot plate • Electronic or triple beam balance • pH paper

Safety Precautions

Follow all normal lab safety rules. Magnesium is a flammable solid, and hydrochloric acid is a corrosive liquid. Your teacher will explain relevant and important information about working with the chemicals associated with this investigation. In addition, take the following safety precautions:

- Wear indirectly vented chemical-splash goggles and chemical-resistant gloves and apron while in the laboratory.

- Do not heat hydrochloric acid directly on a hot plate; rather, heat the hydrochloric acid using a hot bath and keep the temperature of the bath between 20°C and 60°C.

- Use caution when working with hot plates because they can burn skin. Hot plates also need to be kept away from water and other liquids.

- Handle all glassware with care.

- Wash your hands with soap and water before leaving the laboratory.

Investigation Proposal Required?　☐ Yes　　☐ No

Getting Started

The first step in developing your model is to design and carry out two experiments. The goal of the first experiment will be to determine how temperature affects reaction rate. The goal of the second experiment will be to determine how reactant concentration affects reaction rate. For these two experiments, you will focus on the reaction of magnesium and hydrochloric acid. These two chemicals react to form hydrogen gas and magnesium chloride. The equation for this reaction is

$$Mg(s) + 2HCl(aq) \rightarrow H_2(g) + MgCl_2(aq)$$

You can measure the reaction rate by simply timing how long it takes for the solid magnesium to be no longer visible once it is mixed with the hydrochloric acid. You can also measure the reaction rate by timing how long hydrogen gas is produced after the magnesium and hydrochloric acid are mixed. The unit of measurement for a reaction rate is mol/sec. It will therefore be important for you to determine how many moles of each reactant you used for each test.

To design your two experiments, you will need to decide what type of data you need to collect, how you will collect the data, and how you will analyze the data.

To determine *what type of data you need to collect*, think about the following questions:

- How will you determine the number of moles of each reactant at the beginning of the reaction?

LAB 21

- How will you know when the reaction starts and when it is finished?
- What type of measurements will you need to record during each experiment?
- When will you need to make these measurements or observations?

To determine *how you will collect the data*, think about the following questions:

- How much magnesium ribbon will you use in each test?
- How much hydrochloric acid will you need to use to submerge the magnesium ribbon?
- How will you prevent the magnesium ribbon from floating on top of the hydrochloric acid? One way to prevent the magnesium from floating on top of the hydrochloric acid is to use copper wire to create a cage (see Figure L21.2).
- What will serve as your independent variable?
- What types of comparisons will you need to make?
- How will you hold other variables constant?
- What will you do to reduce measurement error?
- How will you keep track of the data you collect and how will you organize it?

To determine *how you will analyze the data*, think about the following questions:

- What type of calculations will you need to make?
- What type of graph could you create to help make sense of your data?

FIGURE L21.2 _____

How to suspend magnesium (Mg) ribbon in the hydrochloric acid (HCl). The copper wire does not react with the HCl.

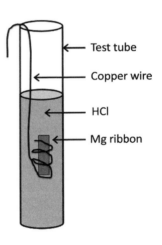

Test tube

Copper wire

HCl

Mg ribbon

Once you have carried out your two experiments, your group will need to develop a conceptual model. This conceptual model must provide an underlying reason for your findings about the effect of temperature and reactant concentration on reaction rate. Your model should also include an explanation of what is happening at the submicroscopic level between and within molecules during a reaction. The molecular-kinetic theory of matter should serve as the theoretical foundation for your model.

The last step in this investigation is to test your model. To accomplish this goal, you can use the same reaction but test different temperatures and concentrations to determine if your model is consistent with the rates of reactions you observe under different conditions. If you can use your model to make accurate predictions about the rate of this reaction under different conditions, then you will be able to generate the evidence you need to convince others that the conceptual model you developed is valid.

Connections to Crosscutting Concepts, the Nature of Science, and the Nature of Scientific Inquiry

As you work through your investigation, be sure to think about

- the importance of developing causal explanations for observations,
- how models are used to help understand natural phenomena,
- the importance of imagination and creativity in science, and
- the role of experiments in science.

Initial Argument

Once your group has finished collecting and analyzing your data, you will need to develop an initial argument. Your argument must include a *claim*, which is your answer to the guiding question. Your argument must also include *evidence* in support of your claim. The evidence is your analysis of the data and your interpretation of what the analysis means. Finally, you must include a *justification* of the evidence in your argument. You will therefore need to use a scientific concept or principle to explain why the evidence that you decided to use is relevant and important. You will create your initial argument on a whiteboard. Your whiteboard must include all the information shown in Figure L21.3.

FIGURE L21.3

Argument presentation on a whiteboard

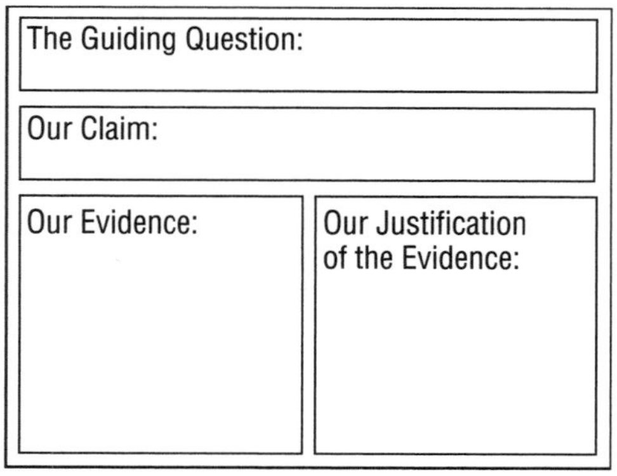

The Guiding Question:	
Our Claim:	
Our Evidence:	Our Justification of the Evidence:

Argumentation Session

The argumentation session allows all of the groups to share their arguments. One member of each group stays at the lab station to share that group's argument, while the other members of the group go to the other lab stations one at a time to listen to and critique the arguments developed by their classmates. The goal of the argumentation session is not to convince others that your argument is the best one; rather, the goal is to identify errors or instances of faulty reasoning in the initial arguments so these mistakes can be fixed. You will therefore need to evaluate the content of the claim, the quality of the evidence used to support the claim, and the strength of the justification of the evidence included in each argument that you see. To critique an argument, you might need more information than what is included on the whiteboard. You might, therefore, need to ask the presenter one or more follow-up questions, such as:

- How did your group collect the data? Why did you use that method?

- What did your group do to make sure the data you collected are reliable? What did you do to decrease measurement error?

- What did your group do to analyze the data, and why did you decide to do it that way? Did you check your calculations?

- Is that the only way to interpret the results of your group's analysis? How do you know that your interpretation of the analysis is appropriate?

- Why did your group decide to present your evidence in that manner?

- What other claims did your group discuss before deciding on that one? Why did you abandon those alternative ideas?

- How confident are you that your group's claim is valid? What could you do to increase your confidence?

Once the argumentation session is complete, you will have a chance to meet with your group and revise your original argument. Your group might need to gather more data or design a way to test one or more alternative claims as part of this process. Remember, your goal at this stage of the investigation is to develop the most valid or acceptable answer to the research question!

Report

Once you have completed your research, you will need to prepare an *investigation report* that consists of three sections that provide answers to the following questions:

1. What question were you trying to answer and why?

2. What did you do during your investigation and why did you conduct your investigation in this way?

3. What is your argument?

Your report should answer these questions in two pages or less. The report must be typed and any diagrams, figures, or tables should be embedded into the document. Be sure to write in a persuasive style; you are trying to convince others that your claim is acceptable or valid!

Checkout Questions

Lab 21. Reaction Rates: Why Do Changes in Temperature and Reactant Concentration Affect the Rate of a Reaction?

Chemists must be able to measure and control the rate of a chemical reaction in order to produce substances in a safe and economical way. Chemists can slow down a reaction rate by lowering the temperature of the reaction or by diluting the concentration of the reactants.

1. Describe the concept of a reaction rate.

2. Describe the molecular-kinetic theory of matter.

3. Use what you know about reaction rates and the molecular-kinetic theory of matter to explain why lowering the temperature of a reaction or diluting the concentration of the reactants in a reaction will decrease the rate of a chemical reaction.

LAB 21

4. Scientists use experiments to prove ideas right or wrong.

   a. I agree with this statement.
   b. I disagree with this statement.

   Explain your answer, using an example from your investigation about reaction rates.

5. Scientists need to be creative and have a good imagination to excel in science.

   a. I agree with this statement.
   b. I disagree with this statement.

   Explain your answer, using an example from your investigation about reaction rates.

6. An important goal in science is to develop causal explanations for observations. Explain what a casual explanation is and why these explanations are important, using an example from your investigation about reaction rates.

7. Scientists often use or develop new models to help them understand natural phenomena. Explain what a model is in science and why models are important, using an example from your investigation about reaction rates.

Teacher Notes

Lab 22. Chemical Equilibrium: Why Do Changes in Temperature, Reactant Concentration, and Product Concentration Affect the Equilibrium Point of a Reaction?

Purpose

The purpose of this lab is to *introduce* students to the concept of chemical equilibrium and some of the factors that affect the equilibrium point of a reaction. This lab gives students an opportunity to use the collision theory of reaction rates to develop a conceptual model that can help them explain why changes in temperature, reactant concentration, and product concentration affect the equilibrium point of a chemical reaction. Students will also learn about the role experiments play in science and why scientists need to be creative and have a good imagination.

The Content

Chemical equilibrium is defined as the state where the concentrations of reactants and products remain constant with time. This condition makes it appear that the reaction has stopped because there is no change in the concentrations of reactants or products at this point. However, on a molecular level, reactants are being converted to products and the products are reverting back to to reactants at the same rate. To illustrate the nature of chemical equilibrium, consider Figure 22.1.

As the reaction proceeds, one can measure both a decrease in the concentration of the reactant(s) and an increase in the concentration of the product(s). As shown in Figure 22.1, there is a decrease in reactants and an increase in products over time. Eventually, however, the concentrations of reactants and products become stable. This point is indicated on the graph as a dashed line. When a system is in equilibrium there is no net change in the concentrations of all chemical species involved in the reaction, even though reactants continue to transform into products and products continue to revert back into reactants over time. There are, as a result, no further apparent changes in the system.

FIGURE 22.1

Concentration of reactant and product over time

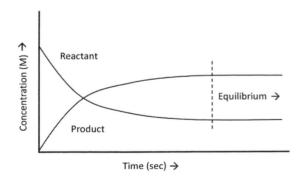

Chemical equilibrium occurs because the forward reaction slows down as the concentration of reactants decreases over time. Initially, no product exists and the reverse reaction cannot occur. As reactants are converted into products, however, the concentration of products increases, which in turn, causes the rate of the reverse reaction to increase. Once the concentrations of reactants and products reach levels where the forward rate equals the reverse rate, the system reaches equilibrium (see Figure 22.2).

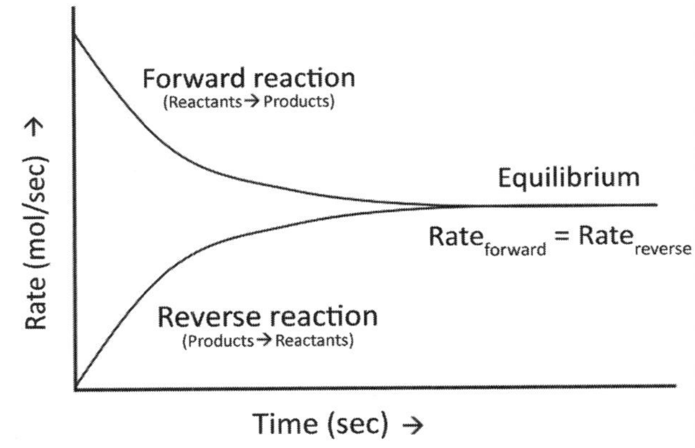

Once equilibrium is reached, it is maintained only if all relevant factors remain the same. A change in reactant or product concentrations, temperature, pressure, or volume will disturb the equilibrium, causing the system to undergo additional net changes to establish a new equilibrium. Le Châtelier's principle predicts how equilibrium will be restored. This principle states:

FIGURE 22.2

Rate of forward and reverse reactions over time

Change to a system includes such things as adding more reactant or product to the system or changing the temperature or pressure of the system. To reduce the effect of a change, one of two things can happen. The reversible reaction can shift in the forward direction or the reverse direction. When the reaction shifts in the forward direction, more reactant is transformed into product. When the reaction shifts in the reverse direction, more products revert into reactants. These shifts, as a result, will cause the reaction rates of the forward and reverse reactions to be the same and the concentration of the products and the reactants in the system to become stable.

> When a change is imposed on a system at equilibrium, the position of the equilibrium shifts in a direction that tends to reduce the effect of that change. (Zumdahl and DeCoste 2010, p. 559)

Timeline

The instructional time needed to complete this lab investigation is 180–250 minutes. Appendix 2 (p. 501) provides options for implementing this lab investigation over several class periods. Option E (250 minutes) should be used if students are unfamiliar with scientific writing because this option provides extra instructional time for scaffolding the writing process. You can scaffold the writing process by modeling, providing examples, and providing hints as students write each section of the report. Option F (180 minutes) should be used if students are familiar with scientific writing and have the skills needed to write an investigation report on their own. In option F, students complete stage 6 (writing the investigation report) and stage 8 (revising the investigation report) as homework.

LAB 22

Materials and Preparation

The materials needed to implement this investigation are listed in Table 22.1. The consumables and equipment can be purchased from a science supply company such as Carolina, Flinn Scientific, or Ward's Science. We recommend that you use a set routine for distributing and collecting the materials during the lab investigation. For example, the consumables and equipment for each group can be set up at each group's lab station before class begins, or one member from each group can collect them from a table or a cart when needed during class.

Prepare the solutions for this investigation as follows:

- *0.1 M iron(III) nitrate (Fe(NO$_3$)$_3$) solution*: Add 4.0 g of ferric nitrate nonahydrate [Fe(NO$_3$)$_3$ • 9H$_2$O] to 50 ml of distilled water. Stir to dissolve and then dilute to 100 ml with distilled water.

- *0.1 M potassium thiocyanate (KSCN) solution*: Add 1.0 g of KSCN to 50 ml of distilled water. Stir to dissolve and then dilute to 100 ml with distilled water.

- *1.5 M copper(II) chloride (CuCl$_2$)*: Add 60.50 g of CuCl$_2$ to 150 ml of distilled water. Stir to dissolve and then dilute to 300 ml with distilled water.

- *4 M sodium chloride (NaCl)*: Add 23.38 g of NaCl to 50 ml of distilled water. Stir to dissolve and then dilute to 100 ml with distilled water.

- *0.05 M silver nitrate (AgNO$_3$)*: Add 5 ml of 1 M AgNO$_3$ to 50 ml of distilled water. Stir and add water up to 100 ml.

Safety Precautions

Remind students to follow all normal lab safety rules. Iron(III) nitrate is a body tissue irritant; it will also stain clothes and skin. Potassium thiocyanate and copper(II) chloride are toxic by ingestion. Silver nitrate is toxic by ingestion, is corrosive to body tissues, and stains clothes and skin. You will therefore need to explain the potential hazards of working with these chemicals and how to work with hazardous chemicals. In addition, tell students to take the following safety precautions:

- Wear indirectly vented chemical-splash goggles and chemical-resistant gloves and aprons when they are collecting their data.

- Use caution when working with hot plates, and keep them away from water and other liquids.

- Handle all glassware (including thermometers) with care.

- Wash their hands with soap and water when they are done collecting the data.

Chemical Equilibrium

*Why Do Changes in Temperature, Reactant Concentration, and Product Concentration Affect
the Equilibrium Point of a Reaction?*

TABLE 22.1

Materials list

Item	Quantity
Consumables	
0.1 M $Fe(NO_3)_3$	10 ml per group
0.1 M KSCN	10 ml per group
1.5 M $CuCl_2$	30 ml per group
4 M NaCl	5 ml per class
0.05 $AgNO_3$	5 ml per group
Distilled water	50 ml per group
Ice (for ice baths)	As needed
Equipment and other materials	
Pyrex test tubes	9 per group
Test tube rack	1 per group
Graduated cylinder, 10 ml	1 per group
Disposable graduated Beral pipettes	6 per group
Beaker, 50 ml	1 per group
Beakers, 250 ml	2 per group
Thermometer (or temperature probe)	1 per group
Hot plate	1 per group
Investigation Proposal A (optional but recommended)	3 per group
Whiteboard, 2' x 3' *	1 per group
Lab handout	1 per student
Peer-review guide and instructor scoring rubric	1 per student

* As an alternative, students can use computer and presentation software such as Microsoft
PowerPoint or Apple Keynote to create their arguments.

LAB 22

Laboratory Waste Disposal

Have students pour solutions containing silver nitrate in a separate waste beaker, then use Flinn laboratory waste disposal method 11 to dispose of it. We recommend following Flinn laboratory waste disposal method 26b for all the other waste solutions. Information about laboratory waste disposal methods is included in the Flinn Catalog and Reference Manual; you can request a free copy at *www.flinnsci.com*.

Topics for the Explicit and Reflective Discussion

Concepts That Can Be Used to Justify the Evidence

To provide an adequate justification of their evidence, students must explain why they included the evidence in their arguments and make the assumptions underlying their analysis and interpretation of the data explicit. In this investigation, students can use the following concepts to help justify their evidence:

- Collision theory of reactions
- Reaction rates
- Chemical equilibrium

We recommend that you discuss these fundamental concepts during the explicit and reflective discussion to help students make this connection.

How to Design Better Investigations

It is important for students to reflect on the strengths and weaknesses of the investigation they designed during the explicit and reflective discussion. Students should therefore be encouraged to discuss ways to eliminate potential flaws, measurement errors, or sources of bias in their investigations. To help students be more reflective about the design of their investigations, you can ask the following questions:

- What were some of the strengths of your investigation? What made it scientific?
- What were some of the weaknesses of your investigation? What made it less scientific?
- If you were to do this investigation again, what would you do to address the weaknesses in your investigation? What could you do to make it more scientific?

Crosscutting Concepts

This investigation is well aligned with two crosscutting concepts found in *A Framework for K–12 Science Education,* and you should review these concepts during the explicit and reflective discussion.

Chemical Equilibrium

Why Do Changes in Temperature, Reactant Concentration, and Product Concentration Affect the Equilibrium Point of a Reaction?

- *Systems and system models:* Scientists often need to use models to understand complex phenomena. In this investigation, students are directed to develop a model to help explain what is happening during a chemical reaction at the submicroscopic level.

- *Stability and change:* It is critical for scientists to understand what makes a system stable or unstable and what controls rates of change in system (e.g., why changes in temperature and reactant concentration affect the equilibrium point of a chemical reaction).

The Nature of Science and the Nature of Scientific Inquiry

This investigation is well aligned with two important concepts related to the *nature of science* (NOS) and the *nature of scientific inquiry* (NOSI), and you should review these concepts during the explicit and reflective discussion.

- *The importance of imagination and creativity in science:* Students should learn that developing explanations for or models of natural phenomena and then figuring out how they can be put to the test of reality is as creative as writing poetry, composing music, or designing skyscrapers. Scientists must also use their imagination and creativity to figure out new ways to test ideas and collect or analyze data.

- *Nature and role of experiments:* Scientists use experiments to test the validity of a hypothesis (i.e., a tentative explanation) for an observed phenomenon. Experiments include a test and the formulation of predictions (expected results) if the test is conducted and the hypothesis is valid. The experiment is then carried out and the predictions are compared with the observed results of the experiment. If the predictions match the observed results, then the hypothesis is supported. If the observed results do not match the prediction, then the hypothesis is not supported. A signature feature of an experiment is the control of variables to help eliminate alternative explanations for observed results.

Hints for Implementing the Lab

- We recommend that students fill out an investigation proposal for this investigation. These proposals provide a way for you to offer students hints and suggestions without telling them how to do it. You can also check the proposals quickly during a class period. We recommend that students fill out one investigation proposal for each of their experiments.

- We recommend that you show students how to prepare a stock solution of $FeSCN^{2+}$ as part of the tool talk. Students can then use the stock solution to fill test tubes, which can be used to make control and treatment groups in all their experiments. You can make a stock solution of $FeSCN^{2+}$ by mixing 40 ml of distilled water with 1 ml of 0.1 M $Fe(NO_3)_3$ and 2 ml of 0.1 M KSCN.

- Students should only need to use 2 ml of the $FeSCN^{2+}$ stock solution for each of their tests. When they use this amount, they only need to add about 10 drops of additional reactant or product to observe a color change.

- Students should only need to use 2 ml of the copper(II) chloride solution for each of their tests. When they use this amount, they only need to add about 5 drops of NaCl or $AgNO_3$ to observe a color change.

- Students will need to test at least three different amounts of additional reactant in the first experiment, three different amounts of additional product in the second experiment, and three different temperatures in the third experiment.

- Remind students to include multiple trials in each experiment and to average their results to help reduce measurement error.

- Show students how to make a hot-water bath and an ice bath as part of the tool talk. Make sure students do not heat any of the solutions directly on the hot plate. The temperature of the hot-water bath should not exceed 60°C.

- Remind students that they will need to test their models.

Topic Connections

Table 22.2 provides an overview of the scientific practices, crosscutting concepts, disciplinary core ideas, and supporting ideas at the heart of this lab investigation. In addition, it lists NOS and NOSI concepts for the explicit and reflective discussion. Finally, it lists literacy and mathematics skills (*CCSS ELA* and *CCSS Mathematics*) that are addressed during the investigation.

Chemical Equilibrium

Why Do Changes in Temperature, Reactant Concentration, and Product Concentration Affect the Equilibrium Point of a Reaction?

TABLE 22.2

Lab 22 alignment with standards

Scientific practices	• Asking questions and defining problems • Developing and using models • Planning and carrying out investigations • Analyzing and interpreting data • Using mathematics and computational thinking • Constructing explanations and designing solutions • Engaging in argument from evidence • Obtaining, evaluating, and communicating information
Crosscutting concepts	• Systems and system models • Stability and change
Core idea	• PS1.B: Chemical reactions
Supporting ideas	• Collision theory of reactions • Reaction rates • Chemical equilibrium
NOS and NOSI concepts	• Imagination and creativity in science • Nature and role of experiments
Literacy connections (*CCSS ELA*)	• *Reading:* Key ideas and details, craft and structure, integration of knowledge and ideas • *Writing:* Text types and purposes, production and distribution of writing, research to build and present knowledge, range of writing • *Speaking and listening:* Comprehension and collaboration, presentation of knowledge and ideas
Mathematics connections (*CCSS Mathematics*)	• Reason abstractly and quantitatively • Model with mathematics

Reference

Zumdahl, S., and D. DeCoste. 2010. *Introductory chemistry: A foundation.* Belmont, CA: Cengage Learning.

LAB 22

Lab Handout

Lab 22. Chemical Equilibrium: Why Do Changes in Temperature, Reactant Concentration, and Product Concentration Affect the Equilibrium Point of a Reaction?

Introduction

It is often useful to think of a reaction as a process that consists of two components acting in opposite directions. From this view, a reaction begins with all reactants and no products. The reactants then begin to interact with each other and transform into products. The rate at which the reactants transform into products will begin to decrease over time as the concentration of the reactant decreases. At this point, some of the products will begin to revert back into reactants. The rate at which the products revert back into reactants will increase as the concentration of the product increases. There is a point, as a result, where the forward and reverse components of a reaction are happening at equal rates. This point is called *chemical equilibrium*. At equilibrium, the rates of the forward and reverse components of the reaction are equal but the concentrations of reactants and products are not. Figure L22.1 illustrates this process.

Chemical equilibrium, therefore, can be defined as the point in a reaction where the rate at which reactants transform into products is equal to the rate at which products revert back into reactants. The *equilibrium point* of a chemical reaction occurs when the amount or concentration of the products and reactants in a closed system is stable. Chemists use a specific property, such as color, concentration, or density, to determine when a reaction is in equilibrium. It is important to note, however, that chemists view the state of chemical equilibrium as dynamic because reactants continue to transform into products and products continue to revert back into reactants even though the amount of reactants and products in the closed system is stable.

The equilibrium point of a reaction can change because chemical equilibrium is not static. There are a number of different factors that can change the equilibrium point of a reaction by changing the rate at which reactants transform into products or by changing the rate at which products revert back into the reactants. These factors include a change in temperature, pressure, reactant concentration, and product concentration. When any of these factors are changed, the equilibrium point of the reaction will move and the concentration of products and reactants in the system at the new equilibrium point will be different.

Chemical Equilibrium

Why Do Changes in Temperature, Reactant Concentration, and Product Concentration Affect the Equilibrium Point of a Reaction?

FIGURE L22.1

The forward and reverse reactions associated with chemical equilibrium

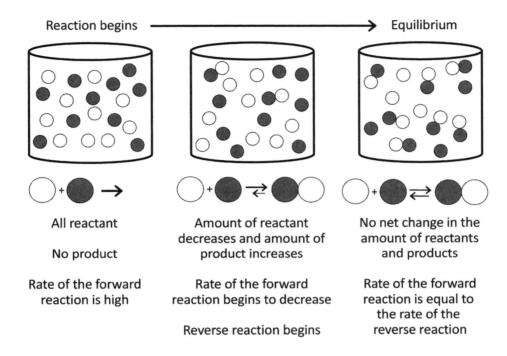

Reaction begins ⟶ Equilibrium

All reactant	Amount of reactant decreases and amount of product increases	No net change in the amount of reactants and products
No product		
Rate of the forward reaction is high	Rate of the forward reaction begins to decrease	Rate of the forward reaction is equal to the rate of the reverse reaction
	Reverse reaction begins	

To control the amount of product or reactant present at the equilibrium point of a reaction in a closed system, chemists need to understand how various factors affect chemical equilibrium and why these various factors change the equilibrium point of a reaction. You will therefore explore how three specific factors affect the equilibrium point of chemical reaction. You will then develop a conceptual model that you can use to explain your observations and predict how the equilibrium point of a different reaction will change when the equilibrium point is disturbed by changing these same three factors.

Your Task

Determine *how* changes in temperature and the addition of extra reactant and product affect the equilibrium point of the reaction between iron(III) nitrate and potassium thiocyanate. Then develop a conceptual model that you can use to explain *why* these factors influence the equilibrium point of a reaction. Once you have developed your conceptual model, you will need to test it to determine if it allows you to predict how the equilibrium point of a different reaction will change under similar conditions.

The guiding question of this investigation is, **Why do changes in temperature, reactant concentration, and product concentration affect the equilibrium point of a reaction?**

LAB 22

Materials

You may use any of the following materials during your investigation:

Consumables	Equipment
• 0.1 M iron(III) nitrate, $Fe(NO_3)_3$ • 0.1 M potassium thiocyanate, KSCN • 1.5 M copper(II) chloride, $CuCl_2$ • 4 M sodium chloride, NaCl • 0.05 M silver nitrate, $AgNO_3$ • Distilled water • Ice	• 9 test tubes • Test tube rack • Graduated cylinder (10 ml) • 6 disposable graduated Beral pipettes • Beaker (50 ml) • 2 beakers (each 250 ml), for hot- and cold-water baths • Thermometer • Hot plate

Safety Precautions

Follow all normal lab safety rules. Iron(III) nitrate is a body tissue irritant; it will also stain clothes and skin. Potassium thiocyanate and copper(II) chloride are toxic by ingestion. Silver nitrate is toxic by ingestion, is corrosive to body tissues, and stains clothes and skin. Your teacher will explain relevant and important information about working with the chemicals associated with this investigation. In addition, take the following safety precautions:

- Wear indirectly vented chemical-splash goggles and chemical-resistant gloves and apron while in the laboratory.
- Use caution when working with hot plates because they can burn skin. Hot plates also need to be kept away from water and other liquids.
- Handle all glassware (including thermometers) with care.
- Wash your hands with soap and water before leaving the laboratory.

Investigation Proposal Required?   ☐ Yes   ☐ No

Getting Started

The first step in developing your model is to design and carry out three experiments. The goal of the first experiment will be to determine how a change in reactant concentration affects the equilibrium point of a reaction. The goal of the second experiment will be to determine how a change in product concentration affects the equilibrium point of a reaction. The goal of the third experiment will be to determine how temperature affects the equilibrium point of a reaction. For these three experiments, you will focus on the reaction of iron(III) nitrate and potassium thiocyanate. Iron(III) ions react with thiocyanate ions to form $FeSCN^{2+}$ complex ions according to the following reaction:

$$Fe^{3+}(aq) + SCN^-(aq) \leftrightarrows FeSCN^{2+}(aq)$$

Yellow   Colorless   Orange-Red

Chemical Equilibrium

Why Do Changes in Temperature, Reactant Concentration, and Product Concentration Affect the Equilibrium Point of a Reaction?

You can prepare a stock solution of $FeSCN^{2+}$ by mixing 40 ml of distilled water with 1 ml of 0.1 M $Fe(NO_3)_3$ and 2 ml of 0.1 M KSCN. You can then add 2 ml of this stock solution to several different test tubes to create a control condition and several treatment conditions for each experiment. You can then change the temperature, reactant concentration, or product concentration in the treatment conditions as needed and leave the control condition alone for comparison purposes. You must, however, determine what type of data you need to collect, how you will collect the data, and how you will analyze the data for each experiment.

To determine *what type of data you need to collect*, think about the following questions:

- What type of measurements or observations will you need to record during each experiment?
- When will you need to make these measurements or observations?

To determine *how you will collect the data*, think about the following questions:

- What will serve as your independent variable in each experiment?
- How will you change the independent variable in each experiment?
- What types of comparisons will you need to make?
- What will you do to reduce measurement error?
- How will you keep track of the data you collect and how will you organize it?

To determine *how you will analyze the data*, think about the following questions:

- What type of calculations will you need to make?
- What type of graph could you create to help make sense of your data?

Once you have carried out your three experiments, your group will need to develop a conceptual model. This conceptual model will need to be able to provide an underlying reason for your findings about the effect of temperature, changes in reactant concentration, and changes in product concentration on the equilibrium point of a reaction. Your model should also include an explanation of what is happening at the submicroscopic level between and within molecules during a reaction. The collision theory of reaction rates and the concept of chemical equilibrium should serve as the theoretical foundation for your model.

The last step in this investigation is to test your model. To accomplish this goal, you can use a different reaction to determine if your model leads to accurate predictions about how the equilibrium point changes in response to different factors. If you can use your model to make accurate predictions about how the equilibrium point of a different reversible reaction changes, then you will be able to generate the evidence you need to convince others that the conceptual model you developed is valid.

You can use the reversible formation of copper(II) complexes to test your model. When copper(II) chloride ($CuCl_2$) is dissolved in water, two different solutes are present in the solution. These solutes include Cu^{2+} ions and Cl^- ions. These solutes interact with water molecules to form two different complex ions. One complex ion is $Cu(H_2O)_6^{2+}$ and the other is $CuCl_4^{2-}$. The reversible equation for the formation of the two complex ions is

$$Cu(H_2O)_6^{2+}(aq) + 4Cl^-(aq) \leftrightarrows CuCl_4^{2-}(aq) + 6H_2O$$

| Blue | Green |

You can change the equilibrium point by adding NaCl, or $AgNO_3$ or by changing the temperature of the solution. To change the concentration of the reactants or the products, simply add 2 ml of the copper(II) chloride solution to a test tube and then add up to eight drops of NaCl or $AgNO_3$. The addition of NaCl will increase the number of Cl^- ions in the system. The addition of $AgNO_3$, in contrast, will decrease the number of Cl^- ions in the system (because the Ag^+ ions react with Cl^- ions to form AgCl). To change the temperature of the system, use a hot-water bath or an ice bath.

Connections to Crosscutting Concepts, the Nature of Science, and the Nature of Scientific Inquiry

As you work through your investigation, be sure to think about

- how models are used to help understand natural phenomena,
- why it is important to understand what makes a system stable or unstable and what controls rates of change in a system,
- the importance of imagination and creativity in science, and
- the role of experiments in science.

Initial Argument

Once your group has finished collecting and analyzing your data, you will need to develop an initial argument. Your argument must include a *claim*, which is your answer to the guiding question. Your argument must also include *evidence* in support of your claim. The evidence is your analysis of the data and your interpretation of what the analysis means. Finally, you must include a *justification* of the evidence in your argument. You will therefore need to use a scientific concept or principle to explain why the evidence that you decided to use is relevant and important. You will create your initial argument on a whiteboard. Your whiteboard must include all the information shown in Figure L22.2.

Argumentation Session

The argumentation session allows all of the groups to share their arguments. One member of each group stays at the lab station to share that group's argument, while the other members of the group go to the other lab stations one at a time to listen to and critique the arguments developed by their classmates. The goal of the argumentation session is not to convince others that your argument is the best one; rather, the goal is to identify errors or instances of faulty reasoning in the initial arguments so these mistakes can be fixed. You will therefore need to evaluate the content of the claim, the quality of the evidence used to support the claim, and the strength of the justification of the evidence included in each argument that you see. To critique an argument, you might need more information than what is included on the whiteboard. You might, therefore, need to ask the presenter one or more follow-up questions, such as:

FIGURE L22.2
Argument presentation on a whiteboard

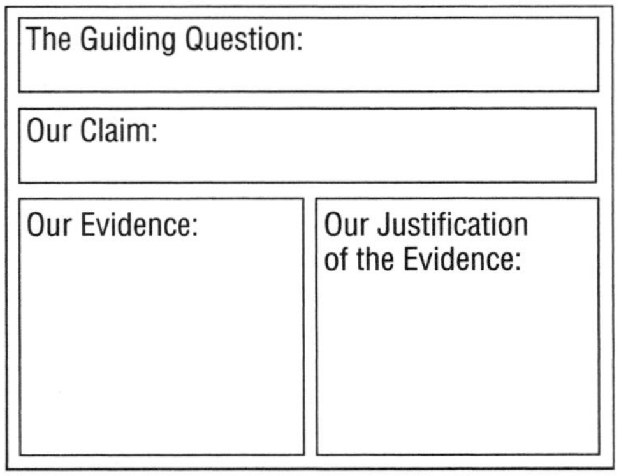

- How did your group collect the data? Why did you use that method?

- What did your group do to make sure the data you collected are reliable? What did you do to decrease measurement error?

- What did your group do to analyze the data, and why did you decide to do it that way? Did you check your calculations?

- Is that the only way to interpret the results of your group's analysis? How do you know that your interpretation of the analysis is appropriate?

- Why did your group decide to present your evidence in that manner?

- What other claims did your group discuss before deciding on that one? Why did you abandon those alternative ideas?

- How confident are you that your group's claim is valid? What could you do to increase your confidence?

Once the argumentation session is complete, you will have a chance to meet with your group and revise your original argument. Your group might need to gather more data or design a way to test one or more alternative claims as part of this process. Remember, your goal at this stage of the investigation is to develop the most valid or acceptable answer to the research question!

Report

Once you have completed your research, you will need to prepare an *investigation report* that consists of three sections that provide answers to the following questions:

1. What question were you trying to answer and why?

2. What did you do during your investigation and why did you conduct your investigation in this way?

LAB 22

3. What is your argument?

Your report should answer these questions in two pages or less. The report must be typed and any diagrams, figures, or tables should be embedded into the document. Be sure to write in a persuasive style; you are trying to convince others that your claim is acceptable or valid!

Checkout Questions

Lab 22. Chemical Equilibrium: Why Do Changes in Temperature, Reactant Concentration, and Product Concentration Affect the Equilibrium Point of a Reaction?

Chemists must understand chemical equilibrium in order to control the amount of product produced as a result of a reaction. Chemists can alter the equilibrium point of a reaction by changing the temperature or the concentration of the reactants or products.

1. What happens to the concentration of the reactants and the products in a system when a reaction is at equilibrium?

2. How is the rate at which the reactants transform into products related to the rate at which the products revert back into reactants when a reaction is at equilibrium?

LAB 22

3. Paper coated with cobalt chloride is sold commercially as test strips for estimating humidity. The following reversible reaction takes place between cobalt chloride and water:

$$CoCl_2(s) + H_2O(g) \leftrightarrows CoCl_2 \bullet 6H_2O(s)$$

Blue   Pink

   a. What color will the test strip be when the humidity is low (20%), and what color will it be when the humidity is high (80%)? Explain your answer.

   b. The test strips come with a color chart to estimate intermediate humidity levels. What color do you think the paper will be when the humidity is about 50%? Explain your answer.

4. Scientists use experiments to test potential explanations for a phenomenon.

   a. I agree with this statement.

   b. I disagree with this statement.

Explain your answer, using an example from your investigation about chemical equilibrium.

5. Scientists do not need to be creative or have a good imagination.

    a. I agree with this statement.
    b. I disagree with this statement.

    Explain your answer, using an example from your investigation about chemical equilibrium.

6. An important goal in science is to understand what makes a system stable or unstable and what controls rates of change in system. Explain why this important, using an example from your investigation about chemical equilibrium.

7. Scientists often use or develop new models to help them understand natural phenomena. Explain what a model is in science and why models are important, using an example from your investigation about chemical equilibrium.

Application Labs

Teacher Notes

Lab 23. Classification of Changes in Matter: Which Changes Are Examples of a Chemical Change, and Which Are Examples of a Physical Change?

Purpose

The purpose of this lab is to allow students to *apply* their understanding of chemical and physical changes to different scenarios to determine what type of change has occurred. Students will also learn about the difference between data and evidence in science and the role of observation and inference in science.

The Content

Matter, the stuff of which the universe is composed, has two characteristics: it has mass and it occupies space. These two characteristics of matter, however, are not sufficient for understanding matter, its various properties, and the many ways that matter can change. It is important to remember that even though there are many types of matter—solids, liquids, gases, and specific substances within these categories—the amount of matter in the universe is constant. The *law of conservation of mass* states that matter is neither created nor destroyed, but it can change forms. Matter may change forms by simply transitioning from a solid to a liquid phase, which represents a *physical change,* or matter may undergo a *chemical change* (also called a chemical reaction) where the composition of a substance is altered by the rearrangement of atoms or molecules.

An example of a chemical change is the burning of a substance like gasoline. Octane (C_8H_{18}) is a large hydrocarbon found in gasoline, which reacts with oxygen gas (O_2) during combustion. The chemical equation for this reaction is

$$C_8H_{18} + O_2 \rightarrow CO_2 + H_2O$$

In the combustion reaction of octane and oxygen, the composition of each substance is changed because the atoms rearrange to form carbon dioxide (CO_2) and water (H_2O). It may seem as though we have violated the law of conservation of mass because we have new substances after the chemical change and the number of atoms is different on both sides of the equation. It is correct that the products of the reaction are different than the starting reactants, but the products are formed from the same original atoms. The following balanced chemical equation shows the relative proportion of each substance that reacts and is produced; by balancing the equation we are able to account for all the matter involved in the process:

$$2C_8H_{18} + 25O_2 \rightarrow 16CO_2 + 18H_2O$$

Understanding chemical and physical processes in science is important because it is an underlying concept for making sense of how matter and energy flow within systems. Distinguishing between physical and chemical processes helps us understand and determine if and how matter has changed as a result of a process.

Physical properties of substances can be measured without changing the identity or composition of the substance. For example, measuring the density or melting point of a substance does not change how it behaves chemically. *Chemical properties*, however, describe the potential for a substance to react with another substance, and measuring such a property involves a chemical change. For instance, to measure if a substance is actually flammable requires subjecting it to a flame in the presence of oxygen. If the substance undergoes combustion, then it exhibits the chemical property of flammability.

Timeline

The instructional time needed to complete this lab investigation is 130–180 minutes. Appendix 2 (p. 501) provides options for implementing this lab investigation over several class periods. Option D (130 minutes) should be used if students are proficient with the lab equipment and general laboratory procedures. This option completes the relatively quick data collection process (stage 2) on day 1 of the investigation. Option H (180 minutes) adds additional time during investigation day 2 to finish data collection during stage 2, which makes this option appropriate for students who are less familiar with the lab equipment and procedures. Each of these options assumes that the students have experience with the scientific writing process and have the skills needed to write an investigation report on their own. In options D and H, students complete stage 6 (writing the investigation report) and stage 8 (revising the investigation report) as homework. If students need additional scaffolding and class time to work on writing their investigation reports, consider Option C (200 minutes).

Materials and Preparation

The materials needed to implement this investigation are listed in Table 23.1 (p. 369). It is recommended that the solutions be prepared prior to lab in large batches and that each group receive a small dropper bottle containing each solution. Scenario 5 requires that students heat paraffin wax in a hot-water bath; this is best accomplished by placing the wax in a heat-resistant test tube (or similar) and then submerging the test tube in the water bath. These wax test tubes should be prepared before class. After students have completed scenario 5, they may return their test tubes to a central cool-water bath so that the wax

LAB 23

solidifies for later class periods. Central hot-water and cool-water baths may be set up for the whole class so that each lab group does not need to construct their own setup.

To minimize time lost due to waiting for chemicals, it is recommended that you provide each lab group with a complete set of equipment and materials. To minimize time lost due to changing equipment setups for the different scenarios, this investigation could be conducted in a station format. Using a station format will also contain the area where Bunsen burners are used in the classroom.

Safety Precautions

Remind students to follow all normal lab safety rules. Hydrochloric acid and sodium hydroxide are corrosive to eyes, skin, and other body tissues. Sodium hydroxide is also toxic by ingestion. Copper(II) nitrate is a tissue irritant. You will therefore need to explain the potential hazards of working with these chemicals and how to work with hazardous chemicals. Be sure to show students how to heat a substance using a crucible. In addition, tell students to take the following safety precautions:

- Wear indirectly vented chemical-splash goggles and chemical-resistant gloves and aprons when they are collecting their data.
- Wipe up any water spilled on the floor from water baths.
- Waft the fumes from a chemical toward their noses with their hands in order to smell it.
- Use caution when working with Bunsen burners. They can burn skin, and combustibles and flammables must be kept away from the open flame. Students with long hair should tie it back behind their heads.
- Inspect the crucible for cracks and exchange it for a new one if it has any cracks. Clean the crucible and lid thoroughly, then dry the crucible and lid by heating them for five minutes before using it.
- Be careful with a crucible after removing it from a flame because it will still be hot.
- Use caution when working with hot plates, hot water, and melted wax because they can burn skin. Keep hot plates away from water or other liquids.
- Handle test tubes placed in the hot-water bath with test tube tongs.
- Handle all glassware (including thermometers) with care.
- Wash their hands with soap and water when they are done collecting the data.

Laboratory Waste Disposal

We recommend following Flinn laboratory waste disposal methods 3 to dispose of the magnesium ribbon and 26a and 26b to dispose of the waste solutions and solids. Information about laboratory waste disposal methods is included in the Flinn Catalog and Reference Manual; you can request a free copy at *www.flinnsci.com*.

TABLE 23.1

Materials list

Item	Quantity
Consumables	
Sodium chloride (NaCl)	5 g per group
1 M sodium hydroxide (NaOH) solution (in a dropper bottle)	2–3 ml per group
1 M copper(II) nitrate ($CuNO_3$) solution (in a dropper bottle)	2–3 ml per group
1 M hydrochloric acid (HCl) (in a dropper bottle)	2–3 ml per group
Sodium bicarbonate ($NaHCO_3$)	2–3 g per group
Paraffin wax, in a test tube (or similar)	1 per group
Magnesium ribbon	2 cm per group
pH paper	As needed
Distilled water (in squirt bottles)	As needed
Equipment and other materials	
Spot (reaction) plate	1 per group
Graduated cylinder, 50 ml	1 per group
Beaker, 150 ml	1 per group
Beaker, 500 ml	1 per group
Hot plate	1 per group
Bunsen burner	1 per group
Ring stand with metal ring	1 per group
Clay triangle	1 per group
Wire gauze square	1 per group
Crucible with lid	1 per group
Crucible tongs	1 per group
Test tube tongs	1 per group
Spatula	1 per group
Thermometer	1 per group
Electronic or triple beam balance	1 per group
Whiteboard, 2' x 3' *	1 per group
Lab handout	1 per student
Peer-review guide and instructor scoring rubric	1 per student

* As an alternative, students can use computer and presentation software such as Microsoft PowerPoint or Apple Keynote to create their arguments.

LAB 23

The paraffin wax samples may be covered, stored, and used again when this investigation is repeated.

Topics for the Explicit and Reflective Discussion

Concepts That Can Be Used to Justify the Evidence

To provide an adequate justification of their evidence, students must explain why they included the evidence in their arguments and make the assumptions underlying their analysis and interpretation of the data explicit. In this investigation, students can use the following concepts to help justify their evidence:

- Matter
- Chemical and physical properties
- Chemical and physical changes

We recommend that you discuss these fundamental concepts during the explicit and reflective discussion to help students make this connection.

How to Design Better Investigations

It is important for students to reflect on the strengths and weaknesses of the investigation they designed during the explicit and reflective discussion. Students should therefore be encouraged to discuss ways to eliminate potential flaws, measurement errors, or sources of bias in their investigations. To help students be more reflective about the design of their investigation, you can ask the following questions:

- What were some of the strengths of your investigation? What made it scientific?
- What were some of the weaknesses of your investigation? What made it less scientific?
- If you were to do this investigation again, what would you do to address the weaknesses in your investigation? What could you do to make it more scientific?

Crosscutting Concepts

This investigation is well aligned with two crosscutting concepts found in *A Framework for K–12 Science Education,* and you should review these concepts during the explicit and reflective discussion.

- *Patterns:* Observed patterns in nature, such as how groups of substances interact with each other, guide the way scientists organize and classify properties of matter and changes in matter. Scientists also explore the relationships between and the underlying causes of the patterns they observe in nature.

- *Energy and matter: Flows, cycles, and conservation*: In science it is important to track how energy and matter move into, out of, and within systems. In chemistry this is particularly important with respect to chemical and physical changes.

The Nature of Science and the Nature of Scientific Inquiry

This investigation is well aligned with two important concepts related to the *nature of science* (NOS) and the *nature of scientific inquiry* (NOSI), and you should review these concepts during the explicit and reflective discussion.

- *The difference between observations and inferences*: An observation is a descriptive statement about a natural phenomenon, whereas an inference is an interpretation of an observation. Students should also understand that current scientific knowledge and the perspectives of individual scientists guide both observations and inferences. Thus, different scientists can have different but equally valid interpretations of the same observations due to differences in their perspectives and background knowledge.

- *The difference between data and evidence in science:* Data are measurements, observations, and findings from other studies that are collected as part of an investigation. Evidence, in contrast, is analyzed data and an interpretation of the analysis.

Hints for Implementing the Lab

- Provide a demonstration of the proper use and setup for a crucible as part of the tool talk. We recommend that students cover the crucible with the lid squarely the whole time it is heated. They can remove the lid briefly every three minutes to allow air to enter. This approach is the simplest and safest way to use a crucible.

- Data collection can take a large portion of class time because of the different scenarios. Both you and the students should keep a close eye on time to ensure that data collection is not rushed due to limited class time.

Topic Connections

Table 23.2 (p. 372) provides an overview of the scientific practices, crosscutting concepts, disciplinary core ideas, and supporting ideas at the heart of this lab investigation. In addition, it lists NOS and NOSI concepts for the explicit and reflective discussion. Finally, it lists literacy skills (*CCSS ELA*) that are addressed during the investigation.

LAB 23

TABLE 23.2

Lab 23 alignment with standards

Scientific practices	• Asking questions and defining problems • Planning and carrying out investigations • Analyzing and interpreting data • Constructing explanations and designing solutions • Engaging in argument from evidence • Obtaining, evaluating, and communicating information
Crosscutting concepts	• Patterns • Energy and matter: Flows, cycles, and conservation
Core idea	• PS1.B: Chemical reactions
Supporting ideas	• Matter • Chemical and physical properties • Chemical and physical changes
NOS and NOSI concepts	• Observations and inferences • Difference between data and evidence
Literacy connections (CCSS ELA)	• *Reading:* Key ideas and details, craft and structure, integration of knowledge and ideas • *Writing:* Text types and purposes, production and distribution of writing, research to build and present knowledge, range of writing • *Speaking and listening:* Comprehension and collaboration, presentation of knowledge and ideas

Lab Handout

Lab 23. Classification of Changes in Matter: Which Changes Are Examples of a Chemical Change, and Which Are Examples of a Physical Change?

Introduction

Matter, the "stuff" of which the universe is composed, has two characteristics: it has mass and it occupies space. Physical properties of matter, such as density, odor, color, melting point, boiling point, state at room temperature (liquid, gas or solid), and magnetism, are often useful for identifying different substances. Matter, however, can also go through changes in both its physical and chemical properties. During *physical changes* the composition of matter does not change; for example, freezing a sample of water results in a change of state (i.e., going from liquid to solid), but the substance is still water (H_2O)—its chemical composition did not change. During *chemical changes* the chemical composition of a substance does change; for example, burning a piece of wood in a fireplace is a chemical change. In this example, the original wood is transformed into ashes and smoke, which both have different chemical properties than the original piece of wood.

Your Task

Create and observe the five scenarios listed below. Using your data and observations, determine if a physical or chemical change has occurred when

1. 100 ml of water (H_2O) is mixed with 5 g of table salt (NaCl),

2. a 2 cm magnesium strip is placed in a crucible and heated,

3. 10 drops of sodium hydroxide (NaOH) and 10 drops of copper(II) nitrate ($CuNO_3$) are mixed,

4. 5 drops of hydrochloric acid (HCl) are added to 2 g of sodium bicarbonate ($NaHCO_3$), and

5. paraffin wax is subjected to heat in a hot-water bath.

The guiding question of this investigation is, **Which changes are examples of a chemical change, and which are examples of a physical change?**

Materials

You may use any of the following materials during your investigation:

Consumables	Equipment
• NaCl	• Spot (reaction) plate
• NaOH solution	• Graduated cylinder (50 ml)
• $CuNO_3$ solution	• Beaker (150 ml)
• HCl solution	• Beaker (500 ml)
• $NaHCO_3$	• Hot plate
• Magnesium strip, 2 cm	• Bunsen burner
• Paraffin wax	• Ring stand with metal ring
• pH paper	• Clay triangle
• Distilled water (in squirt bottles)	• Wire gauze square
	• Crucible with lid
	• Crucible tongs
	• Test tube tongs
	• Spatula
	• Thermometer
	• Electronic or triple beam balance

Safety Precautions

Follow all normal lab safety rules. Your teacher will explain relevant and important information about working with the chemicals associated with this investigation. In addition, take the following safety precautions:

- Wear indirectly vented chemical-splash goggles and chemical-resistant gloves and apron while in the laboratory.

- Wipe up any water spilled on the floor from water baths.

- When investigating the odor associated with chemicals, never inhale with your nose directly over a tube, beaker or bottle; your instructor will demonstrate wafting the fumes toward your nose with your hand.

- Use caution when working with Bunsen burners. They can burn skin, and combustibles and flammables must be kept away from the open flame. If you have long hair, tie it back behind your head.

- Inspect the crucible for cracks. If it is cracked, exchange it for a new one. Clean the crucible and lid thoroughly before using them.

- Be careful with a crucible after removing it from a flame because it will still be hot.

- Use caution when working with hot plates, hot water, and melted wax because they can burn skin. Hot plates also need to be kept away from water and other liquids.

- Handle test tubes placed in the hot-water bath ONLY with test tube tongs.

- Handle all glassware (including thermometers) with care.

- Wash your hands with soap and water before leaving the laboratory.

Investigation Proposal Required? ☐ Yes ☐ No

Getting Started

Create each of the scenarios listed on the previous page and record what happens. Then conduct additional tests as needed to determine if a chemical or physical change took place.

Connections to Crosscutting Concepts, the Nature of Science, and the Nature of Scientific Inquiry

As you work through your investigation, be sure to think about

- the importance of patterns within science,
- the flow of energy and matter within a system,
- the difference between observations and inferences in science, and
- the difference between data and evidence in science.

Initial Argument

Once your group has finished collecting and analyzing your data, you will need to develop an initial argument. Your argument must include a *claim*, which is your answer to the guiding question. Your argument must also include *evidence* in support of your claim. The evidence is your analysis of the data and your interpretation of what the analysis means. Finally, you must include a *justification* of the evidence in your argument. You will therefore need to use a scientific concept or principle to explain why the evidence that you decided to use is relevant and important. You will create your initial argument on a whiteboard. Your whiteboard must include all the information shown in Figure L23.1.

FIGURE L23.1 _____

Argument presentation on a whiteboard

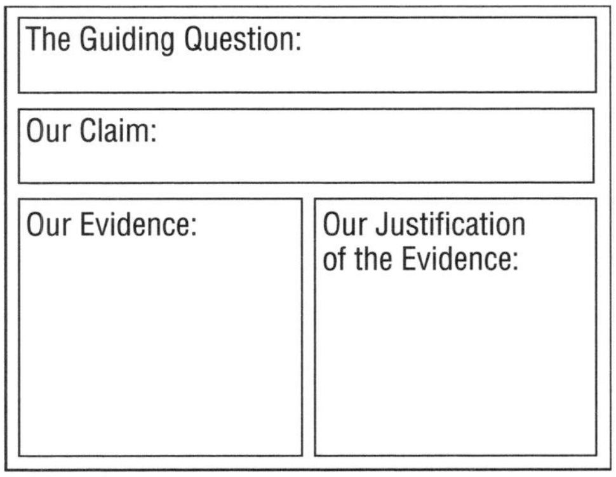

The Guiding Question:	
Our Claim:	
Our Evidence:	Our Justification of the Evidence:

Argumentation Session

The argumentation session allows all of the groups to share their arguments. One member of each group stays at the lab station to share that group's argument, while the other members of the group go to the other lab stations one at a time to listen to and critique the arguments developed by their classmates. The goal of the argumentation session is not to convince others that your argument is the best one; rather, the goal is to identify errors or instances of faulty reasoning in the initial arguments so these mistakes can be fixed. You will therefore need to evaluate the content of the claim, the quality of the evidence used to support the claim, and the strength of the justification of the evidence included in each argument that you see. To critique an argument, you might need more information than what is included on the whiteboard. You might, therefore, need to ask the presenter one or more follow-up questions, such as:

- What did your group do to analyze the data, and why did you decide to do it that way?
- Is that the only way to interpret the results of your group's analysis? How do you know that your interpretation of the analysis is appropriate?
- Why did your group decide to present your evidence in that manner?
- What other claims did your group discuss before deciding on that one? Why did you abandon those alternative ideas?
- How confident are you that your group's claim is valid? What could you do to increase your confidence?

Once the argumentation session is complete, you will have a chance to meet with your group and revise your original argument. Your group might need to gather more data or design a way to test one or more alternative claims as part of this process. Remember, your goal at this stage of the investigation is to develop the most valid or acceptable answer to the research question!

Report

Once you have completed your research, you will need to prepare an *investigation report* that consists of three sections that provide answers to the following questions:

1. What question were you trying to answer and why?
2. What did you do during your investigation and why did you conduct your investigation in this way?
3. What is your argument?

Your report should answer these questions in two pages or less. The report must be typed and any diagrams, figures, or tables should be embedded into the document. Be sure to write in a persuasive style; you are trying to convince others that your claim is acceptable or valid!

Checkout Questions

Lab 23. Classification of Changes in Matter: Which Changes Are Examples of a Chemical Change, and Which Are Examples of a Physical Change?

1. What are the characteristics of a physical change? What are the characteristics of a chemical change?

2. One night for dinner Jaxon decided to make baked potatoes for himself and his sister, Jade. He placed two potatoes in the oven set at 350°F for about 45 minutes. When Jaxon removed the potatoes from the oven, he noticed that they were soft and had a different texture than before they were cooked. Jade pointed out that the potatoes had undergone a chemical change. Jaxon did not believe that the potatoes had changed chemically; he thought they only experienced a physical change.

   Do you agree with Jade or with Jaxon? Use what you know about chemical and physical changes to provide a supporting argument.

3. Data and evidence are interchangeable in science.

    a. I agree with this statement.

    b. I disagree with this statement.

Explain your answer, using an example from your investigation about classification of changes in matter.

4. In science, observations are objective, but inferences are subjective.

    a. I agree with this statement.

    b. I disagree with this statement.

Explain your answer, using an example from your investigation about classification of changes in matter.

5. All things in the universe are made of matter. Understanding how matter moves within and between systems is important within science. Explain why understanding this is important, using an example from your investigation about classification of changes in matter.

6. When scientists observe events, often they are trying to recognize and identify patterns. Describe why patterns are important in science, using an example from your investigation about classification of changes in matter.

LAB 24

Teacher Notes

Lab 24. Identification of Reaction Products: What Are the Products of the Chemical Reactions?

Purpose

The purpose of this lab is to give students an opportunity to *apply* their understanding of chemical reactions and solubility rules to determine the products of a chemical reaction and then identify a precipitate. Students will also learn about the role of observation and inference in science and the difference between laws and theories.

The Content

Predicting and identifying the products of a chemical reaction is possible because chemical reactions follow two key principles in chemistry: the *law of conservation of mass* and the *law of definite proportions*. The law of conservation of mass states that matter is neither created nor destroyed during chemical processes. The law of definite proportions states that a compound always contains the same elements combined together in the same proportion by mass. When these two laws are applied together in the context of chemical reactions, it is possible to predict the outcome of reactions as well as be confident that a particular reaction will consistently result in the same products.

For example, consider a double replacement (or precipitation) reaction between a solution of potassium iodide (KI) and a solution of lead nitrate ($Pb(NO_3)_2$). The balanced chemical equation for this reaction is shown below. In this example, the ions within each solution will rearrange to form the insoluble compound lead iodide (PbI_2) and potassium nitrate (KNO_3) will remain in solution:

$$2KI_{(aq)} + Pb(NO_3)_{2(aq)} \rightarrow 2KNO_{3(aq)} + PbI_{2(s)}$$

This reaction is predictable based on the principles that no mass can be lost during a reaction, that we must have the same numbers of total atoms before and after the reaction, and that the ions in the compounds will arrange in consistent proportions (i.e., $1Pb^{2+}:2I^-$). Additionally, if we consider the role of ion charges in this example, we know that the ions will not rearrange to form a lead-potassium compound because both ions have a positive charge. In the case of ionic compounds it is necessary for the sum of the positive and negative changes to equal zero, resulting in a neutral substance; likewise, there is no iodine-nitrate compound formed because both ions carry a negative charge.

The example reaction above represents a precipitation reaction, due to the formation of a solid compound that is not soluble in water. The insoluble compound, lead iodide, precipitates or "falls out" out of solution. Understanding how *cations* (positively charged ions) and *anions* (negatively charged ions) rearrange during a chemical reaction is only part of the information needed to predict the potential products of a reaction and especially the identification of the precipitate. Knowing if and when a precipitate will form is based on an understanding of general solubility rules. Table 24.1 identifies some common ions and their solubility when paired with other common ions.

TABLE 24.1

Solubility rules for ionic compounds in water

Ion	Soluble?	Exceptions
NO_3^-	Yes	None
ClO_4^-	Yes	None
Cl^-	Yes	Ag^+, Hg_2^{2+}, Pb^{2+}
I^-	Yes	Ag^+, Hg_2^{2+}, Pb^{2+}
SO_4^{2-}	Yes	Ca^{2+}, Ba^{2+}, Sr^{2+}, Ag^+, Hg^{2+}, Pb^{2+}
CO_3^{2-}	No	Group IA and NH_4^+
PO_4^{3-}	No	Group IA and NH_4^+
OH^-	No	Group IA, Ca^{2+} (slightly soluble), Ba^{2+}, Sr^{2+}
S^{2-}	No	Groups IA and IIA and NH_4^+
Na^+	Yes	None
NH_4^+	Yes	None
K^+	Yes	None

Timeline

The instructional time needed to complete this lab investigation is 130–200 minutes. Appendix 2 (p. 501) provides options for implementing this lab investigation over several class periods. Typical data collection for this investigation will take less than a single class period, therefore option D (130 minutes) is a good timeline, provided that your students have experience with the scientific writing process and have the skills needed to write an investigation report on their own. In option D students complete stage 6 (writing the investigation report) and stage 8 (revising the investigation report) as homework. If your students need more scaffolding with the writing process, option C (200 minutes) allows class time to complete stages 6 and 8. You can scaffold the writing process by modeling, providing examples, and providing hints as students write each section of the report.

LAB 24

Materials and Preparation

The materials needed to implement this investigation are listed in Table 24.2. The consumables and equipment can be purchased from a science supply company such as Carolina, Flinn Scientific, or Ward's Science.

Prepare 1.0 M solutions using the salts identified in the four reactions the students will test (see Table 24.2). Place the solutions into dropper bottles so that each lab group has its own set of reagents to work with. Students can mix the solutions in test tubes or in the well plate; advise them to use as little solution as necessary to observe the reaction. Upon generating the precipitate, students should be able to use what they know about chemical reactions and the solubility rules to determine the identity of the precipitate. If students wish to go further and generate additional evidence from stoichiometric calculations, they may need to isolate and weigh their precipitate. The vacuum filtration kit will be needed for groups that choose this approach; the kit should include a side-arm flask, a Buchner funnel with stopper, filter paper, and a sink aspirator with hose.

Safety Precautions

Remind students to follow all normal lab safety rules. Silver nitrate is toxic by ingestion, is corrosive to body tissues, and stains clothes and skin. You will therefore need to explain the potential hazards of working with silver nitrate and how to work with hazardous chemicals. In addition, tell students to take the following safety precautions:

- Wear indirectly vented chemical-splash goggles and chemical-resistant gloves and aprons when they are collecting their data.
- Wash their hands with soap and water when they are done collecting the data.

Laboratory Waste Disposal

Given the small amounts of solution the students will use, it is best to simply dump the waste from the well plates onto paper towels and discard the towels as solid waste. Clean the well plates with damp paper towels or other swabs and dispose the towels as solid waste also. For additional disposal methods and information, consult the Flinn Catalog and Reference Manual (available at *www.flinnsci.com*). Retain the stock solutions for use when the investigation is repeated.

TABLE 24.2

Materials list

Item	Quantity
Consumables	
Calcium chloride, $CaCl_2$, in 30 ml dropper bottle	1 per group
1 M calcium nitrate, $Ca(NO_3)_2$, in 30 ml dropper bottle	1 per group
1 M nickel(II) chloride, $NiCl_2$, in 30 ml dropper bottle	1 per group
1 M silver nitrate, $AgNO_3$, in 30 ml dropper bottle	1 per group
1 M sodium chloride, NaCl, in 30 ml dropper bottle	1 per group
1 M sodium chromate, Na_2CrO_4, in 30 ml dropper bottle	1 per group
1 M sodium hydroxide, NaOH, in 30 ml dropper bottle	1 per group
1 M sodium phosphate, Na_3PO_4, in 30 ml dropper bottle	1 per group
Distilled water	As needed
Equipment and other materials	
Toothpicks	As needed
Filter paper	As needed
Test tubes	4 per group
Well plate	1 per group
Electronic or triple beam balance	1 per group
Graduated cylinder, 10 ml	1 per group
Vacuum filtration kit	1 per group
Whiteboard, 2' x 3' *	1 per group
Lab handout	1 per student
Peer-review guide and instructor scoring rubric	1 per student

* As an alternative, students can use computer and presentation software such as Microsoft PowerPoint or Apple Keynote to create their arguments.

Topics for the Explicit and Reflective Discussion

Concepts That Can Be Used to Justify the Evidence

To provide an adequate justification of their evidence, students must explain why they included the evidence in their arguments and make the assumptions underlying their analysis and interpretation of the data explicit. In this investigation, students can use the following concepts to help justify their evidence:

- Chemical equations
- Conservation of matter
- Solubility

We recommend that you discuss these fundamental concepts during the explicit and reflective discussion to help students make this connection.

How to Design Better Investigations

It is important for students to reflect on the strengths and weaknesses of the investigation they designed during the explicit and reflective discussion. Students should therefore be encouraged to discuss ways to eliminate potential flaws, measurement errors, or sources of bias in their investigations. To help students be more reflective about the design of their investigation, you can ask the following questions:

- What were some of the strengths of your investigation? What made it scientific?
- What were some of the weaknesses of your investigation? What made it less scientific?
- If you were to do this investigation again, what would you do to address the weaknesses in your investigation? What could you do to make it more scientific?

Crosscutting Concepts

This investigation is well aligned with two crosscutting concepts found in *A Framework for K–12 Science Education,* and you should review these concepts during the explicit and reflective discussion.

- *Patterns:* Observed patterns in nature, such as how groups of substances interact with each other, guide the way scientists organize and classify properties of matter and changes in matter. Scientists also explore the relationships between and the underlying causes of the patterns they observe in nature.
- *Energy and matter: Flows, cycles, and conservation*: In science it is important to track how energy and matter move into, out of, and within systems. In chemistry this is particularly important with respect to chemical and physical changes.

The Nature of Science and the Nature of Scientific Inquiry

This investigation is well aligned with two important concepts related to the *nature of science* (NOS) and the *nature of scientific inquiry* (NOSI), and you should review these concepts during the explicit and reflective discussion.

- *The difference between observations and inferences*: An observation is a descriptive statement about a natural phenomenon, whereas an inference is an interpretation of an observation. Students should also understand that current scientific knowledge and the perspectives of individual scientists guide both observations and inferences. Thus, different scientists can have different but equally valid interpretations of the same observations due to differences in their perspectives and background knowledge.

- *The difference between laws and theories in science:* A scientific law describes the behavior of a natural phenomenon or a generalized relationship under certain conditions; a scientific theory is a well-substantiated explanation of some aspect of the natural world. Theories do not become laws even with additional evidence; they explain laws. However, not all scientific laws have an accompanying explanatory theory. It is also important for students to understand that scientists do not discover laws or theories; the scientific community develops them over time.

Hints for Implementing the Lab

- The time needed for data collection and analysis for this lab is very short, so be prepared to accomplish multiple stages of the ADI model within a single class period.

- It is possible to conduct this investigation without actually reacting any chemicals as described. Given the solubility rules, the students will have all the information they need to identify the precipitates. To increase the rigor, consider requiring students to determine how much product should be generated and use stoichiometric calculations to determine if the anticipated precipitate was generated.

Topic Connections

Table 24.3 (p. 386) provides an overview of the scientific practices, crosscutting concepts, disciplinary core ideas, and supporting ideas at the heart of this lab investigation. In addition, it lists NOS and NOSI concepts for the explicit and reflective discussion. Finally, it lists literacy and mathematics skills (*CCSS ELA* and *CCSS Mathematics*) that are addressed during the investigation.

LAB 24

TABLE 24.3

Lab 24 alignment with standards

Scientific practices	• Asking questions and defining problems • Planning and carrying out investigations • Analyzing and interpreting data • Constructing explanations and designing solutions • Engaging in argument from evidence • Obtaining, evaluating, and communicating information
Crosscutting concepts	• Patterns • Energy and matter: Flows, cycles, and conservation
Core idea	• PS1.B: Chemical reactions
Supporting ideas	• Conservation of matter • Chemical equations • Solubility • Precipitates
NOS and NOSI concepts	• Observations and inferences • Scientific laws and theories
Literacy connections (*CCSS ELA*)	• *Reading:* Key ideas and details, craft and structure, integration of knowledge and ideas • *Writing:* Text types and purposes, production and distribution of writing, research to build and present knowledge, range of writing • *Speaking and listening:* Comprehension and collaboration, presentation of knowledge and ideas
Mathematics connections (*CCSS Mathematics*)	• Reason abstractly and quantitatively • Look for and express regularity in repeated reasoning

Lab Handout

Lab 24. Identification of Reaction Products: What Are the Products of the Chemical Reactions?

Introduction

Chemical reactions are the result of a rearrangement of the molecular or ionic structure of a substance. It is important to remember that the *law of conservation of mass* states that mass is conserved in ordinary chemical changes. The total amount of mass before and after the reaction is therefore the same, even though there are new substances with different properties than the original substances. Additionally, the *law of definite proportions* states that atoms combine in specific ways when they form compounds; therefore, a given compound always contains the same proportion of elements by mass. These two laws allow us to predict the rearrangement of atoms during chemical reactions, with no atoms being destroyed and no new atoms being produced. Balanced chemical equations are used to show the relative amounts of substances that react with each other and how the structures are rearranged during a chemical reaction.

An example of a precipitate reaction

One specific type of chemical reaction is a double replacement reaction or a *precipitation reaction*. Precipitation reactions typically occur when two solutions are mixed together and a nonsoluble product—the precipitate—is formed. Figure L24.1 shows an example of a precipitation reaction involving potassium iodide and lead nitrate. The balanced chemical equation is

$$2KI(aq) + Pb(NO_3)_2(aq) \rightarrow 2KNO_3(aq) + PbI_2(s)$$

In this example clear potassium iodide (KI) and clear lead nitrate ($Pb(NO_3)_2$) solutions are mixed together, producing a bright yellow precipitate, lead iodide (PbI_2). The other product, potassium nitrate (KNO_3), is soluble and remains dissolved in the solution.

To predict the products during a precipitation reaction, you must know the ion charges for the substances dissolved into the solutions and understand which types of substances are soluble in water. There are some general rules that can help you determine if an ionic compound will dissolve in water. Table L24.1 (p. 388) lists some basic solubility rules for ionic compounds; the table shows common *anions* (negatively charged ions) and *cations* (positively charged ions) along with their solubility. General solubility rules do not hold true in every case; therefore, exceptions to the solubility rules are also noted in Table L24.1 .

LAB 24

Solubility rules for ionic compounds in water

Ion	Soluble?	Exceptions
NO_3^-	Yes	None
ClO_4^-	Yes	None
Cl^-	Yes	Ag^+, Hg_2^{2+}, Pb^{2+}
I^-	Yes	Ag^+, Hg_2^{2+}, Pb^{2+}
SO_4^{2-}	Yes	Ca^{2+}, Ba^{2+}, Sr^{2+}, Ag^+, Hg_2^{2+}, Pb^{2+}
CO_3^{2-}	No	Group IA and NH_4^+
PO_4^{3-}	No	Group IA and NH_4^+
OH^-	No	Group IA, Ca^{2+} (slightly soluble), Ba^{2+}, Sr^{2+}
S^{2-}	No	Groups IA and IIA and NH_4^+
Na^+	Yes	None
NH_4^+	Yes	None
K^+	Yes	None

Understanding how ions may rearrange during a chemical reaction and understanding the general solubility rules will go a long way in helping you predict the products of a precipitation reaction and identify the actual precipitate. However, depending on the chemicals involved, there may be no obvious way to identify the precipitate using qualitative observations. In those cases it may be necessary to use stoichiometric procedures to determine the precipitate based on a balanced chemical equation.

Your Task

Four partial chemical equations are provided below. Your task is to identify the products of the four chemical reactions, including the precipitate in each reaction.

$$AgNO_{3(aq)} + NaCl_{(aq)} \rightarrow ?$$

$$Na_2CrO_{4(aq)} + Ca(NO_3)_{2(aq)} \rightarrow ?$$

$$CaCl_{2(aq)} + Na_3PO_{4(aq)} \rightarrow ?$$

$$NaOH_{(aq)} + NiCl_{2(aq)} \rightarrow ?$$

The guiding question of this investigation is, **What are the products of the chemical reactions?**

Materials

You may use any of the following materials during your investigation:

Consumables	Equipment
• Calcium chloride, $CaCl_2$	• Toothpicks
• 1 M calcium nitrate, $Ca(NO_3)_2$	• Filter paper
• 1 M nickel(II) chloride, $NiCl_2$	• 4 test tubes
• 1 M silver nitrate, $AgNO_3$	• Well plate
• 1 M sodium chloride, NaCl	• Electronic or triple beam balance
• 1 M sodium chromate, Na_2CrO_4	• Graduated cylinder (10 ml)
• 1 M sodium hydroxide, NaOH	• Vacuum filtration kit
• 1 M sodium phosphate, Na_3PO_4	
• Distilled water	

Safety Precautions

Follow all normal lab safety rules. Silver nitrate is toxic by ingestion, is corrosive to body tissues, and stains clothes and skin. Your teacher will explain relevant and important information about working with the chemicals associated with this investigation. In addition, take the following safety precautions:

- Wear indirectly vented chemical-splash goggles and chemical-resistant gloves and apron while in the laboratory.

- Handle all glassware with care.

- Wash your hands with soap and water before leaving the laboratory.

Investigation Proposal Required?   ☐ Yes   ☐ No

Getting Started

To answer the guiding question, you will need to determine what type of data you need to collect, how you will collect the data, and how you will analyze the data.

To determine *what type of data you need to collect*, think about the following questions:

- How much of each chemical will you need to use?

- What masses will you need to measure during the investigation?

- What observations will you need to make?

To determine *how you will collect the data*, think about the following questions:

- How long will you need to allow the chemicals to react?

- How will you reduce error?

To determine *how you will analyze the data*, think about the following questions:

- What type of calculations will you need to make (if any)?
- How will you determine the precipitate in each reaction?

Connections to Crosscutting Concepts, the Nature of Science, and the Nature of Scientific Inquiry

As you work through your investigation, be sure to think about

- the importance of patterns in science,
- the importance of the flow of matter and energy within systems,
- the difference between observations and inferences in science, and
- the difference between laws and theories in science.

Initial Argument

Once your group has finished collecting and analyzing your data, you will need to develop an initial argument. Your argument must include a *claim*, which is your answer to the guiding question. Your argument must also include *evidence* in support of your claim. The evidence is your analysis of the data and your interpretation of what the analysis means. Finally, you must include a *justification* of the evidence in your argument. You will therefore need to use a scientific concept or principle to explain why the evidence that you decided to use is relevant and important. You will create your initial argument on a whiteboard. Your whiteboard must include all the information shown in Figure L24.2.

FIGURE L24.2

Argument presentation on a whiteboard

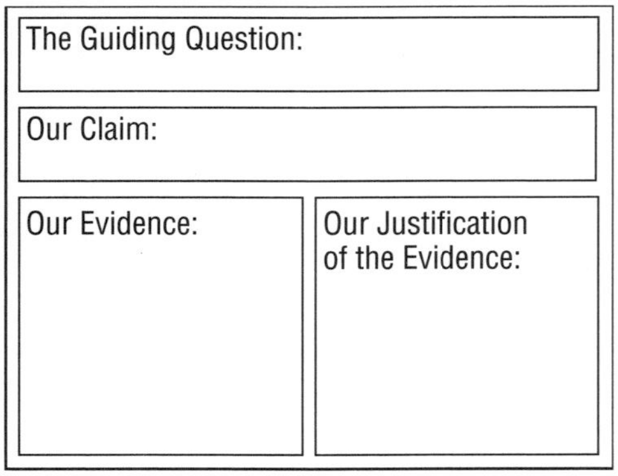

The Guiding Question:	
Our Claim:	
Our Evidence:	Our Justification of the Evidence:

Argumentation Session

The argumentation session allows all of the groups to share their arguments. One member of each group stays at the lab station to share that group's argument, while the other members of the group go to the other lab stations one at a time to listen to and critique the arguments developed by their classmates. The goal of the argumentation session is not to convince others that your argument is the best one; rather, the goal is to identify errors or instances of faulty reasoning in the initial arguments so these mistakes can be fixed. You will therefore need to evaluate the content of the claim, the quality of the evidence used to support the claim, and the strength of the justification of the evidence included in each argument that you see. To critique an argument, you might need more information than what is included on the whiteboard. You might, therefore, need to ask the presenter one or more follow-up questions, such as:

- What did your group do to analyze the data, and why did you decide to do it that way?

- Is that the only way to interpret the results of your group's analysis? How do you know that your interpretation of the analysis is appropriate?

- Why did your group decide to present your evidence in that manner?

- What other claims did your group discuss before deciding on that one? Why did you abandon those alternative ideas?

- How confident are you that your group's claim is valid? What could you do to increase your confidence?

Once the argumentation session is complete, you will have a chance to meet with your group and revise your original argument. Your group might need to gather more data or design a way to test one or more alternative claims as part of this process. Remember, your goal at this stage of the investigation is to develop the most valid or acceptable answer to the research question!

Report

Once you have completed your research, you will need to prepare an *investigation report* that consists of three sections that provide answers to the following questions:

1. What question were you trying to answer and why?

2. What did you do during your investigation and why did you conduct your investigation in this way?

3. What is your argument?

Your report should answer these questions in two pages or less. The report must be typed and any diagrams, figures, or tables should be embedded into the document. Be sure to write in a persuasive style; you are trying to convince others that your claim is acceptable or valid!

LAB 24

Lab 24. Identification of Reaction Products: What Are the Products of the Chemical Reactions?

1. Describe why a precipitate forms during some double displacement reactions and not others.

2. Stephanie and Tamara are conducting an investigation in chemistry class. They mix a solution of barium nitrate with a solution of sodium sulfate. When they mix the solutions, a reaction occurs and a precipitate is formed. Stephanie thinks the precipitate is sodium nitrate, but Tamara thinks it is barium sulfate.

   Use what you know about chemical reactions and solubility to provide an argument in support of either Stephanie or Tamara.

3. In science, observations are more important than inferences.

   a. I agree with this statement.
   b. I disagree with this statement.

   Explain your answer, using an example from your investigation about identification of reaction products.

4. In science, there is a hierarchy of ideas that builds from hypothesis to theory to law, with each idea becoming more certain than the other.

    a. I agree with this statement.
    b. I disagree with this statement.

    Explain your answer, using an example from your investigation about identification of reaction products.

5. Scientists find it important to identify patterns in their observations. Explain why this is important, using an example from your investigation about identification of reaction products.

6. One of the main foci of chemistry is investigating the flow of matter and energy within systems. Explain why understanding the flow of matter and energy is important, using an example from your investigation about identification of reaction products.

LAB 25

Teacher Notes

Lab 25. Acid-Base Titration and Neutralization Reactions: What Is the Concentration of Acetic Acid in Each Sample of Vinegar?

Purpose

The purpose of this lab is for students to *apply* what they know about the characteristics of acids and bases, neutralization reactions, and stoichiometry to answer a practical question. This lab gives students an opportunity to develop the skills they need to perform an acid-base titration. Students will also learn about the difference between observations and inferences in science and how science is influenced by society and the culture in which it is practiced.

The Content

Acid-base titrations can be used to measure the concentration of an acid or base in a solution. An acid-base titration is based on the premise that acids and bases neutralize each other when mixed in an exact stoichiometric ratio. A chemist, as a result, can use this proportional relationship to determine how many moles of an acid or a base are present in an unknown solution. To accomplish this task, the chemist must first determine how many moles of an acid or base need to be added to the unknown solution in order to neutralize it. The chemist can then use the stoichiometric ratio that exists between the reactants to determine the number of moles of the acid or base in the unknown solution. Finally, the chemist can calculate the concentration of the solution based on the moles of acid or base found in the sample and the volume of the sample.

The objective of an acid-base titration is to measure the volume of one reactant of known concentration (the *titrant*) that is required to neutralize another reactant of unknown concentration (the *analyte*). The titrant is gradually added to the analyte in small amounts until an end point is reached. The end point is represented by a color change in an indicator that is added to the analyte. If the indicator is chosen well, the end point will correspond with the *equivalence point* of reaction, which is the point at which the amount of titrant is stoichiometrically equivalent to the amount of the analyte.

In this investigation, the students need to determine the acetic acid concentration of three different samples of vinegar. The analyte in this case is vinegar (a solution of acetic acid) and the titrant is sodium hydroxide (NaOH). When the sodium hydroxide is added to acetic acid, the sodium hydroxide will react with the acetic acid based on the following neutralization reaction:

$$CH_3COOH(aq) + NaOH(aq) \rightarrow NaCH_3COO(aq) + H_2O(l)$$

This balanced chemical equation indicates that 1 mole of NaOH is needed to completely neutralize 1 mole of CH_3COOH. The volume of the NaOH that was added to the vinegar can therefore be used to calculate the molarity of the solution given this stoichiometric relationship. The steps to calculate the molarity of the unknown solution are as follows:

1. Determine the moles of NaOH used in the reaction from the known molarity of the titrant and the volume of titrant needed to reach the end point. The general equation for this calculation is *moles of titrant = molarity of titrant × liters of titrant.* For this investigation, it is *moles of NaOH = molarity of NaOH × liters of NaOH.*

2. Determine the moles of the analyte reactant using the stoichiometric ratio of the reactants provided in the neutralization equation and the number of moles of titrant used. In the case, the stoichiometric ratio is 1:1. The general equation for this calculation is *moles of analyte reactant = moles of titrant × (1 mole analyte reactant / 1 mole of titrant).* For this investigation, it would be *moles of CH₃COOH = moles of NaOH × (1 mole CH₃COOH / 1 mole of NaOH).*

3. Determine the molarity of the unknown solution from the moles of the analyte reactant and the volume of the unknown sample. The general equation for this calculation is *molarity of sample = moles of analyte reactant / liters of the sample.* For this investigation, it would be *molarity of vinegar = moles of acetic acid / liters of vinegar.*

The concentration of the analyte can also be expressed as a mass percent. To determine the mass percent of the reactant in the analyte, perform the following calculations:

1. Use the moles of the analyte reactant and the molar mass of the analyte reactant to determine the mass of the reactant in the sample. The general equation for this calculation is *mass of analyte reactant = moles of analyte reactant × molar mass of analyte reactant.* For this investigation, it would be *mass of acetic acid = moles of acetic acid × molar mass of acetic acid.*

2. Determine the total mass of the sample. The general equation for this calculation is *mass of sample = density of sample / volume of sample.* For this investigation, it would be *mass of vinegar = density of vinegar / volume of vinegar.* (The density of vinegar is 1.005 g/ml.)

3. Determine the mass percent of the analyte reactant from the mass of the analyte reactant in the sample and the total mass of the sample. The general equation for this calculation is *mass percent of analyte reactant = mass of analyte reactant / mass of the sample.* For this investigation, it would be *mass percent of acetic acid = mass of acetic acid / mass of vinegar.*

LAB 25

Timeline

The instructional time needed to complete this lab investigation is 180–250 minutes. Appendix 2 (p. 501) provides options for implementing this lab investigation over several class periods. Option E (250 minutes) should be used if students are unfamiliar with scientific writing because this option provides extra instructional time for scaffolding the writing process. You can scaffold the writing process by modeling, providing examples, and providing hints as students write each section of the report. Option F (180 minutes) should be used if students are familiar with scientific writing and have the skills needed to write an investigation report on their own. In option F, students complete stage 6 (writing the investigation report) and stage 8 (revising the investigation report) as homework.

Materials and Preparation

The materials needed to implement this investigation are listed in Table 25.1. The consumables and equipment can be purchased from a science supply company such as Carolina, Flinn Scientific, or Ward's Science.

It is important to standardize the ~ 0.1 M NaOH solution before it is used as a titrant because solid NaOH absorbs moisture from the air and solutions of NaOH absorb carbon dioxide from the air. To standardize the NaOH solution, you can follow these steps:

1. Weigh out ~ 0.5 g of potassium hydrogen phthalate (KHP).

2. Transfer the KHP to a 250 ml Erlenmeyer flask. Dissolve the KHP in 50 ml of distilled water.

3. Add a few drops of phenolphthalein.

4. Titrate to end point with the ~ 0.1 M NaOH solution.

5. Record the volume of NaOH and determine the exact molarity of the NaOH. (The neutralization reaction for KHP by NaOH is in a 1:1 mole ratio.)

You can standardize the NaOH solution before the lab begins and tell students the exact molarity of it, or you can standardize it as part of your tool talk. You can also require the students to standardize it as part of their investigation. If you decide to have the students standardize the NaOH solution, be sure to supply them with the supplemental information sheet.

We recommend that you use a set routine for distributing and collecting the materials during the lab investigation. For example, the consumables and equipment for each group can be set up at each group's lab station before class begins, or one member from each group can collect them from a table or a cart when needed during class.

TABLE 25.1

Materials list

Item	Quantity
Consumables	
KHP	1 g per group
~0.1 M NaOH solution (standardized)	100 ml per group
Vinegar sample A	10 ml per group
Vinegar sample B	10 ml per group
Vinegar sample C	10 ml per group
Distilled water	1 wash bottle per group
Bromthymol blue (in a dropper bottle)	1 per group
Cresol red (in a dropper bottle)	1 per group
Phenolphthalein (in a dropper bottle)	1 per group
Thymol blue (in a dropper bottle)	1 per group
Equipment and other materials	
Volumetric pipette, 5 ml	1 per group
Pipette bulb	1 per group
Burette, 50 ml	1 per group
Burette clamp	1 per group
Test tube clamp	1 per group
Ring stand	1 per group
Beakers (250 ml)	2 per group
Erlenmeyer flask, 250 ml	2 per group
Graduated cylinder, 25 ml	1 per group
Magnetic stirrer or glass stirring rod	1 per group
Funnel	1 per group
Electronic or triple beam balance	2 per class
Lab Reference Sheet on standardization of NaOH	1 per group
Investigation Proposal C (optional but recommended)	1 per group
Whiteboard, 2' x 3' *	1 per group
Lab handout	1 per student
Peer-review guide and instructor scoring rubric	1 per student

*As an alternative, students can use computer and presentation software such as Microsoft PowerPoint or Apple Keynote to create their arguments.

LAB 25

Safety Precautions

Remind students to follow all normal lab safety rules. Acetic acid and sodium hydroxide are corrosive to eyes, skin, and other body tissues. Sodium hydroxide is also toxic by ingestion. Phenolphthalein is an alcohol-based solution and is flammable, and it is moderately toxic by ingestion. You will therefore need to explain the potential hazards of working with these chemicals and how to work with hazardous chemicals. In addition, tell students to take the following safety precautions:

- Wear indirectly vented chemical-splash goggles and chemical-resistant gloves and aprons when they are collecting their data.
- Handle all glassware with care.
- Wash their hands with soap and water when they are done collecting the data.

Laboratory Waste Disposal

We recommend following Flinn laboratory waste disposal method 10 to dispose of excess NaOH solution. The neutralized solutions of CH_3COOH and NaOH can be flushed down a drain with excess water following Flinn laboratory waste disposal method 26b. Information about laboratory waste disposal methods is included in the Flinn Catalog and Reference Manual; you can request a free copy at *www.flinnsci.com*.

Topics for the Explicit and Reflective Discussion

Concepts That Can Be Used to Justify the Evidence

To provide an adequate justification of their evidence, students must explain why they included the evidence in their arguments and make the assumptions underlying their analysis and interpretation of the data explicit. In this investigation, students can use the following concepts to help justify their evidence:

- Characteristics of acids and bases
- Neutralization reactions
- Stoichiometry and mole ratios
- Molarity

We recommend that you discuss these fundamental concepts during the explicit and reflective discussion to help students make this connection.

How to Design Better Investigations

It is important for students to reflect on the strengths and weaknesses of the investigation they designed during the explicit and reflective discussion. Students should therefore be encouraged to discuss ways to eliminate potential flaws, measurement errors, or sources of bias in their investigations. To help students be more reflective about the design of their investigation, you can ask the following questions:

- What were some of the strengths of your investigation? What made it scientific?
- What were some of the weaknesses of your investigation? What made it less scientific?
- If you were to do this investigation again, what would you do to address the weaknesses in your investigation? What could you do to make it more scientific?

Crosscutting Concepts

This investigation is well aligned with two crosscutting concepts found in *A Framework for K–12 Science Education,* and you should review these concepts during the explicit and reflective discussion.

- *Scale, proportion, and quantity:* It is critical for scientists to be able to recognize what is relevant at different sizes, time frames, and scales. Scientists must also be able to recognize proportional relationships between categories or quantities. In this investigation, students need to recognize and use proportional relationships to determine the concentration of acetic acid in vinegar samples.
- *Energy and matter: Flows, cycles, and conservation:* In science it is important to track how energy and matter move into, out of, and within systems. In this investigation, students need to track how much of one type of matter they add to a system to determine the amount of a different type of matter that was already present in the system.

The Nature of Science and the Nature of Scientific Inquiry

This investigation is well aligned with two important concepts related to the *nature of science* (NOS) and the *nature of scientific inquiry* (NOSI), and you should review these concepts during the explicit and reflective discussion.

- *The difference between observations and inferences*: An observation is a descriptive statement about a natural phenomenon, whereas an inference is an interpretation of an observation. Students should also understand that current scientific knowledge and the perspectives of individual scientists guide both observations and inferences. Thus, different scientists can have different but equally valid interpretations of the same observations due to differences in their perspectives and background knowledge.
- *The influence of society and culture on science*: Science is influenced by the society and culture in which it is practiced because science is a human endeavor. Cultural values and expectations determine what scientists choose to investigate, how investigations are conducted, how research findings are interpreted, and

what people see as implications. People also view some research as being more important than others because of cultural values and current events. For example, many viewed the development of a method for determining the exact concentration of a solution and the equipment used to do it as being extremely valuable because of its implications for business and future research.

Hints for Implementing the Lab

- Allowing students to design their own procedures for collecting data gives them an opportunity to try, to fail, and to learn from their mistakes. However, you can scaffold students as they develop their procedure for conducting a titration by having them fill out an investigation proposal. These proposals provide a way for you to offer students hints and suggestions without telling them how to do it. You can also check the proposals quickly during a class period.

- Students should decide which type of indicator to use. This will lead to some diversity in the methods and will promote more discussion during the argumentation sessions.

- There are three options for standardizing the NaOH. First, you can standardize it before class. This option saves valuable class time. Second, you can standardize it as part of your tool talk. This is a good option because it allows you to illustrate the proper way to use the titration equipment. Finally, you can require students to standardize the solution as part of the investigation. This is a good option if you want to make the laboratory more challenging for advanced students or if students have performed a titration in the past.

Topic Connections

Table 25.2 provides an overview of the scientific practices, crosscutting concepts, disciplinary core ideas, and supporting ideas at the heart of this lab investigation. In addition, it lists NOS and NOSI concepts for the explicit and reflective discussion. Finally, it lists literacy and mathematics skills (*CCSS ELA* and *CCSS Mathematics*) that are addressed during the investigation.

TABLE 25.2

Lab 25 alignment with standards

Scientific practices	• Asking questions and defining problems • Planning and carrying out investigations • Analyzing and interpreting data • Using mathematics and computational thinking • Engaging in argument from evidence • Obtaining, evaluating, and communicating information
Crosscutting concepts	• Scale, proportion, and quantity • Energy and matter: Flows, cycles, and conservation
Core ideas	• PS1.A: Structure and properties of matter • PS1.B: Chemical reactions
Supporting ideas	• Acids and bases • Stoichiometry • Titration • Molarity
NOS and NOSI Concepts	• Observations and inferences • Social and cultural influences
Literacy connections (*CCSS ELA*)	• *Reading:* Key ideas and details, craft and structure, integration of knowledge and ideas • *Writing:* Text types and purposes, production and distribution of writing, research to build and present knowledge, range of writing • *Speaking and listening:* Comprehension and collaboration, presentation of knowledge and ideas
Mathematics connections (*CCSS Mathematics*)	• Make sense of problems and persevere in solving them • Reason abstractly and quantitatively • Model with mathematics • Use appropriate tools strategically • Attend to precision • Look for and make use of structure • Look for and express regularity in repeated reasoning

LAB 25

Lab Handout

Lab 25. Acid-Base Titration and Neutralization Reactions: What Is the Concentration of Acetic Acid in Each Sample of Vinegar?

Introduction

Vinegar is basically a solution of acetic acid (CH_3COOH). It is commonly used as an ingredient in salad dressing and marinades. People also use it as a cleaning agent because it dissolves mineral deposits that often build up on appliances. The acetic acid found in vinegar is produced through the oxidation of ethanol (CH_3CH_2OH) by bacteria from the genus *Acetobacter* that is added to alcoholic liquids such as wine, apple cider, and beer. The overall chemical reaction that is facilitated by these bacteria is

$$CH_3CH_2OH + O_2 \rightarrow CH_3COOH + H_2O$$

The acetic acid concentration of vinegar differs depending on its intended use. The acetic acid concentration in table vinegars, such as red wine, apple cider or balsamic, ranges from 4% to 8%. When used for pickling, the acetic acid concentration of vinegar can be as high as 12%. Companies that produce vinegar therefore need to be able to determine the exact concentration of acetic acid in a sample of vinegar in order to ensure consistency between batches and to maintain quality control over their product. One way to determine the exact concentration of acid in a solution is to use a technique called an acid-base titration.

An acid-base titration is based on the premise that acids and bases neutralize each other when mixed in an exact stoichiometric ratio. For example, when sodium hydroxide is added to acetic acid, the sodium hydroxide will react with and consume the acetic acid based on the following neutralization reaction:

$$CH_3COOH(aq) + NaOH(aq) \rightarrow NaCH_3COO(aq) + H_2O(l)$$

This balanced chemical equation indicates that 1 mole of NaOH is needed to completely neutralize 1 mole of CH_3COOH. A chemist, as a result, can use this proportional relationship to determine how many moles of acetic acid are present in a solution. To accomplish this task, the chemist must first determine how many moles of sodium hydroxide need to be added to the solution in order to neutralize the acetic acid in it. The chemist can then use the stoichiometric ratio that exists between the reactants to determine the number of moles of acetic acid in the solution. Finally, the chemist can calculate the concentration of acetic acid in the solution based on the moles of acetic acid found in the sample and the volume of the sample.

In this investigation you will use an acid-base titration to determine the concentration of acetic acid in several different types of vinegar. This is important for you to be able to do because a common question that chemists have to answer is how much acid or base is present in a given solution. It is also a key aspect of doing acid-base chemistry.

Your Task

Determine the concentration of acetic acid in three different samples of vinegar.

The guiding question for this investigation is, **What is the concentration of acetic acid in each sample of vinegar?**

Materials

You may use any of the following materials during this investigation:

Consumables	Equipment
• Potassium hydrogen phthalate, KHP	• Volumetric pipette (5 ml)
• 0.1 M NaOH solution	• Pipette bulb
• Vinegar sample A	• Burette (50 ml)
• Vinegar sample B	• Burette clamp
• Vinegar sample C	• Test tube clamp
• Distilled water (in a wash bottle)	• Ring stand
Indicators	• 2 beakers (each 250 ml)
• Bromthymol blue	• 2 Erlenmeyer flasks (each 250 ml)
• Cresol red	• Graduated cylinder (25 ml)
• Phenolphthalein	• Magnetic stirrer or glass stirring rod
• Thymol blue	• Funnel
	• Electronic or triple beam balance

Safety Precautions

Follow all normal lab safety rules. Acetic acid and sodium hydroxide are corrosive to eyes, skin, and other body tissues. Sodium hydroxide is also toxic by ingestion. Phenolphthalein is an alcohol-based solution and is flammable, and it is moderately toxic by ingestion. Keep away from flames. Your teacher will explain relevant and important information about working with the chemicals associated with this investigation. In addition, take the following safety precautions:

- Wear indirectly vented chemical-splash goggles and chemical-resistant gloves and apron while in the laboratory.

- Handle all glassware with care.

- Wash your hands with soap and water before leaving the laboratory.

LAB 25

Getting Started

The objective of an acid-base titration is to measure the volume of one reactant of known concentration that is required to neutralize another reactant of unknown concentration. The reactant with the known concentration is called the *titrant*, and the reactant with the unknown concentration is called the *analyte*. The titrant is gradually added to the analyte in small amounts until an end point is reached. The end point is represented by a color change in an indicator that is added to the analyte. If the indicator is chosen well, the end point will correspond with the equivalence point of reaction. The equivalence point is the point at which the amount of titrant is stoichiometrically equivalent to the amount of the analyte.

To perform an acid-base titration, you first need to understand how to measure the volume of the titrant needed to neutralize a specific volume of the analyte. The equipment setup for the process is illustrated in Figure L25.1. The basic steps are as follows:

1. Use a 5 ml volumetric pipette to add exactly 5 ml of the analyte (solution of unknown concentration) to a 250 ml beaker or 250 ml Erlenmeyer flask.

2. Add about 20 ml of deionized water to the analyte.

3. Add a few drops of indicator solution to the analyte.

4. Fill the burette with the titrant and record the initial burette reading.

5. Add the titrant drop by drop until the end point is reached. Be sure to stir or swirl the analyte as the titrant is added to make sure the chemicals are completely mixed.

6. Record the final burette reading.

Once the end point is reached, the volume of the titrant that was added to the analyte is used to calculate the molarity of the sample, as follows:

1. Determine the moles of titrant used in the reaction from the known molarity of the titrant and the volume of titrant needed to reach the end point.

2. Determine the moles of the analyte reactant using the stoichiometric ratio of the reactants provided in the neutralization equation and the number of moles of titrant used.

3. Determine the molarity of the unknown solution from the moles of the analyte reactant and the volume of the unknown sample (molarity = moles/L).

The concentration of the analyte can also be expressed as a mass percent. To determine the mass percent of the reactant in the analyte, perform the following calculations:

1. Use the moles of the analyte reactant and the molar mass of the analyte reactant to determine the mass of the reactant in the sample.

2. Determine the total mass of the sample.

3. Determine the mass percent of the analyte reactant from the mass of the analyte reactant in the sample and the total mass of the sample.

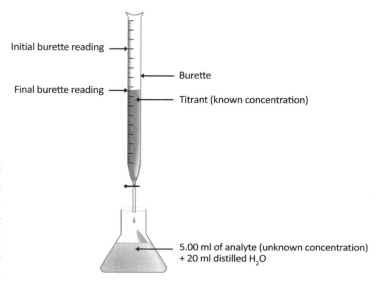

FIGURE L25.1

Equipment setup for a titration

Initial burette reading

Burette

Final burette reading

Titrant (known concentration)

5.00 ml of analyte (unknown concentration) + 20 ml distilled H_2O

When performing a titration, it is important to know the exact concentration of the titrant. Solid NaOH, for example, tends to absorb moisture from the air. It is therefore often difficult to determine the exact mass of NaOH that is added to solution, which leads to inaccurate molarity calculations. In addition, solutions of NaOH tend to absorb carbon dioxide from the air, which neutralizes some of the base. NaOH should therefore be standardized before it is used as the titrant in a titration. You instructor will either tell you the standardized concentration of the NaOH solution that you will use during this investigation or have you standardize your NaOH solution using a primary standard called potassium hydrogen phthalate (KHP).

It is also important to choose an indicator solution that will signal the end point of the titration that is as close as possible to the equivalence point of the reaction. Acetic acid is a weak acid. The equivalence point of the reaction between acetic acid and sodium hydroxide is at a pH of 8.72. KHP is also a weak acid. The equivalence point of the reaction between KHP and sodium hydroxide is at a pH of 8.50. Table L25.1 provides a list of the potential indicators that you can use for this investigation.

TABLE L25.1

Potential indicators

Indicator	pH range	Low pH color	High pH color
Bromthymol blue	6.0–7.6	Yellow	Blue
Cresol red	7.2–8.8	Orange	Red
Phenolphthalein	8.0–9.6	Clear	Pink
Thymol blue	8.0–9.6	Yellow	Blue

Now that you understand how to perform a titration, you will need to determine what type of data you need to collect, how you will collect the data, and how you will analyze the data to answer the guiding question.

To determine *what type of data you need to collect*, think about the following questions:

- What type of measurements will you need to record during your investigation?
- When will you need to take your measurements?

To determine *how you will collect the data*, think about the following questions:

- What samples of vinegar will you need to titrate?
- Which indicator will you use and why?
- How will you eliminate confounding variables?
- How will you reduce measurement error? (Hint: You can run multiple trials and then determine if the results are significantly close. You will also need to average your results.)
- How will you keep track of the data you collect?

To determine *how you will analyze the data*, think about the following questions:

- What types of calculations will you need to make to determine the concentration of acetic acid in each sample of vinegar?
- How will you determine which sample of vinegar has the greatest concentration?
- What type of graph could you create to help make sense of your data?

Connections to Crosscutting Concepts, the Nature of Science, and the Nature of Scientific Inquiry

As you work through your investigation, be sure to think about

- the importance of recognizing and using proportional relationships,
- the importance of tracking how matter moves into and within a system,
- the difference between observations and inferences in science, and
- how science is influenced by the society and culture in which it is practiced.

Initial Argument

Once your group has finished collecting and analyzing your data, you will need to develop an initial argument. Your argument must include a *claim*, which is your answer to the guiding question. Your argument must also include *evidence* in support of your claim. The evidence is your analysis of the data and your interpretation of what the analysis means. Finally, you must include a *justification* of the evidence in your argument. You will therefore need to use a scientific concept or principle to explain why the evidence that you decided to use is relevant and important. You will create your initial argument on a whiteboard. Your whiteboard must include all the information shown in Figure L25.2.

Argumentation Session

The argumentation session allows all of the groups to share their arguments. One member of each group stays at the lab station to share that group's argument, while the other members of the group go to the other lab stations one at a time to listen to and critique the arguments developed by their classmates. The goal of the argumentation session is not to convince others that your argument is the best one; rather, the goal is to identify errors or instances of faulty reasoning in the initial arguments so these mistakes can be fixed. You will therefore need to evaluate the content of the claim, the quality of the evidence used to support the claim, and the strength of the justification of the evidence included in each argument that you see. To critique an argument, you might need more information than what is included on the whiteboard. You might, therefore, need to ask the presenter one or more follow-up questions, such as:

FIGURE L25.2

Argument presentation on a whiteboard

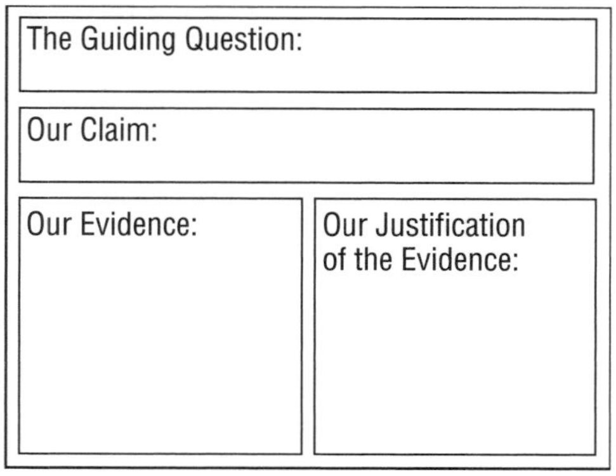

The Guiding Question:	
Our Claim:	
Our Evidence:	Our Justification of the Evidence:

- What did your group do to make sure the data you collected are reliable? What did you do to decrease measurement error?

- What did your group do to analyze the data, and why did you decide to do it that way? Did you check your calculations?

- Is that the only way to interpret the results of your group's analysis? How do you know that your interpretation of the analysis is appropriate?

- Why did your group decide to present your evidence in that manner?

- What other claims did your group discuss before deciding on that one? Why did you abandon those alternative ideas?

- How confident are you that your group's claim is valid? What could you do to increase your confidence?

Once the argumentation session is complete, you will have a chance to meet with your group and revise your original argument. Your group might need to gather more data or design a way to test one or more alternative claims as part of this process. Remember, your goal at this stage of the investigation is to develop the most valid or acceptable answer to the research question!

LAB 25

Report

Once you have completed your research, you will need to prepare an *investigation report* that consists of three sections that provide answers to the following questions:

1. What question were you trying to answer and why?

2. What did you do during your investigation and why did you conduct your investigation in this way?

3. What is your argument?

Your report should answer these questions in two pages or less. The report must be typed and any diagrams, figures, or tables should be embedded into the document. Be sure to write in a persuasive style; you are trying to convince others that your claim is acceptable or valid!

Lab 25 Reference Sheet

Standardization of a NaOH Solution

A solution of potassium hydrogen phthalate ($KHC_8H_4O_4$), often called KHP, can be used to standardize a NaOH solution. KHP serves as a primary standard, which is a stable solid that can be weighed out and used to standardize a titrant solution, such as NaOH, that has a concentration that may be slightly different from the reported value.

The balanced chemical equation for the neutralization of KHP by NaOH is

$$KHC_8H_4O_4(aq) + NaOH(aq) \rightarrow KNaC_8H_4O_4(aq) + H_2O(l)$$

This equation indicates that the stoichiometric ratio of NaOH and KHP is 1:1. Therefore if the number of grams of solid KHP used to make the solution of KHP is known, it is easy to determine the number of moles of NaOH used in the titration and the exact concentration of the NaOH solution. The basic steps are as follows:

1. Weigh out about 0.5 g of KHP and record the exact mass.

2. Transfer the KHP to a 250 ml Erlenmeyer flask and add exactly 50 ml of distilled water.

3. Add a few drops of indicator solution to the KHP solution.

4. Fill the burette with the NaOH solution and record the initial burette reading.

5. Add the titrant drop by drop until the end point is reached. Be sure to stir or swirl the KHP solution as the titrant is added to make sure the chemicals are completely mixed.

6. Record the final burette reading.

Once the end point is reached, the volume of the NaOH that was added to the KHP solution can be used to calculate the molarity of the NaOH, as follows:

1. Determine the moles of KHP used (molar mass of KHP is 204.23 g/mol).

2. Determine the moles of NaOH used (moles of NaOH = moles of KHP).

3. Determine the molarity of the NaOH solution (molarity = moles of NaOH used in the titration / liters of NaOH used in the titration).

Be sure to complete at least three trials to determine the amount of measurement error associated with the titration. If the results are sufficiently close to each other, use the average value for the calculations.

LAB 25

Lab 25. Acid-Base Titration and Neutralization Reactions: What Is the Concentration of Acetic Acid in Each Sample of Vinegar?

1. Why can chemists use an acid-base titration to determine the concentration of an unknown acid or base?

2. If you titrate 20.0 ml of 5 M sodium hydroxide with 5.0 ml of hydrochloric acid, what is the molarity of the acid? Hydrochloric acid reacts with sodium hydroxide as follows: $HCl + NaOH \rightarrow NaCl + H_2O$.

3. All scientists will make the same observations during an investigation.

    a. I agree with this statement.
    b. I disagree with this statement.

    Explain your answer, using an example from your investigation about acid-base titration and neutralization reactions.

4. The research done by a scientist is often influenced by what is important in society.

   a. I agree with this statement.

   b. I disagree with this statement.

   Explain your answer, using an example from your investigation about acid-base titration and neutralization reactions.

5. Scientists often need to track how matter moves into and within a system. Explain why this is important, using an example from your investigation about the concentration of acetic acid in vinegar.

6. Scientists often focus on proportional relationships. Explain what a proportional relationship is and why these relationships are useful, using an example from your investigation about acid-base titration and neutralization reactions.

LAB 26

Lab 26. Composition of Chemical Compounds: What Is the Empirical Formula of Magnesium Oxide?

Purpose

The purpose of this lab is to allow students to *apply* their understanding of the law of constant composition and the law of conservation of mass to determine the percent composition and then the empirical formula of magnesium oxide. Students will also learn about the different methods used in scientific investigations and how scientific knowledge can change over time in light of new evidence.

The Content

Chemists can determine the composition of a compound using two key principles: the *law of definite proportions* and the *law of conservation of mass*. The law of constant composition states that a compound is always made up of the exact same proportion of elements by mass. The law of conservation of mass states that matter is neither created nor destroyed during a chemical reaction. When these two laws are applied together, it is possible to determine both the *percent composition* and the *empirical formula* of a compound. The percent composition is the percent of each element found in the compound by mass. The empirical formula of a compound indicates the lowest whole-number ratio of the atoms found within one unit of that compound. The empirical formula and the molar mass of a compound can then be used to determine the *molecular formula*. The molecular formula indicates the actual number of atoms that are found in a single unit of a compound.

In this lab investigation, students must determine the empirical formula for magnesium oxide. To create magnesium oxide, students will need to react magnesium with oxygen from the air by heating it in a crucible. They can then measure the mass of the magnesium before and after oxidation. The resulting masses are used to calculate the empirical formula of magnesium oxide. The following steps can be used to determine the empirical formula from the masses of magnesium and magnesium oxide:

1. Use the law of conservation of mass to calculate the mass of the oxygen that combined with the magnesium.

   mass of Mg + mass of O = mass of magnesium oxide

   OR

   mass of O = mass of magnesium oxide − mass of Mg

2. Calculate the percent composition of magnesium oxide.

$$\% \ Mg = (grams \ of \ Mg \div grams \ of \ magnesium \ oxide) \times 100$$

$$\% \ O = (grams \ of \ O \div grams \ of \ magnesium \ oxide) \times 100$$

3. Use the molar masses of magnesium and oxygen to calculate the number of moles of each reactant in the product.

$$moles \ of \ Mg = grams \ of \ Mg \div 24.31 \ g/mol$$

$$moles \ of \ O = grams \ of \ O \div 16.0 \ g/mol$$

4. Determine the empirical formula of magnesium oxide. The empirical formula is the lowest whole-number ratio between the moles of Mg and O in the product.

The empirical formula for magnesium oxide is MgO.

Timeline

The instructional time needed to complete this lab investigation is 130–200 minutes. Appendix 2 (p. 501) provides options for implementing this lab investigation over several class periods. Option D (130 minutes) is appropriate if your students have experience with the scientific writing process and have the skills needed to write an investigation report on their own. In option D, students complete stage 6 (writing the investigation report) and stage 8 (revising the investigation report) as homework. Option C (200 minutes) incorporates additional class time for students to work on the written aspects of the investigation, such as their investigation report. Data collection for this investigation does not take a significant amount of time, but students will need ample time to work through the calculations to determine the empirical formula of magnesium oxide.

Materials and Preparation

The materials needed to implement this investigation are listed in Table 26.1 (p. 414). There is minimal preparation needed to complete this activity. If possible, each lab group should have its own set of equipment to facilitate data collection. Having the materials in place at each lab station will help reduce setup time; additionally, having the crucible heating apparatus set up will help ensure that students use the necessary materials and use them correctly. Providing magnesium ribbon in precut pieces will help limit waste and also decrease individual group setup times.

LAB 26

TABLE 26.1

Materials list

Item	Quantity
Consumable	
Magnesium ribbon	3–4 cm per group
Equipment and other materials	
Bunsen burner	1 per group
Striker	1 per group
Crucible with lid	1 per group
Clay triangle	1 per group
Crucible tongs	1 per group
Ring stand with metal ring	1 per group
Wire gauze square	1 per group
Electronic or triple beam balance	1 per group
Periodic table	1 per group
Investigation Proposal A	1 per group
Whiteboard, 2' x 3' *	1 per group
Lab handout	1 per student
Peer-review guide and instructor scoring rubric	1 per student

* As an alternative, students can use computer and presentation software such as Microsoft PowerPoint or Apple Keynote to create their arguments.

Safety Precautions

Remind students to follow all normal lab safety rules. Magnesium is a flammable metal that burns with an intense flame. Students will also be working with crucibles and Bunsen burners. You will therefore need to explain the potential hazards of burning magnesium, that it is important for them to not look at magnesium as it burns, and how to burn magnesium in a crucible in a safe way. In addition, tell students to take the following safety precautions:

- Wear indirectly vented chemical-splash goggles and chemical-resistant gloves and aprons when they are collecting their data.

- Use caution when working with Bunsen burners. They can burn skin, and combustibles and flammables must be kept away from the open flame. Students with long hair should tie it back behind their heads.

- Inspect the crucible for cracks and exchange it for a new one if it has any cracks. Clean the crucible and lid thoroughly, then dry the crucible and lid by heating them for five minutes before using it.

- Be careful with a crucible after removing it from a flame because it will still be hot.

- Wash their hands with soap and water when they are done collecting the data.

Laboratory Waste Disposal

We recommend following Flinn laboratory waste disposal method 3 to dispose of the magnesium ribbon used in this lab. Information about laboratory waste disposal methods is included in the *Flinn Science Catalog Reference Manual*; you can request a free copy at *www.flinnsci.com*.

Topics for the Explicit and Reflective Discussion

Concepts That Can Be Used to Justify the Evidence

To provide an adequate justification of their evidence, students must explain why they included the evidence in their arguments and make the assumptions underlying their analysis and interpretation of the data explicit. In this investigation, students can use the following concepts to help justify their evidence:

- Percent composition
- Empirical formula
- Law of conservation of mass
- Law of definite proportions

We recommend that you discuss these fundamental concepts during the explicit and reflective discussion to help students make this connection.

How to Design Better Investigations

It is important for students to reflect on the strengths and weaknesses of the investigation they designed during the explicit and reflective discussion. Students should therefore be encouraged to discuss ways to eliminate potential flaws, measurement errors, or sources of bias in their investigations. To help students be more reflective about the design of their investigation, you can ask the following questions:

- What were some of the strengths of your investigation? What made it scientific?

LAB 26

- What were some of the weaknesses of your investigation? What made it less scientific?
- If you were to do this investigation again, what would you do to address the weaknesses in your investigation? What could you do to make it more scientific?

Crosscutting Concepts

This investigation is well aligned with two crosscutting concepts found in *A Framework for K–12 Science Education,* and you should review these concepts during the explicit and reflective discussion.

- *Scale, proportion, and quantity*: It is critical for scientists to be able to recognize what is relevant at different sizes, time frames, and scales. Scientists must also be able to recognize proportional relationships between categories or quantities.
- *Energy and matter: Flows, cycles, and conservation:* In science it is important to track how energy and matter move into, out of, and within systems.

The Nature of Science and the Nature of Scientific Inquiry

This investigation is well aligned with two important concepts related to the *nature of science* (NOS) and the *nature of scientific inquiry* (NOSI), and you should review these concepts during the explicit and reflective discussion.

- *Changes in scientific knowledge over time:* A person can have confidence in the validity of scientific knowledge but must also accept that scientific knowledge may be abandoned or modified in light of new evidence or because existing evidence has been reconceptualized by scientists. There are many examples in the history of science of both evolutionary changes (i.e., the slow or gradual refinement of ideas) and revolutionary changes (i.e., the rapid abandonment of a well-established idea) in scientific knowledge.
- *Methods used in scientific investigations*: Examples of methods include experiments, systematic observations of a phenomenon, literature reviews, and analysis of existing data sets; the choice of method depends on the objectives of the research. There is no universal step-by step scientific method that all scientists follow; rather, different scientific disciplines (e.g., chemistry vs. physics) and fields within a discipline (e.g., organic vs. physical chemistry) use different types of methods, use different core theories, and rely on different standards to develop scientific knowledge.

Hints for Implementing the Lab

- We recommend that students fill out an investigation proposal for this lab. Students will need to heat their crucibles and magnesium ribbons to very high temperatures, so it will be important for you to check their procedure before they begin. The investigation proposal also encourages students to think about the type of data they will need to collect and how they will analyze it.

- Provide a demonstration of the proper use and setup for a crucible as part of the tool talk. We recommend that students cover the crucible with the lid squarely the whole time it is heated. They can remove the lid briefly every three minutes to allow air to enter. This approach is the simplest and safest way to use a crucible.

- Weighing a hot crucible can affect the accuracy of measurements, so be sure that students allow ample time for the crucible to cool prior to weighing.

- A major source of error in this investigation is the reaction not going to completion. Through questions, prompt the students to think about ways to account for this issue.

- Another major source of error is a side reaction of magnesium with nitrogen in the air to produce magnesium nitride (Mg_3N_2). This side reaction reduces the apparent yield of magnesium oxide. The magnesium nitride can be removed with the addition of water, which converts the nitride to magnesium hydroxide and ammonia gas. Heating the product again causes the loss of water and conversion of the hydroxide to the oxide.

- Emphasize to your students that it is not enough to simply suggest which empirical formula is correct; they need to show why the alternatives are not acceptable. This approach will generate high-quality arguments.

- Students may claim that other empirical formulas, such as Mg_5O_4, are valid based on the data they collected. The difference in findings will likely be due to measurement error. Students should therefore be encouraged to discuss potential sources of error and ways to correct them as part of the argumentation session. You can then give your students an opportunity to return to lab after the argumentation session to re-collect data and see if they can eliminate the sources of error that they identified.

Topic Connections

Table 26.2 (p. 418) provides an overview of the scientific practices, crosscutting concepts, disciplinary core ideas, and supporting ideas at the heart of this lab investigation. In addition, it lists NOS and NOSI concepts for the explicit and reflective discussion. Finally, it lists literacy and mathematics skills (*CCSS ELA* and *CCSS Mathematics*) that are addressed during the investigation.

LAB 26

TABLE 26.2 _____

Lab 26 alignment with standards

Scientific practices	• Asking questions and defining problems • Planning and carrying out investigations • Analyzing and interpreting data • Using mathematical and computational thinking • Constructing explanations and designing solutions • Engaging in argument from evidence • Obtaining, evaluating, and communicating information
Crosscutting concepts	• Scale, proportion, and quantity • Energy and matter: Flows, cycles, and conservation
Core idea	• PS1.B: Chemical reactions
Supporting ideas	• Percent composition • Empirical formula • Law of conservation of mass • Law of definite proportions
NOS and NOSI concepts	• Changes in scientific knowledge over time • Methods used in scientific investigations
Literacy connections (CCSS ELA)	• *Reading:* Key ideas and details, craft and structure, integration of knowledge and ideas • *Writing:* Text types and purposes, production and distribution of writing, research to build and present knowledge, range of writing • *Speaking and listening:* Comprehension and collaboration, presentation of knowledge and ideas
Mathematics connections (CCSS Mathematics)	• Reason abstractly and quantitatively • Construct viable arguments and critique the reasoning of others • Attend to precision • Look for and express regularity in repeated reasoning

Lab Handout

Lab 26. Composition of Chemical Compounds: What Is the Empirical Formula of Magnesium Oxide?

Introduction

Chemists can describe the composition of a chemical compound in at least three different ways. The first way is to define the *percent composition* of a compound; this is the percent of each element found in the compound by mass. For example, a chemical compound called acetylene is composed of 92.25% carbon and 7.75% hydrogen by mass. The second way to describe the composition of a compound is to provide its *empirical formula*, which indicates the lowest whole-number ratio of the atoms found within one unit of that compound. The empirical formula of acetylene, for example, is CH. The empirical formula indicates the ratio of atoms in a compound but does not always represent the actual number of each kind of atom found in one unit of that compound. The compound benzene, for example, has the same empirical formula (CH) as acetylene because both acetylene and benzene contain one carbon atom for every atom of hydrogen. The third way to define the percent composition of a compound is to use a *molecular formula*, which indicates the actual number of atoms that are found in a single unit of that compound. The molecular formulas of acetylene and benzene are different even though they share the same empirical formula because each compound contains a different number of carbon and hydrogen atoms. The molecular formula of acetylene is C_2H_2, whereas the molecular formula for benzene is C_6H_6.

Chemists rely on two important principles when they attempt to determine the composition of an unknown compound. The first principle is the *law of definite proportions*, which indicates that a compound is always made up of the exact same proportion of elements by mass. The percent composition of a compound is therefore a constant and does not depend on the amount of a sample. The second principle is the *law of conservation of mass*, which states that mass is neither created nor destroyed during a chemical reaction. These two principles enable chemists to determine the percent composition of an unknown compound. Chemists can then use the percent composition of the compound and some simple mathematics to determine its empirical formula. In this investigation, you will have an opportunity to use these two principles to determine the empirical formula of a compound that you create inside the lab by heating magnesium in the presence of oxygen.

Your Task

Magnesium oxide is a compound that consists of magnesium and oxygen. It is produced when magnesium metal is heated. The heat causes the magnesium to combine with molecules of oxygen found in the air. The magnesium and oxygen may combine in a number of different ratios during this reaction. These ratios include, but are not limited to, MgO, Mg_2O, Mg_3O_2, and Mg_5O_4. Your goal is to determine the percent composition of magnesium oxide and then use this information to calculate its empirical formula.

The guiding question for this lab is, **What is the empirical formula of magnesium oxide?**

Materials

You may use any of the following materials during your investigation:

Consumables	Equipment
• Magnesium ribbon (3–4 cm long)	• Bunsen burner
	• Striker
	• Crucible with lid
	• Clay triangle
	• Crucible tongs
	• Ring stand with metal ring
	• Wire gauze square
	• Electronic or triple beam balance
	• Periodic table

Safety Precautions

Follow all normal lab safety rules. Your teacher will explain relevant and important information about working with the chemicals associated with this investigation. In addition, take the following safety precautions:

- Wear indirectly vented chemical-splash goggles and chemical-resistant gloves and apron while in the laboratory.
- Use caution when working with Bunsen burners. They can burn skin, and combustibles and flammables must be kept away from the open flame. If you have long hair, tie it back behind your head.
- Inspect the crucible for cracks. If it is cracked, exchange it for a new one. Clean the crucible and lid thoroughly before using them.
- Be careful with a crucible after removing it from a flame because it will still be hot.
- Handle all glassware with care.
- Wash your hands with soap and water before leaving the laboratory.

Investigation Proposal Required?   ☐ Yes   ☐ No

Getting Started

The first step in your investigation is to determine the percent composition of magnesium oxide. As noted earlier, when magnesium is heated it, combines with oxygen to form magnesium oxide. The law of conservation of mass suggests that the total mass of the products of a chemical reaction must equal the mass of the reactants. The mass of the oxygen in a sample of magnesium oxide will therefore equal the mass of the magnesium oxide minus the mass of the original piece of magnesium that was used in the reaction. You can use this information to determine the percent of oxygen by mass found in magnesium oxide. You will, however, need to first determine what type of data you need to collect and how you will collect the data to be able to make this calculation.

To determine *what type of data you need to collect*, think about the following questions:

- What type of measurements or observations will you need to record during your investigation?
- When will you need to make these measurements or observations?

To determine *how you will collect the data*, think about the following questions:

- What equipment will you need to use to make magnesium oxide?
- How much magnesium will you use to make magnesium oxide?
- How will you know when all the magnesium has been converted to magnesium oxide?
- How will you make sure that your data are of high quality (i.e., how will you reduce error)?
- How will you keep track of the data you collect and how will you organize it?

The second step in your investigation is to calculate the empirical formula for magnesium oxide from its percent composition. This calculation requires two steps. First, you will need to calculate the number of moles of magnesium and oxygen in your sample of magnesium oxide. Next, you will need to use the ratio between the number of moles of magnesium and the number of moles of oxygen to calculate the empirical formula of magnesium oxide. Keep in mind that fractions of atoms do not exist.

Connections to Crosscutting Concepts, the Nature of Science, and the Nature of Scientific Inquiry
As you work through your investigation, be sure to think about

- how scale, proportion, and quantity play a role in science;
- the flow of energy and matter within systems;
- how scientific knowledge can change over time in light of new evidence; and
- the different methods used in scientific investigations.

LAB 26

Initial Argument

Once your group has finished collecting and analyzing your data, you will need to develop an initial argument. Your argument must include a *claim*, which is your answer to the guiding question. Your argument must also include *evidence* in support of your claim. The evidence is your analysis of the data and your interpretation of what the analysis means. Finally, you must include a *justification* of the evidence in your argument. You will therefore need to use a scientific concept or principle to explain why the evidence that you decided to use is relevant and important. You will create your initial argument on a whiteboard. Your whiteboard must include all the information shown in Figure L26.1.

FIGURE L26.1

Argument presentation on a whiteboard

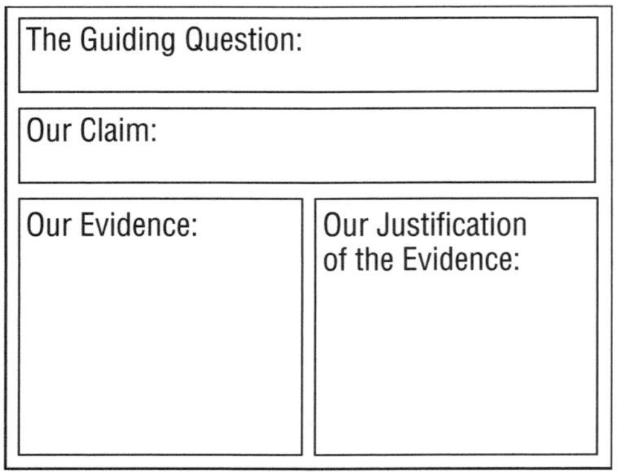

The Guiding Question:	
Our Claim:	
Our Evidence:	Our Justification of the Evidence:

Argumentation Session

The argumentation session allows all of the groups to share their arguments. One member of each group stays at the lab station to share that group's argument, while the other members of the group go to the other lab stations one at a time to listen to and critique the arguments developed by their classmates. The goal of the argumentation session is not to convince others that your argument is the best one; rather, the goal is to identify errors or instances of faulty reasoning in the initial arguments so these mistakes can be fixed. You will therefore need to evaluate the content of the claim, the quality of the evidence used to support the claim, and the strength of the justification of the evidence included in each argument that you see. To critique an argument, you might need more information than what is included on the whiteboard. You might, therefore, need to ask the presenter one or more follow-up questions, such as:

- How did your group collect the data? Why did you use that method?
- Is that the only way to interpret the results of your group's analysis? How do you know that your interpretation of the analysis is appropriate?
- Why did your group decide to present your evidence in that manner?
- What other claims did your group discuss before deciding on that one? Why did you abandon those alternative ideas?
- How confident are you that your group's claim is valid? What could you do to increase your confidence?

Once the argumentation session is complete, you will have a chance to meet with your group and revise your original argument. Your group might need to gather more data or design a way to test one or more alternative claims as part of this process. Remember, your goal at this stage of the investigation is to develop the most valid or acceptable answer to the research question!

Report

Once you have completed your research, you will need to prepare an *investigation report* that consists of three sections that provide answers to the following questions:

1. What question were you trying to answer and why?

2. What did you do during your investigation and why did you conduct your investigation in this way?

3. What is your argument?

Your report should answer these questions in two pages or less. The report must be typed and any diagrams, figures, or tables should be embedded into the document. Be sure to write in a persuasive style; you are trying to convince others that your claim is acceptable or valid!

LAB 26

Checkout Questions

1. Lab 26. Composition of Chemical Compounds: What Is the Empirical Formula of Magnesium Oxide? ⊚ Describe how the law of definite proportions and the law of conservation of mass are useful for determining the composition of a compound.

2. Felicity and Dawson are conducting an investigation to determine the empirical formula of iron oxide. They start with an 85.65 g piece of iron metal and burn it in air. The mass of the iron oxide produced is 118.37. Felicity thinks the empirical formula of iron oxide is Fe_3O_4 and Dawson thinks it is FeO. Use what you know about how to determine the empirical formula of a compound to provide an argument in support of Felicity or Dawson.

3. If multiple scientists are investigating the same question in their own labs, each scientist will use the same method to try and answer the question.

   a. I agree with this statement.
   b. I disagree with this statement.

   Explain your answer, using an example from your investigation about the composition of chemical compounds.

4. In science, once an idea has been established it will not be changed because it has been investigated so thoroughly.

   a. I agree with this statement.
   b. I disagree with this statement.

Explain your answer, using an example from your investigation about the composition of chemical compounds.

5. Scale, proportion, and quantity play a central role in science and chemistry. Explain why these concepts are important, using an example from your investigation about the composition of chemical compounds.

6. One of the main foci of chemistry is investigating the flow of matter and energy within systems. Explain how understanding the flow of matter and energy can be useful, using an example from your investigation about the composition of chemical compounds.

LAB 27

Teacher Notes

Lab 27. Stoichiometry and Chemical Reactions: Which Balanced Chemical Equation Best Represents the Thermal Decomposition of Sodium Bicarbonate?

Purpose

The purpose of this lab is to allow students to *apply* their understanding of stoichiometry as it relates to the reactants and products of a chemical reaction. This lab is designed for students to generate empirical evidence to determine the appropriate chemical reaction that represents the thermal decomposition of sodium bicarbonate. Students will also learn about the difference between data and evidence and the nature and role of experiments in science.

The Content

Predicting and identifying the products of a chemical reaction is possible because chemical reactions follow two key principles: the *law of conservation of mass* and the *law of definite proportions*. The law of conservation of mass states that matter is neither created nor destroyed during chemical processes. The law of definite proportions states that a compound is always made up of the exact same proportion of elements by mass. When these two laws are applied together, in the context of chemical reactions, it is possible to predict the outcome of reactions as well as be confident that a particular reaction will consistently result in the same products.

Predicting and identifying the products of a chemical reaction is considerably easier if the reactants are known. When the actual reactants are not known, the difficulty of predicting the products with any certainty is increased. Stoichiometry, however, is useful for generating empirical evidence regarding the products of a chemical reaction when there are several possibilities. Stoichiometry is the study of the quantitative relationships between the reactants and products of a chemical reaction. Through the use of stoichiometry and a balanced chemical equation, it is possible to determine how many grams of a product will be generated from a particular amount of reactants. Using the ideal amounts, or theoretical yields, of product that would be produced during a reaction also gives us a strategy for identifying the product generated during a chemical reaction.

To further explain the usefulness of stoichiometry, consider the problem posed for this activity related to the thermal decomposition of sodium bicarbonate (baking soda). The main task of this investigation is to determine which possible decomposition reaction represents what actually happens when sodium bicarbonate is heated and decomposes.

To approach this problem, each of the four reactions must be evaluated, with each representing potentially different products generated during the decomposition. If there is a known mass of reactant (i.e., sodium bicarbonate), then a balanced chemical equation will allow for a calculation of the theoretical yield of product. The reaction whose theoretical yield most closely aligns with the actual yield and empirical evidence is best supported as the decomposition reaction of sodium bicarbonate.

The following procedures demonstrate how to calculate the theoretical yield for the decomposition reaction of sodium bicarbonate that results in solid sodium carbonate (Na_2CO_3), carbon dioxide (CO_2) gas, and water vapor (H_2O) and assess the degree to which this reaction is empirically supported as the decomposition reaction of sodium bicarbonate.

1. Begin with the balanced chemical reaction that is being assessed. The balanced chemical reaction equation for the decomposition of sodium bicarbonate is

$$2NaHCO_{3(s)} \rightarrow Na_2CO_{3(s)} + CO_{2(g)} + H_2O_{(g)}$$

The balanced chemical reaction equation above represents the decomposition reaction for solid sodium bicarbonate. Reading the equation as a "sentence" it would be interpreted as "Two moles of solid sodium bicarbonate decomposes to one mole of solid sodium carbonate, one mole of carbon dioxide gas, and one mole of water vapor." The coefficients of the balanced chemical equation tell us the relative amounts (in moles) of each substance involved in the reaction. Therefore, for every 2 moles of sodium bicarbonate decomposed, 1 mole of each product is produced; the mole ratio for the solids would be 2 moles $NaHCO_3$:1 mole Na_2CO_3.

2. Determine how much product (in grams) would be produced based on the starting mass of sodium bicarbonate. Assume that the above reaction is conducted by starting with 3.0 g $NaHCO_3$.

   a. Determine how many moles of sodium bicarbonate were decomposed by converting grams to moles using the molar mass as a conversion. From the periodic table, the molar mass of sodium bicarbonate is the sum of the molar masses of the atoms in the compound. The molar mass of sodium bicarbonate is approximately 84.0 g/mol.

   Calculation: (3.0 g $NaHCO_3$) × (1 mol $NaHCO_3$/84.0 g) = 0.036 moles of $NaHCO_3$ reacted

   b. Determine how many moles of product would be produced from this starting amount using the mole ratio of reactant to product from the balanced chemical equation. In each reaction there are multiple products, any of which could be used for this calculation; however, the solid product will be the easiest to measure and will be the focus of this analysis.

Calculation: 2 moles of $NaHCO_3$ produce 1 mole of Na_2CO_3, mole ratio = 2:1 (0.036 mol $NaHCO_3$) × (1 mol Na_2CO_3/2 mol $NaHCO_3$) = 0.018 mol Na_2CO_3

    c. Determine the theoretical yield of Na_2CO_3, in grams, if 0.018 moles of Na_2CO_3 are produced by using the molar mass of Na_2CO_3 as a conversion. The molar mass of Na_2CO_3 is the sum of the molar mass of the atoms in the compound. The molar mass of sodium carbonate is approximately 106.0 g/mol.

Calculation: (0.018 mol Na_2CO_3) × (106.0g/1mol Na_2CO_3) = 1.91 g of Na_2CO_3 produced

Steps 2a–2c can be combined into one calculation: (3.0g $NaHCO_3$) × (1 mol $NaHCO_3$/84.0g Mg) × (1 mol Na_2CO_3/2 mol $NaHCO_3$) × (106.0 g $NaHCO_3$ /1 mol $NaHCO_3$) = 1.91 g $NaHCO_3$

3. Determine the agreement between the empirical results of the investigation and the theoretical results of the decomposition of sodium bicarbonate to sodium carbonate. Assume that the final mass of product was 2.1 g. Agreement between the empirical results and theoretical results can be calculated following a typical percent error calculation.

Calculation: [(Actual yield – Theoretical yield)/Theoretical yield] × 100% = [(2.1 g – 1.9 g)/1.91 g] × 100% = (0.2 g/1.91 g) × 100% = 10% error

This investigation will require students to determine which of the hypothesized reactions best aligns with the empirical results that they generate in the laboratory. To accomplish this task, students will need to compare the actual and theoretical yields of each reaction and then identify the reaction that has the lowest percent error. This type of analysis, however, does not confirm which reaction is "correct," nor does it eliminate the possibility of yet another alternative reaction being more closely aligned with the empirical results.

Timeline

The instructional time needed to complete this lab investigation is 130–200 minutes. Appendix 2 (p. 501) provides options for implementing this lab investigation over several class periods. Option D (130 minutes) is appropriate if your students have experience with the scientific writing process and have the skills needed to write an investigation report on their own. In option D, students complete stage 6 (writing the investigation report) and stage 8 (revising the investigation report) as homework. Option C (200 minutes) incorporates additional class time for students to work on the written aspects of the investigation, such as their investigation report. Data collection for this investigation does not take a significant amount of time, but students will need ample time to work through the stoichiometric calculations to determine which potential reaction is best supported by their data.

Materials and Preparation

The materials needed to implement this investigation are listed in Table 27.1. There is minimal preparation needed before this activity. If possible, each lab group should have its own set of equipment to facilitate data collection. Having the materials in place at each lab station will help reduce setup time; additionally, having the crucible heating apparatus set up as an example will help ensure that students use the necessary materials and use them correctly.

TABLE 27.1 _____

Materials list

Item	Quantity
Consumable	
Sodium bicarbonate, $NaHCO_3$	5–10 g per group
Equipment and other materials	
Bunsen burner	1 per group
Striker	1 per group
Ring stand with metal ring	1 per group
Crucible with lid	1 per group
Crucible tongs	1 per group
Pipe-stem triangle	1 per group
Wire gauze square	1 per group
Electronic or triple beam balance	1 per group
Periodic table	1 per group
Investigation Proposal A	1 per group
Whiteboard, 2' x 3' *	1 per group
Lab handout	1 per student
Peer-review guide and instructor scoring rubric	1 per student

* As an alternative, students can use computer and presentation software such as Microsoft PowerPoint or Apple Keynote to create their arguments.

LAB 27

Safety Precautions

Remind students to follow all normal lab safety rules. Students will be working with Bunsen burners and crucibles. You will therefore need to explain the potential hazards of using this equipment and how to burn sodium bicarbonate in a crucible in a safe way. In addition, tell students to take the following safety precautions:

- Wear indirectly vented chemical-splash goggles and chemical-resistant gloves and aprons when they are collecting their data.
- Use caution when working with Bunsen burners. They can burn skin, and combustibles and flammables must be kept away from the open flame. Students with long hair should tie it back behind their heads.
- Inspect the crucible for cracks and exchange it for a new one if it has any cracks. Clean the crucible and lid thoroughly, then dry the crucible and lid by heating them for five minutes before using it.
- Be careful with a crucible after removing it from a flame because it will still be hot.
- Wash their hands with soap and water when they are done collecting the data.

Laboratory Waste Disposal

We recommend following Flinn laboratory waste disposal method 26a to dispose of the waste products generated during this lab. Information about laboratory waste disposal methods is included in the Flinn Catalog and Reference Manual; you can request a free copy at *www.flinnsci.com*.

Topics for the Explicit and Reflective Discussion

Concepts That Can Be Used to Justify the Evidence

To provide an adequate justification of their evidence, students must explain why they included the evidence in their arguments and make the assumptions underlying their analysis and interpretation of the data explicit. In this investigation, students can use the following concepts to help justify their evidence:

- Stoichiometry
- Law of conservation of mass
- Law of definite proportions

We recommend that you discuss these fundamental concepts during the explicit and reflective discussion to help students make this connection.

How to Design Better Investigations

It is important for students to reflect on the strengths and weaknesses of the investigation they designed during the explicit and reflective discussion. Students should therefore be encouraged to discuss ways to eliminate potential flaws, measurement errors, or sources of bias in their investigations. To help students be more reflective about the design of their investigation, you can ask the following questions:

- What were some of the strengths of your investigation? What made it scientific?

- What were some of the weaknesses of your investigation? What made it less scientific?

- If you were to do this investigation again, what would you do to address the weaknesses in your investigation? What could you do to make it more scientific?

Crosscutting Concepts

This investigation is well aligned with two crosscutting concepts found in *A Framework for K–12 Science Education,* and you should review these concepts during the explicit and reflective discussion.

- *Cause and effect: Mechanism and explanation:* One of the main objectives of science is to identify and establish relationships between a cause and an effect.

- *Systems and system models*: Defining a system under study and making a model of it are tools for developing a better understanding of natural phenomena in science.

The Nature of Science and the Nature of Scientific Inquiry

This investigation is well aligned with two important concepts related to the *nature of science* (NOS) and the *nature of scientific inquiry* (NOSI), and you should review these concepts during the explicit and reflective discussion.

- *The difference between data and evidence in science:* Data are measurements, observations, and findings from other studies that are collected as part of an investigation. Evidence, in contrast, is analyzed data and an interpretation of the analysis.

- *The nature and role of experiments:* Scientists use experiments to test the validity of a hypothesis (i.e., a tentative explanation) for an observed phenomenon. Experiments include a test and the formulation of predictions (expected results) if the test is conducted and the hypothesis is valid. The experiment is then carried out and the predictions are compared with the observed results of the experiment. If the predictions match the observed results, then the hypothesis is supported. If the observed results do not match the prediction, then the hypothesis is not supported. A signature feature of an experiment is the control of variables to help eliminate alternative explanations for observed results.

LAB 27

Hints for Implementing the Lab

- Allowing students to design their own procedures for collecting data gives them an opportunity to try, to fail, and to learn from their mistakes. However, you can scaffold students as they develop their procedure by having them fill out an investigation proposal. These proposals provide a way for you to offer students hints and suggestions without telling them how to do it. You can also check the proposals quickly during a class period.

- Provide a demonstration of the proper use and setup for a crucible as part of the tool talk.

- The decomposition of sodium bicarbonate generates gas as a product; do not seal the crucible while heating.

- Weighing a hot crucible can affect the accuracy of measurements, so be sure that students allow ample time for their crucible to cool prior to weighing.

- A major source of error in this investigation is the reaction not going to completion. Through questions, prompt the students to reheat the product and then reweigh to ensure that the reaction has gone to completion. The sodium bicarbonate does not show any obvious signs of decomposition; it will remain a white powder while being heated in the crucible.

- Encourage students to calculate the percent error for each reaction as a way to help determine which explanation is more valid or acceptable.

- Emphasize to your students that it is not enough to simply suggest which chemical equation is correct; they need to show why the alternatives are not acceptable. This approach will generate high-quality arguments. This approach will also provide them more opportunity to practice stoichiometric calculations.

Topic Connections

Table 27.2 provides an overview of the scientific practices, crosscutting concepts, disciplinary core ideas, and supporting ideas at the heart of this lab investigation. In addition, it lists NOS and NOSI concepts for the explicit and reflective discussion. Finally, it lists literacy and mathematics skills (*CCSS ELA* and *CCSS Mathematics*) that are addressed during the investigation.

TABLE 27.2

Lab 27 alignment with standards

Scientific practices	• Asking questions and defining problems • Planning and carrying out investigations • Analyzing and interpreting data • Using mathematics and computational thinking • Constructing explanations and designing solutions • Engaging in argument from evidence • Obtaining, evaluating, and communicating information
Crosscutting concepts	• Cause and effect: Mechanism and explanation • Systems and system models
Core ideas	• PS1.A: Structure and properties of matter • PS1.B: Chemical reactions
Supporting ideas	• Stoichiometry • Conservation of mass • Law of definite proportions
NOS and NOSI concepts	• Difference between data and evidence • Nature and role of experiments
Literacy connections (*CCSS ELA*)	• *Reading:* Key ideas and details, craft and structure, integration of knowledge and ideas • *Writing:* Text types and purposes, production and distribution of writing, research to build and present knowledge, range of writing • *Speaking and listening:* Comprehension and collaboration, presentation of knowledge and ideas
Mathematics connections (*CCSS Mathematics*)	• Reason abstractly and quantitatively • Construct viable arguments and critique the reasoning of others • Attend to precision • Look for and express regularity in repeated reasoning

LAB 27

Lab 27. Stoichiometry and Chemical Reactions: Which Balanced Chemical Equation Best Represents the Thermal Decomposition of Sodium Bicarbonate?

Introduction

The *law of conservation of mass* states that mass is conserved during a chemical reaction. The *law of definite proportions* states that a compound is always made up of the exact same proportion of elements by mass. John Dalton was able to explain these two fundamental laws of chemistry with his *atomic theory*, which states that a chemical reaction is simply the rearrangement of atoms with no atoms being destroyed and no new atoms being produced during the process. Chemists use a balanced chemical equation to represent what happens on the submicroscopic level during a chemical reaction.

The *stoichiometric coefficient* is the number written in front of atoms, ions, or molecules in a chemical equation. These numbers are used to balance the number of each type of atom found on both the reactant and product sides of the equation. Stoichiometric coefficients are also useful because they identify the mole ratio between reactants and products. The mole ratio is important because it allows chemists to determine how many moles of a product will be produced from a specific number of moles of a reactant or how many moles of reactant are needed to produce a specific amount of a product.

Molar mass serves as a bridge between the number of moles of a substance and the mass of a substance. The molar mass is the mass of a given substance divided by one mole of the substance. The molar mass of a given substance can be calculated by summing the atomic mass for each atom found in a molecule of that substance. For example, the atomic mass of hydrogen is 1.01 g/mol, and the atomic mass of oxygen is 15.99 g/mol, so the molar mass of H_2O is 18.01 g/mol (1.01 g/mol + 1.01 g/mol + 15.99 g/mol). Once the molar mass of a substance is known, the mass of a sample can be used to determine the number of moles of a substance or the moles of substance can be used to determine the mass of a sample. For example, a 40-gram sample of H_2O consists of 2.2 mol of H_2O (40 g of H_2O ÷ 18.01 g/mol = 2.2 mol of H_2O) and a 3.0 mol sample of H_2O has a mass of 54.03 g (3.0 mol of H_2O x 18.01 g/mol = 54.03 g of H_2O). In this investigation, you will have an opportunity to use atomic theory, molar mass, and stoichiometry to determine how atoms are rearranged during a chemical reaction.

Your Task

There are at least four different balanced chemical equations that could explain how atoms are rearranged during the thermal decomposition of sodium bicarbonate ($NaHCO_3$). The first potential explanation is that the sodium bicarbonate decomposes into sodium hydroxide ($NaOH$) and carbon dioxide (CO_2) when it is heated. The balanced chemical equation for this reaction is

$$NaHCO_3(s) \rightarrow NaOH(s) + CO_2(g)$$

The second potential explanation is that the sodium bicarbonate decomposes into sodium carbonate (Na_2CO_3), carbon dioxide (CO_2), and water when it is heated. The balanced chemical equation for this reaction is

$$2NaHCO_3(s) \rightarrow Na_2CO_3(s) + CO_2(g) + H_2O(g)$$

The third potential explanation is that the sodium bicarbonate decomposes into sodium oxide (Na_2O), carbon dioxide, and water when it is heated. The balanced chemical equation for this potential reaction is

$$2NaHCO_3(s) \rightarrow Na_2O(s) + 2CO_2(g) + H_2O(g)$$

The fourth potential explanation is that the sodium bicarbonate decomposes into sodium hydride (NaH), carbon monoxide (CO), and oxygen when it is heated. The balanced chemical equation for this potential reaction is

$$NaHCO_3(s) \rightarrow NaH(s) + CO(g) + O_2(g)$$

Your goal is to determine which of these four balanced chemical equations best represents how atoms are rearranged during the thermal decomposition of sodium bicarbonate.

The guiding question of this investigation is, **Which balanced chemical equation best represents the thermal decomposition of sodium bicarbonate?**

Materials

You may use any of the following materials during your investigation:

Consumable	Equipment
Solid $NaHCO_3$	• Bunsen burner
	• Striker
	• Ring stand with metal ring
	• Crucible with lid
	• Crucible tongs
	• Pipe-stem triangle
	• Wire gauze square
	• Electronic or triple beam balance
	• Periodic table

LAB 27

Safety Precautions

Follow all normal lab safety rules. Your teacher will explain relevant and important information about working with the chemicals associated with this investigation. In addition, take the following safety precautions:

- Wear indirectly vented chemical-splash goggles and chemical-resistant gloves and apron while in the laboratory.
- Use caution when working with Bunsen burners. They can burn skin, and combustibles and flammables must be kept away from the open flame. If you have long hair, tie it back behind your head.
- Inspect the crucible for cracks. If it is cracked, exchange it for a new one. Clean the crucible and lid thoroughly before using them.
- Be careful with a crucible after removing it from a flame because it will still be hot.
- Wash your hands with soap and water before leaving the laboratory.

Investigation Proposal Required?   ☐ Yes   ☐ No

Getting Started

As part of your investigation, you will need to use a Bunsen burner and a crucible (see Figure L27.1) to increase the temperature of sodium bicarbonate enough for it to decompose. The thermal decomposition of sodium bicarbonate will occur rapidly at 200°C, but the product of the decomposition reaction will begin to decompose at temperatures over 850°C.

FIGURE L27.1 _____
How to heat sodium bicarbonate using a crucible and a Bunsen burner

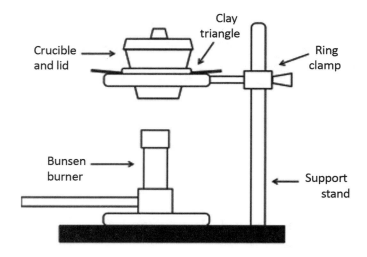

To answer the guiding question, you will also need to determine what type of data you will need to collect during your investigation, how you will collect the data, and how you will analyze the data.

To determine *what type of data to collect*, think about the following questions:

- How much $NaHCO_3$ will you need to use?
- What will you need to measure?

To determine *how you will collect the data*, think about the following questions:

- How long will you need to heat the $NaHCO_3$?
- How will you empirically determine when the decomposition of the $NaHCO_3$ is complete?
- How will you reduce error?

To determine *how you will analyze the data*, think about the following questions:

- What type of calculations will you need to make?
- How will your group take into account the precision of the balance in your analysis?

Connections to Crosscutting Concepts, the Nature of Science, and the Nature of Scientific Inquiry

As you work through your investigation, be sure to think about

- the importance of identifing the underlying cause for observed phenomena,
- how models are used to study natural phenomena,
- the difference between data and evidence in science, and
- the nature and role of experiments in science.

Initial Argument

Once your group has finished collecting and analyzing your data, you will need to develop an initial argument. Your argument must include a *claim*, which is your answer to the guiding question. Your argument must also include *evidence* in support of your claim. The evidence is your analysis of the data and your interpretation of what the analysis means. Finally, you must include a *justification* of the evidence in your argument. You will therefore need to use a scientific concept or principle to explain why the evidence that you decided to use is relevant and important. You will create your initial argument on a whiteboard. Your whiteboard must include all the information shown in Figure L27.2 (p. 438).

LAB 27

FIGURE L27.2 _____

Argument presentation on a whiteboard

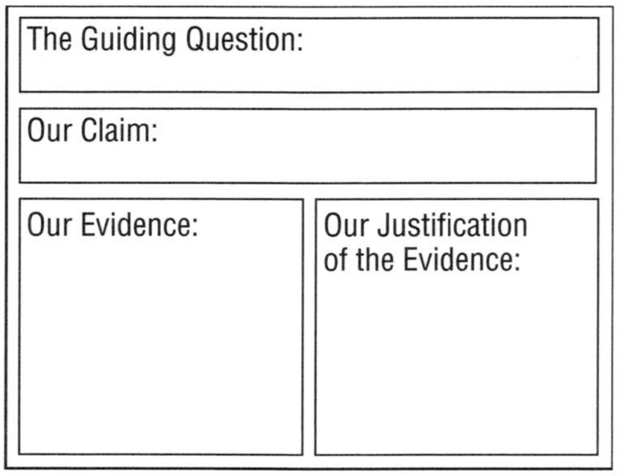

The Guiding Question:	
Our Claim:	
Our Evidence:	Our Justification of the Evidence:

Argumentation Session

The argumentation session allows all of the groups to share their arguments. One member of each group stays at the lab station to share that group's argument, while the other members of the group go to the other lab stations one at a time to listen to and critique the arguments developed by their classmates. The goal of the argumentation session is not to convince others that your argument is the best one; rather, the goal is to identify errors or instances of faulty reasoning in the initial arguments so these mistakes can be fixed. You will therefore need to evaluate the content of the claim, the quality of the evidence used to support the claim, and the strength of the justification of the evidence included in each argument that you see. To critique an argument, you might need more information than what is included on the whiteboard. You might, therefore, need to ask the presenter one or more follow-up questions, such as:

- How did your group collect the data? Why did you use that method?
- What did your group do to make sure the data you collected are reliable? What did you do to decrease measurement error?
- What did your group do to analyze the data, and why did you decide to do it that way? Did you check your calculations?
- Is that the only way to interpret the results of your group's analysis? How do you know that your interpretation of the analysis is appropriate?
- Why did your group decide to present your evidence in that manner?
- What other claims did your group discuss before deciding on that one? Why did you abandon those alternative ideas?
- How confident are you that your group's claim is valid? What could you do to increase your confidence?

Once the argumentation session is complete, you will have a chance to meet with your group and revise your original argument. Your group might need to gather more data or design a way to test one or more alternative claims as part of this process. Remember, your goal at this stage of the investigation is to develop the most valid or acceptable answer to the research question!

Report

Once you have completed your research, you will need to prepare an *investigation report* that consists of three sections that provide answers to the following questions:

438

1. What question were you trying to answer and why?

2. What did you do during your investigation and why did you conduct your investigation in this way?

3. What is your argument?

Your report should answer these questions in two pages or less. The report must be typed and any diagrams, figures, or tables should be embedded into the document. Be sure to write in a persuasive style; you are trying to convince others that your claim is acceptable or valid!

Checkout Questions

Lab 27. Stoichiometry and Chemical Reactions: Which Balanced Chemical Equation Best Represents the Thermal Decomposition of Sodium Bicarbonate?

1. Describe how the law of definite proportions is useful in understanding chemical reactions and predicting their products.

2. Alex and Sam are conducting an investigation to determine the products of the decomposition of barium carbonate ($BaCO_3$). They started with 8.0 g of $BaCO_3$, which was heated and decomposed, leaving 6.5 g of solid product.

   Alex thinks the correct chemical reaction is

   $$BaCO_3(s) \rightarrow BaO(s) + CO_2(g)$$

   But Sam thinks the correct reaction is

   $$BaCO_3(s) \rightarrow BaCO(s) + O_2(g)$$

   Use what you know about chemical reactions to provide an argument in support of Alex or Sam, based on the data provided.

3. Experiments are the best way that scientists can learn about the natural world.

    a. I agree with this statement.
    b. I disagree with this statement.

    Explain your answer, using an example from your investigation about stoichiometry and chemical reactions.

4. In science, evidence is more important than data.

    a. I agree with this statement.
    b. I disagree with this statement.

    Explain your answer, using an example from your investigation about stoichiometry and chemical reactions.

5. Scientists make many observations, and they also propose causal mechanisms that may underlie their observations. Explain why understanding causal mechanisms is important, using an example from your investigation about stoichiometry and chemical reactions.

6. A chemical equation can be considered a model for a chemical reaction. Explain what a model is and why models are important in science, using an example from your investigation about stoichiometry and chemical reactions.

LAB 28

Lab 28. Designing a Cold Pack: Which Salt Should Be Used to Make an Effective but Economical Cold Pack?

Purpose

The purpose of this lab is to allow students to *apply* their understanding of endothermic and exothermic processes and engineering practices to design a new product. This lab gives students an opportunity to learn how to use findings from an investigation and a cost-benefit analysis to determine which salt should be used to develop a new ice pack. Students will also learn about the roles that social and cultural values and creativity play in science.

The Content

Solutions are an important aspect of chemistry and everyday life. Solutions are formed when two or more substances are mixed together in a uniform or homogeneous manner. A common example of a solution is salt water. In this example the salt is the *solute,* or the substance being dissolved, and the water serves as the *solvent,* or the substance doing the dissolving. When an ionic compound is dissolved in water, the enthalpy change is the net result of two processes. First, an energy input breaks the attractive forces between the ions (ΔH_1) and disrupts the hydrogen bonds that exist between the molecules of water (ΔH_2). Second, energy is released as ion-dipole attractive forces form between the dissociated ions and the water molecules (ΔH_3). The enthalpy change of solution can be written as an equation:

$$\Delta H_{soln} = \Delta H_1 + \Delta H_2 + \Delta H_3$$

The dissolution process is exothermic ($\Delta H_{soln} < 0$) when the amount of energy released during the formation of the hydrated ions (ΔH_3) is greater than the amount of energy required to separate the solute particles and the water molecules ($\Delta H_1 + \Delta H_2$). The dissolution process is endothermic ($\Delta H_{soln} > 0$) when the amount of energy released in the formation of the hydrated ions (ΔH_3) is less than the amount of energy required to separate the solute particles and water molecules ($\Delta H_1 + \Delta H_2$). Some salts, like calcium chloride, have an exothermic heat of solution and others, such as ammonium nitrate, have an endothermic heat of solution.

The molar ΔH_{soln} for an ionic compound can be measured using a calorimeter. A calorimeter is designed to prevent or at least reduce heat loss to the atmosphere outside the reaction vessel. An ionic compound can be mixed with water directly in the calorimeter, and the temperature is recorded before and after the dissolution process. The amount of heat change occurring in the calorimeter can then be calculated using the following equation:

$$q = m \times s \times \Delta T$$

where q = heat energy change (in joules), m = total mass of the solution (solute plus solvent), s = the specific heat of the solution (4.18 J/g•°C), and ΔT = the observed temperature change. The molar heat of solution (H_{soln}) is the heat energy change (q) per mole of solute (n).

Timeline

The instructional time needed to complete this lab investigation is 180–250 minutes. Appendix 2 (p. 501) provides options for implementing this lab investigation over several class periods. Option G (250 minutes) and Option H (180 minutes) provide two viable timelines for this investigation. Both options provide students with extended data collection time during stage 2 as well as ample time to analyze the data they generate and consider the cost-benefit analysis aspect of this investigation. The difference between these two options is the amount of class time devoted to the scientific writing process. If your students need additional scaffolding and class time to work on their investigation reports, then Option G is more appropriate. If your students have experience with the scientific writing process and have the skills needed to write an investigation report on their own, then consider Option H, where students complete stage 6 (writing the investigation report) and stage 8 (revising the investigation report) as homework.

Materials and Preparation

The materials needed to implement this investigation are listed in Table 28.1 (p. 444). A temperature sensor kit includes an electronic temperature probe and any associated software or hardware such as a data collection interface or laptop computer. Temperature sensors are available for purchase from a variety of lab supply companies (e.g., Pasco and Vernier). It is recommended that the sensors and interface be positioned at lab stations before class to save time due to technical issues that may arise during setup. A calorimeter can be as simple as a single Styrofoam cup or several cups nested inside one another. There are also more elaborate calorimeters available for purchase from lab supply companies. A container of each salt should be kept in a central location so that students may retrieve the salts as needed for their investigation; be mindful of cross-contamination between containers.

TABLE 28.1

Materials list

Item	Quantity
Consumables	
Ammonium chloride, NH_4Cl	3–5 g per group
Ammonium nitrate, NH_4NO_3	3–5 g per group
Magnesium sulfate, $MgSO_4$	3–5 g per group
Sodium thiosulfate, $Na_2S_2O_3$	3–5 g per group
Distilled water	As needed
Equipment and other materials	
Graduated cylinder, 100 ml	1 per group
Spatula	1 per group
Calorimeter	1 per group
Temperature probe with sensor interface	1 per group
Electronic or triple beam balance	1 per group
Investigation Proposal C	1 per group
Whiteboard, 2' x 3' *	1 per group
Lab handout	1 per student
Peer-review guide and instructor scoring rubric	1 per student

* As an alternative, students can use computer and presentation software such as Microsoft PowerPoint or Apple Keynote to create their arguments.

Safety Precautions

Remind students to follow all normal lab safety rules. Ammonium chloride, ammonium nitrate, sodium thiosulfate, and magnesium sulfate are all tissue irritants and moderately toxic by ingestion. You will therefore need to explain the potential hazards of working with these chemicals and how to work with hazardous chemicals. In addition, tell students to take the following safety precautions:

- Wear indirectly vented chemical-splash goggles and chemical-resistant gloves and aprons when they are collecting their data.
- Handle all glassware with care.
- Wash their hands with soap and water when they are done collecting the data.

Laboratory Waste Disposal

We recommend following Flinn laboratory waste disposal methods 12b for sodium thiosulfate and 26a and 26b for magnesium sulfate, ammonium chloride, and ammonium nitrate. Information about laboratory waste disposal methods is included in the Flinn Catalog and Reference Manual; you can request a free copy at *www.flinnsci.com.*

Topics for the Explicit and Reflective Discussion

Concepts That Can Be Used to Justify the Evidence

To provide an adequate justification of their evidence, students must explain why they included the evidence in their arguments and make the assumptions underlying their analysis and interpretation of the data explicit. In this investigation, students can use the following concepts to help justify their evidence:

- Endothermic and exothermic processes
- Enthalpy change
- Heat of solution

We recommend that you discuss these fundamental concepts during the explicit and reflective discussion to help students make this connection.

How to Design Better Investigations

It is important for students to reflect on the strengths and weaknesses of the investigation they designed during the explicit and reflective discussion. Students should therefore be encouraged to discuss ways to eliminate potential flaws, measurement errors, or sources of bias in their investigations. To help students be more reflective about the design of their investigation, you can ask the following questions:

- What were some of the strengths of your investigation? What made it scientific?
- What were some of the weaknesses of your investigation? What made it less scientific?
- If you were to do this investigation again, what would you do to address the weaknesses in your investigation? What could you do to make it more scientific?

Crosscutting Concepts

This investigation is well aligned with two crosscutting concepts found in *A Framework for K–12 Science Education,* and you should review these concepts during the explicit and reflective discussion.

- *Patterns:* Observed patterns in nature guide the way scientists organize and classify substances. Scientists also explore the relationships between and the underlying causes of the patterns they observe in nature.

- *Energy and matter: Flows, cycles, and conservation:* It is important in science to track how energy and matter move into, out of, and within systems.

The Nature of Science and the Nature of Scientific Inquiry

This investigation is well aligned with two important concepts related to the *nature of science* (NOS) and the *nature of scientific inquiry* (NOSI), and you should review these concepts during the explicit and reflective discussion.

- *The influence of society and culture on science:* Science is influenced by the society and culture in which it is practiced because science is a human endeavor. Cultural values and expectations determine what scientists choose to investigate, how investigations are conducted, how research findings are interpreted, and what people see as implications. People also view some research as being more important than others because of cultural values and current events.

- *The importance of imagination and creativity in science:* Students should learn that developing explanations for or models of natural phenomena and then figuring out how they can be put to the test of reality is as creative as writing poetry, composing music, or designing skyscrapers. Scientists must also use their imagination and creativity to figure out new ways to test ideas and collect or analyze data.

Hints for Implementing the Lab

- Learn how to use the temperature sensor kit and associated software before the lab begins. You must know how to use the equipment so you can help students when technical issues arise.

- Allow the students to become familiar with the temperature sensor kit and software as part of the tool talk before they begin to design their investigation. This gives students a chance to see what they can and cannot do with the equipment.

- Be sure that students record actual values (e.g., temperature readings and changes in temperature) and are not just attempting to hand draw what they see on the computer screen. Alternatively, have them save any graphs generated by the sensor software to analyze or print.

- We suggest using approximately 100 ml of water per trial along with about 3 g of salt per trial.

Topic Connections

Table 28.2 provides an overview of the scientific practices, crosscutting concepts, disciplinary core ideas, and supporting ideas at the heart of this lab investigation. In addition, it lists NOS and NOSI concepts for the explicit and reflective discussion. Finally, it lists literacy and mathematics skills (*CCSS ELA* and *CCSS Mathematics*) that are addressed during the investigation.

TABLE 28.2

Lab 28 alignment with standards

Scientific practices	• Asking questions and defining problems • Planning and carrying out investigations • Analyzing and interpreting data • Using mathematics and computational thinking • Constructing explanations and designing solutions • Engaging in argument from evidence • Obtaining, evaluating, and communicating information
Crosscutting concepts	• Patterns • Energy and matter: Flows, cycles, and conservation
Core idea	• PS1.B: Chemical reactions
Supporting ideas	• Endothermic and exothermic processes • Enthalpy change • Heat of solution
NOS and NOSI concepts	• Social and cultural influences • Imagination and creativity in science
Literacy connections (*CCSS ELA*)	• *Reading:* Key ideas and details, craft and structure, integration of knowledge and ideas • *Writing:* Text types and purposes, production and distribution of writing, research to build and present knowledge, range of writing • *Speaking and listening:* Comprehension and collaboration, presentation of knowledge and ideas
Mathematics connections (*CCSS Mathematics*)	• Reason abstractly and quantitatively • Construct viable arguments and critique the reasoning of others • Model with mathematics • Look for and express regularity in repeated reasoning

LAB 28

Lab Handout

Lab 28. Designing a Cold Pack: Which Salt Should Be Used to Make an Effective but Economical Cold Pack?

Introduction

An instant cold pack is a first aid device that is used to treat injuries. Most commercial instant cold packs contain two plastic bags. One bag contains an ionic compound, and the other bag contains water. When the instant cold pack is squeezed hard enough, the bag containing the water breaks and the ionic compound and water mix. The dissolution of the ionic compound in the water results in an enthalpy change and a decrease in the overall temperature of the cold pack. In this investigation, you will explore the enthalpy changes that are associated with common salts and then apply what you have learned about these enthalpy changes to design an effective but economical instant cold pack.

The enthalpy change associated with the dissolution process is called the *heat of solution* (ΔH_{soln}). At constant pressure, the ΔH_{soln} is equal in magnitude to heat (q) lost to or gained from the surroundings. In the case of a salt dissolving in water, the overall enthalpy change is the net result of two key processes. First, an input of energy is required to break the attractive forces that hold the ions in the salt together and to disrupt the intermolecular forces that hold the water molecules in the solvent together. The system *gains* energy during this process. Second, energy is released from the system as attractive forces form between the dissociated ions and the molecules of water. The system *loses* energy during this process. The ΔH_{soln} can therefore be either endothermic or exothermic depending on the net energy change in the system. The ΔH_{soln} is exothermic when the system releases more energy into the surroundings than it absorbs and endothermic when the system absorbs more energy than it releases.

A chemist can determine the molar ΔH_{soln} for a specific salt by mixing a sample of it with water inside a *calorimeter*. A calorimeter is an insulated container that is designed to prevent or at least reduce heat loss to the atmosphere (see Figure L28.1). Once the salt and water are mixed, the chemist can record the temperature change that occurs inside the calorimeter as a result of the dissolution process. The magnitude of the heat energy change is then calculated using the following equation:

$$q = m \times s \times \Delta T$$

where q = heat energy change (in joules), m = total mass of the solution (solute plus solvent), s = the specific heat of the solution (4.18 J/g•°C), and ΔT = the observed temperature change. The chemist can then calculate the molar ΔH_{soln} for the salt by dividing q by the number of moles of the salt (n) that he or she mixed with the water.

Your Task

Investigate different salts for potential use in a cold pack. Using the empirical data you collect along with the cost data provided in Table L28.1 (p. 451), determine which salt in what quantity should be used to produce an effective but economical cold pack.

The guiding question of this investigation is, **Which salt should be used to make an effective but economical cold pack?**

Materials

You may use any of the following materials during your investigation:

Consumables	Equipment
• Ammonium chloride, NH_4Cl • Ammonium nitrate, NH_4NO_3 • Magnesium sulfate, $MgSO_4$ • Sodium thiosulfate, $Na_2S_2O_3$ • Distilled water	• Graduated cylinder (100 ml) • Spatula • Calorimeter • Temperature probe with sensor interface • Electronic or triple beam balance

Safety Precautions

Follow all normal lab safety rules. Ammonium chloride, ammonium nitrate, sodium thiosulfate, and magnesium sulfate are all tissue irritants and moderately toxic by ingestion. Your teacher will explain relevant and important information about working with the chemicals associated with this investigation. In addition, take the following safety precautions:

- Wear indirectly vented chemical-splash goggles and chemical-resistant gloves and apron while in the laboratory.

- Handle all glassware with care.

- Wash your hands with soap and water before leaving the laboratory.

Investigation Proposal Required? ☐ Yes ☐ No

FIGURE L28.1 _____

Example of a calorimeter

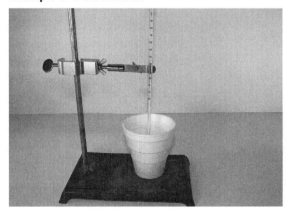

Getting Started

The first step in your investigation is to determine the heat energy change associated with each salt. To accomplish this task, you will need to determine what type of data to collect, how you will collect the data, and how you will analyze the data.

To determine *what type of data you need to collect*, think about the following questions:

- What type of measurements or observations will you need to make during your investigation?
- Is it important to know the change in temperature of the solution or just its final temperature?
- How does the amount of salt or the amount of water influence your potential results?

To determine *how you will collect the data*, think about the following questions:

- What will serve as your independent and dependent variables?
- How often will you collect data and when will you do it?
- How will you make sure that your data are of high quality (i.e., how will you reduce error)?
- How will you keep track of the data you collect and how will you organize it?

To determine *how you will analyze the data*, think about the following questions:

- How will you calculate the heat energy change associated with the formation of a solution?
- How will you calculate the molar ΔH_{soln} for each compound?
- What type of graph could you create to help make sense of your data?

The second step of your investigation will be to determine which salt should be used to make the instant cold pack. The company wants to produce small instant cold packs that will easily fit in a portable first aid kit. The instant cold pack they are planning to make will consist of two bags: one containing water and the other containing one of the salts. The bag of water will be placed inside the bag that contains the salt so when the bag of water is ruptured, the salt and water can mix. The company is planning on using 60 ml of water in this cold pack. For the instant ice pack to be effective, its temperature needs to fall to about 2°C once the salt and water are mixed. The company, however, wants to spend as little as possible to produce the instant cold packs. You will therefore need to conduct a complete cost-benefit analysis for each salt. This will require you to determine how much of each type of salt you will need to use and how much it will cost per instant cold pack. The price of each salt is given in Table L28.1.

TABLE L28.1

Prices of salts

Salt	Amount (in grams)	Price
NH_4Cl	1,000	$13.90
NH_4NO_3	500	$8.95
$MgSO_4$	100	$1.17
$Na_2S_2O_3$	500	$8.55

Connections to Crosscutting Concepts, the Nature of Science, and the Nature of Scientific Inquiry

As you work through your investigation, be sure to think about

- the importance of recognizing and analyzing patterns,
- how energy and matter flow within a system,
- the role of culture and values in science, and
- the importance of creativity in science.

Initial Argument

Once your group has finished collecting and analyzing your data, you will need to develop an initial argument. Your argument must include a *claim*, which is your answer to the guiding question. Your argument must also include *evidence* in support of your claim. The evidence is your analysis of the data and your interpretation of what the analysis means. Finally, you must include a *justification* of the evidence in your argument. You will therefore need to use a scientific concept or principle to explain why the evidence that you decided to use is relevant and important. You will create your initial argument on a whiteboard. Your whiteboard must include all the information shown in Figure L28.2.

FIGURE L28.2

Argument presentation on a whiteboard

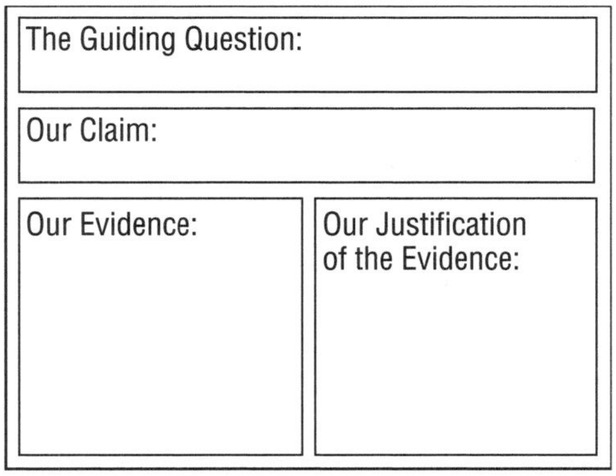

The Guiding Question:	
Our Claim:	
Our Evidence:	Our Justification of the Evidence:

Argumentation Session

The argumentation session allows all of the groups to share their arguments. One member of each group stays at the lab station to share that group's argument, while the other members of the group go to the other lab stations one at a time to listen to and critique the arguments developed by their classmates. The goal of the argumentation session is not to convince others that your argument is the best one; rather, the goal is to identify errors or instances of faulty reasoning in the initial arguments so these mistakes can be fixed. You will therefore need to evaluate the content of the claim, the quality of the evidence used to support the claim, and the strength of the justification of the evidence included in each argument that you see. To critique an argument, you might need more information than what is included on the whiteboard. You might, therefore, need to ask the presenter one or more follow-up questions, such as:

- How did your group collect the data? Why did you use that method?

- What did your group do to analyze the data, and why did you decide to do it that way?

- Is that the only way to interpret the results of your group's analysis? How do you know that your interpretation of the analysis is appropriate?

- Why did your group decide to present your evidence in that manner?

- What other claims did your group discuss before deciding on that one? Why did you abandon those alternative ideas?

- How confident are you that your group's claim is valid? What could you do to increase your confidence?

Once the argumentation session is complete, you will have a chance to meet with your group and revise your original argument. Your group might need to gather more data or design a way to test one or more alternative claims as part of this process. Remember, your goal at this stage of the investigation is to develop the most valid or acceptable answer to the research question!

Report

Once you have completed your research, you will need to prepare an *investigation report* that consists of three sections that provide answers to the following questions:

1. What question were you trying to answer and why?

2. What did you do during your investigation and why did you conduct your investigation in this way?

3. What is your argument?

Your report should answer these questions in two pages or less. The report must be typed and any diagrams, figures, or tables should be embedded into the document. Be sure to write in a persuasive style; you are trying to convince others that your claim is acceptable or valid!

Checkout Questions

Lab 28. Designing a Cold Pack: Which Salt Should Be Used to Make an Effective but Economical Cold Pack?

1. Describe the nature of the attractive forces that must be broken or disrupted when a polar solute is dissolved into a polar solvent.

2. Dissolving calcium chloride ($CaCl_2$) into water is an exothermic process. However, there are intermediate steps in the dissolving process that are endothermic. Use what you know about the process of dissolving to explain how it is possible for an exothermic process to actually involve some smaller endothermic processes.

3. Cultural and societal values have a great influence on science.

    a. I agree with this statement.

    b. I disagree with this statement.

Explain your answer, using an example from your investigation about designing a cold pack.

4. Creativity is not valued in science; it is better to do things the way they have always been done.

    a. I agree with this statement.

    b. I disagree with this statement.

Explain your answer, using an example from your investigation about designing a cold pack.

5. Understanding how energy and matter flow within systems is important in science, particularly when it comes to technological applications like designing a cold pack. Explain why understanding this aspect of science is important, using an example from your investigation.

6. Recognizing consistent patterns across several investigations is important in science. Explain the benefits of recognizing patterns, using an example from your investigation about designing a cold pack.

LAB 29

Teacher Notes

Lab 29. Rate Laws: What Is the Rate Law for the Reaction Between Hydrochloric Acid and Sodium Thiosulfate?

Purpose

The purpose of this lab is for students to *apply* what they know about reaction rates and how concentration affects a reaction to determine the rate law for a specific chemical reaction using the differential method. Students will also learn about the differences between observations and inferences in science and the relationship between scientific laws and theories.

The Content

The reaction rate for a given chemical reaction is a measure of the change in concentration of the reactants or the change in the concentration of the products in a given unit of time. Reaction rates are inversely proportional to time; the greater the rate, the less time that is needed for the reactants to be converted to products. The *collision theory of reactions* suggests that the rate of a chemical reaction depends on three important factors: (1) the number of collisions that take place between molecules during a reaction, (2) the average energy of these collisions, and (3) the orientation of the molecules at the time of the collision. Reaction rates, as a result, are affected by both the temperature and concentration of the reactants. A change in temperature results in a change in the kinetic energy of the reactants. Reaction rates therefore tend to double with every 10°C increase in temperature. A change in the concentration of one or more of the reactants will also speed up or slow down the rate of a reaction. A change in reactant concentration changes the frequency of collisions between molecules, but it does not affect the likelihood that any given collision between any two molecules will be effective and result in a reaction. The likelihood of an effective collision depends on how much kinetic energy the molecules have and the orientation of the molecules when they collide with each other.

The relationship between the rate of a reaction and the concentration of reactants is expressed in a mathematical equation called a rate law. For a general reaction of the form A + B → C, the rate law is written as

$$Rate = k[A]^n[B]^m$$

In this equation, k is the rate constant, [A] is the molar concentration of the reactant A, [B] is the molar concentration of the reactant B, and the exponents n and m define how the reaction rate depends on the concentration of each reactant. The n and m exponents are described as the *reaction order*. In this example, the reaction is to the nth order of A and the *m*th order of B. The reaction order indicates how the concentration of a reactant affects the overall reaction rate and which reactant affects the rate the most.

The reaction order cannot be determined from a chemical equation. One way to determine the reaction order is to use the *differential method*. This approach allows chemists to determine how a rate changes when concentration of the reactants change by making systematic comparisons. The following procedure illustrates how to use the differential method to determine the reaction order and the rate law for the reaction $A + B \rightarrow C$:

1. The concentration of reactant A is held constant while the concentration of reactant B is varied, and the reaction time is measured in each condition. Then the concentration of reactant B is held constant while the concentration of reactant A is varied, and the reaction time is measured in each condition. The rate for each condition is then calculated by taking the inverse of the reaction time. Consider the following data as an example:

Condition	[A] (moles)	[B] (moles)	Reaction time (sec)	Rate (1/time)
1	0.02	0.01	500	0.002
2	0.01	0.01	1,000	0.001
3	0.01	0.02	250	0.004

2. Determine the reaction order with respect to A (n) by comparing conditions 1 and 2. To do this, use the following equation:

$$Rate_{condition1}/Rate_{condition2} = ([A]_{condition1}/[A]_{condition2})^n$$

In this example,

$$0.002/0.001 = (0.02/0.01)^n$$

$$2 = 2^n$$

$$n = 1$$

3. Determine the reaction order with respect to B (m) by comparing conditions 2 and 3. To do this, use the following equation:

$$Rate_{condition3}/Rate_{condition2} = ([B]_{condition3}/[B]_{condition2})^m$$

In this example,

$$0.004/0.001 = (0.02/0.01)^m$$

$$4 = 2^m$$

$$m = 2$$

LAB 29

4. Write out the full rate law.

In this example,

$$Rate = k[A][B]^2$$

Timeline

The instructional time needed to complete this lab investigation is 180–250 minutes. Appendix 2 (p. 501) provides options for implementing this lab investigation over several class periods. Option E (250 minutes) should be used if students are unfamiliar with scientific writing because this option provides extra instructional time for scaffolding the writing process. You can scaffold the writing process by modeling, providing examples, and providing hints as students write each section of the report. Option F (180 minutes) should be used if students are familiar with scientific writing and have the skills needed to write an investigation report on their own. In option F, students complete stage 6 (writing the investigation report) and stage 8 (revising the investigation report) as homework.

Materials and Preparation

The materials needed to implement this investigation are listed in Table 29.1. The consumables and equipment can be purchased from a science supply company such as Carolina, Flinn Scientific, or Ward's Science. We recommend that you use a set routine for distributing and collecting the materials during the lab investigation. For example, the consumables and equipment for each group can be set up at each group's lab station before class begins, or one member from each group can collect them from a table or a cart when needed during class.

Safety Precautions

Remind students to follow all normal lab safety rules. Hydrochloric acid is moderately toxic by ingestion and inhalation, and it is also corrosive to the eyes and skin. Sodium thiosulfate is a body tissue irritant. The sulfur precipitate and aqueous sulfur dioxide, which are products of the reaction between hydrochloric acid and sodium thiosulfate, are skin and eye irritants. You will therefore need to explain the potential hazards of working with these substances and how to work with hazardous chemicals. In addition, tell students to take the following safety precautions:

- Wear indirectly vented chemical-splash goggles and chemical-resistant gloves and aprons when they are collecting their data.

- Handle all glassware with care.

- Wash their hands with soap and water when they are done collecting the data.

TABLE 29.1

Materials list

Item	Quantity
Consumables	
1 ml hydrochloric acid, HCl	50 ml per group
0.3 M sodium thiosulfate, $Na_2S_2O_3$	50 ml per group
Distilled water	50 ml per group
Equipment and other materials	
6-well reaction plate	2 per group
Beakers, 50 ml	3 per group
Graduated cylinders, 10 ml	3 per group
Disposable pipettes	3 per group
Stopwatch or timer	1 per group
Overhead projector or document camera	1 per class
Investigation Proposal A (optional but recommended)	2 per group
Whiteboard, 2' x 3' *	1 per group
Lab handout	1 per student
Lab Reference Sheet on rate law (optional but recommended)	1 per group
Peer-review guide and instructor scoring rubric	1 per student

* As an alternative, students can use computer and presentation software such as Microsoft PowerPoint or Apple Keynote to create their arguments.

Laboratory Waste Disposal

We recommend following Flinn laboratory waste disposal methods 26a for solid waste and 24b for solutions. Information about laboratory waste disposal methods is included in the Flinn Catalog and Reference Manual; you can request a free copy at *www.flinnsci.com*. To separate the solid waste and the solutions before disposal, have students empty their 6-well reaction plates into one large collection container and then filter out the solid material.

Topics for the Explicit and Reflective Discussion

Concepts That Can Be Used to Justify the Evidence

To provide an adequate justification of their evidence, students must explain why they included the evidence in their arguments and make the assumptions underlying their analysis and interpretation of the data explicit. In this investigation, students can use the following concepts to help justify their evidence:

- Atomic theory
- Molecular-kinetic theory of matter
- Collision theory of reactions
- Reaction rates

We recommend that you discuss these fundamental concepts during the explicit and reflective discussion to help students make this connection.

How to Design Better Investigations

It is important for students to reflect on the strengths and weaknesses of the investigation they designed during the explicit and reflective discussion. Students should therefore be encouraged to discuss ways to eliminate potential flaws, measurement errors, or sources of bias in their investigations. To help students be more reflective about the design of their investigations, you can ask the following questions:

- What were some of the strengths of your investigation? What made it scientific?
- What were some of the weaknesses of your investigation? What made it less scientific?
- If you were to do this investigation again, what would you do to address the weaknesses in your investigation? What could you do to make it more scientific?

Crosscutting Concepts

This investigation is well aligned with two crosscutting concepts found in *A Framework for K–12 Science Education*, and you should review these concepts during the explicit and reflective discussion.

- *Scale, proportion, and quantity:* It is critical for scientists to be able to recognize what is relevant at different sizes, time frames, and scales. Scientists must also be able to recognize proportional relationships between categories or quantities.
- *Energy and matter: Flows, cycles, and conservation*: It is important in science to track how energy and matter move into, out of, and within systems.

The Nature of Science and the Nature of Scientific Inquiry

This investigation is well aligned with two important concepts related to the *nature of science* (NOS) and the *nature of scientific inquiry* (NOSI), and you should review these concepts during the explicit and reflective discussion.

- *The difference between observations and inferences*: An observation is a descriptive statement about a natural phenomenon, whereas an inference is an interpretation of an observation. Students should also understand that current scientific knowledge and the perspectives of individual scientists guide both observations and inferences. Thus, different scientists can have different but equally valid interpretations of the same observations due to differences in their perspectives and background knowledge.

- *The difference between laws and theories in science:* A scientific law describes the behavior of a natural phenomenon or a generalized relationship under certain conditions; a scientific theory is a well-substantiated explanation of some aspect of the natural world. Theories do not become laws even with additional evidence; they explain laws. However, not all scientific laws have an accompanying explanatory theory. It is also important for students to understand that scientists do not discover laws or theories; the scientific community develops them over time.

Hints for Implementing the Lab

- This is an application lab, so before beginning the investigation students should be familiar with reaction rates, reaction orders, and rate laws and should understand how to calculate reaction orders from reaction rate data. The Lab Reference Sheet provides a basic overview of how to calculate reaction orders using the differential method.

- We recommend that students fill out an investigation proposal for each experiment associated with this investigation. These proposals provide a way for you to offer students hints and suggestions without telling them how to do it. You can also check the proposals quickly during a class period.

- Students will need to test at least two different concentrations of hydrochloric acid in one experiment and two different concentrations of sodium thiosulfate in the other experiment.

- Make sure students use at least 1 ml of sodium thiosulfate in each reaction. At lower concentrations, the reaction time is more difficult to measure because the onset of turbidity is more gradual.

- Make sure that the total volume of hydrochloric acid, sodium thiosulfate, and water that students add to each well during their experiments is equal to 5 ml each time. For example, if they use 1 ml of 1 M HCl and 2 ml of $Na_2S_2O_3$, then they will need to add 2 ml of water to the well to bring the total volume up to 5 ml.

- Remind students to include multiple trials in each experiment and to average their results to help reduce measurement error.

LAB 29

- Be sure that students know the number of moles of each reactant they are using for each reaction. Remind them how to use the dilution equation ($M_1V_1 = M_2V_2$) to calculate initial molar concentration as part of the tool talk.

- We recommend that you show students how to determine the end of the reaction using visual clues (i.e., not being able to see a "+" on a piece of paper through the precipitate) as part of the tool talk. To do this, use an overhead projector or a document camera to show how the turbidity of the reaction mixture increases over time due to the formation of the sulfur precipitate (see Figure 29.1).

FIGURE 29.1

Students can determine the end of a reaction using visual cues.

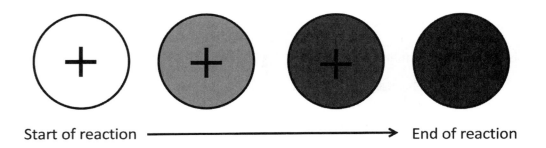

Start of reaction ————————————————▶ End of reaction

Topic Connections

Table 29.2 provides an overview of the scientific practices, crosscutting concepts, disciplinary core ideas, and supporting ideas at the heart of this lab investigation. In addition, it lists NOS and NOSI concepts for the explicit and reflective discussion. Finally, it lists literacy and mathematics skills (*CCSS ELA* and *CCSS Mathematics*) that are addressed during the investigation.

TABLE 29.2

Lab 29 alignment with standards

Scientific practices	• Asking questions and defining problems • Planning and carrying out investigations • Analyzing and interpreting data • Using mathematics and computational thinking • Engaging in argument from evidence • Obtaining, evaluating, and communicating information
Crosscutting concepts	• Scale, proportion, and quantity • Energy and matter: Flows, cycles, and conservation
Core idea	• PS1.B: Chemical reactions
Supporting ideas	• Atomic theory • Molecular-kinetic theory of matter • Moles and molar mass • Reaction rates • Reaction order • Rate laws
NOS and NOSI concepts	• Observations and inferences • Scientific laws and theories
Literacy connections (CCSS ELA)	• *Reading:* Key ideas and details, craft and structure, integration of knowledge and ideas • *Writing:* Text types and purposes, production and distribution of writing, research to build and present knowledge, range of writing • *Speaking and listening:* Comprehension and collaboration, presentation of knowledge and ideas
Mathematics connections (CCSS Mathematics)	• Reason abstractly and quantitatively • Construct viable arguments and critique the reasoning of others • Model with mathematics

LAB 29

Lab Handout

Lab 29. Rate Laws: What Is the Rate Law for the Reaction Between Hydrochloric Acid and Sodium Thiosulfate?

Introduction

The *collision theory of reactions* suggests that the rate of a reaction depends on three important factors. The first is the number of collisions that take place between molecules during a reaction. This factor is important because molecules must collide with each other for a reaction to take place. The second factor is the average energy of these collisions. This factor is important because colliding molecules must have enough kinetic energy to overcome the repulsive and bonding forces of the reactants. The minimum amount of energy needed for the reactants to transform into products is called the *activation energy* of the reaction. The third, and final, factor is the orientation of the molecules at the time of the collision. This factor is important because the reactant molecules must collide with each other in a specific orientation for the atoms in these molecules to rearrange (see Figure L29.1). Chemists can therefore alter the rate of a specific chemical reaction by manipulating one or more of these factors inside the laboratory.

FIGURE L29.1 _____

The collision theory of reactions

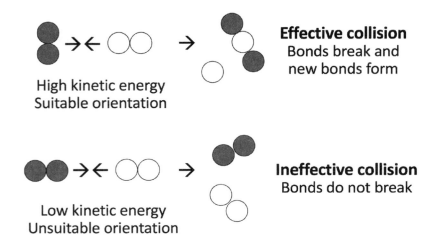

Chemists, for example, can change the temperature of the reactants to speed up or slow down a reaction. Temperature affects the rate of a reaction because the proportion of molecules with enough kinetic energy to overcome the activation energy barrier of a reaction goes up when the temperature increases and goes down when temperature decreases. Chemists can also change the concentration of the reactants to speed up or slow down the rate of a reaction. A change in reactant concentration changes the frequency of collisions between molecules but does not affect the likelihood that any given collision between any two molecules will be effective and result in a reaction. The likelihood of an effective collision, as noted earlier, depends on how much kinetic energy the molecules have and the orientation of the molecules when they collide with each other.

Chemists use an equation called a rate law to describe how the concentration of each reactant affects the overall reaction rate. For a general reaction of the form A + B → C, the rate law is written as

$$Rate = k[\text{A}]^n[\text{B}]^m$$

In this equation, k is the rate constant, [A] is the molar concentration of the reactant A, [B] is the molar concentration of the reactant B, and the exponents n and m define how the reaction rate depends on the concentration of each reactant. The rate constant for a reaction is affected by temperature but not by concentration. The n and m exponents are often described as the reaction order. In this example, the reaction is to the nth order of A and the mth order of B. Typical values of n and m are 0, 1, or 2.

The reaction order for each reactant in the rate law indicates how much the reaction rate will change in response to a change in the concentration of each reactant. If a reactant has an order of 0, then the reaction rate is not affected by the concentration of this reactant. Reactants with an order of 0 are therefore not included in the rate law. When a reactant has an order of 1, the change in reaction rate is directly proportional to the change in reactant concentration. The rate of a chemical reaction, for example, will double when the concentration of a first-order reactant is doubled and will triple when the concentration of the first-order reactant is tripled. When a reactant has an order of 2, the rate of reaction will change twice as much as the change in the reactant concentration. The rate of a chemical reaction, for example, will quadruple when the concentration of a second-order reactant is doubled.

It is important for chemists to understand how changing the concentration of one or more reactants will affect the overall rate of a specific chemical reaction so they can make a specific product in a safe and time-efficient manner. Unfortunately, chemists cannot determine the reaction order for each reactant in a specific reaction by simply looking at a balanced chemical equation; they must determine it through a process called the *differential method*. This method allows chemists to determine a rate law by varying the concentration of the reactants in a reaction in a systematic manner and then measuring how these changes affect the reaction rate. A description of the differential method can be found in the Lab Reference Sheet (p. 470).

LAB 29

Your Task

Use the differential method to determine the rate law for the reaction between hydrochloric acid (HCl) and sodium thiosulfate ($Na_2S_2O_3$). The balanced chemical equation for this reaction is

$$2HCl(aq) + Na_2S_2O_3(aq) \rightarrow S(s) + SO_2(aq) + H_2O(l) + 2NaCl(aq)$$

The guiding question of this investigation is, **What is the rate law for the reaction between hydrochloric acid and sodium thiosulfate?**

Materials

You may use any of the following materials during your investigation:

Consumables	Equipment
• 1.0 M HCl	• 2 6-well reaction plates
• 0.3 M $Na_2S_2O_3$	• 3 beakers (each 50 ml)
• Distilled water	• 3 graduated cylinders (each 10 ml)
	• 3 disposable pipettes
	• Stopwatch or timer

Safety Precautions

Follow all normal lab safety rules. Hydrochloric acid is moderately toxic by ingestion and inhalation, and it is also corrosive to the eyes and skin. Sodium thiosulfate is a body tissue irritant. The sulfur precipitate and aqueous sulfur dioxide, which are products of the reaction between hydrochloric acid and sodium thiosulfate, are skin and eye irritants. Your teacher will explain relevant and important information about working with the chemicals associated with this investigation. In addition, take the following safety precautions:

- Wear indirectly vented chemical-splash goggles and chemical-resistant gloves and apron while in the laboratory.

- Handle all glassware with care.

- Wash your hands with soap and water before leaving the laboratory.

Investigation Proposal Required? ☐ Yes ☐ No

Getting Started

The first step in the differential method is to carry out two experiments. The goal of the first experiment is to determine how a change in concentration of hydrochloric acid affects the reaction rate. The goal of the second experiment is to determine how a change in the concentration of sodium thiosulfate affects reaction rate. In each of your experiments, you will be able to measure reaction time by monitoring the appearance of the sulfur precipitate. The solution will change from clear to cloudy over time.

You will need to mix the reactants together in different concentrations in each of your experiments. You can change the concentration of each reactant by changing the amount of reactant you add to a well in a well plate. You can add between 1 and 4 ml of a reactant to a well, but the total volume of liquid added to each well should always equal 5 ml to keep volume constant across conditions. You can use water to bring the total volume of the liquid in a well up to 5 ml. For example, if you add 1 ml of HCl to 3 ml of $Na_2S_2O_3$, you will need to add 1 ml of water to bring the total volume of liquid in the well up to 5 ml. If you add 2 ml of HCl to 3 ml of $Na_2S_2O_3$, you will not need to add any water because the total volume of the liquid in the well is already 5 ml. Keep this information in mind as you design your two experiments. You will also need to determine what type of data you need to collect, how you will collect the data, and how you will analyze the data.

To determine *what type of data you need to collect*, think about the following questions:

- What types of measurements or observation will you need to make during each experiment?
- When will you need to make these measurements or observations?

To determine *how you will collect the data*, think about the following questions:

- What will serve as your independent variable in each experiment?
- What will serve as your dependent variable in each experiment?
- What comparisons will you need to make in each experiment?
- How will you determine when a reaction is finished?
- How will you hold other variables constant across each comparison?
- What will you do to reduce measurement error?
- How will you keep track of the data you collect and how will you organize it?

To determine *how you will analyze the data*, think about the following questions:

- How will you determine the concentration of each reactant at the beginning of the reaction?
- What type of calculations will you need to make?
- What type of graph could you create to help make sense of your data?

The second step in this investigation is to determine the rate law for this reaction. To determine the rate law, you will need to first calculate the reaction order for each reactant using the data you collected in each experiment. The Lab Reference Sheet describes how to make these calculations. Next, you will need to calculate the value of k for the rate law. To calculate the rate constant, you simply need to substitute the appropriate values for the concentration of each reactant, the reaction order for each reactant, and the reaction rate into the rate law equation and solve for *k*.

LAB 29

The third and final step in the investigation is to test your rate law. To accomplish this task, you will need to determine if you can use your rate law to make accurate predictions about reaction times. If you can use the rate law you developed to make accurate predictions about the time it takes for the sulfur precipitate to appear when using different concentrations of hydrochloric acid and sodium thiosulfate, then you will have evidence that suggests that the rate law you developed is a valid one.

Connections to Crosscutting Concepts, the Nature of Science, and the Nature of Scientific Inquiry

As you work through your investigation, be sure to think about

- why it is important to look for and use proportional relationships;
- why it is important to track how energy and matter move into, out of, and within systems;
- the difference between observations and inferences in science; and
- the difference between laws and theories in science.

Initial Argument

Once your group has finished collecting and analyzing your data, you will need to develop an initial argument. Your argument must include a *claim*, which is your answer to the guiding question. Your argument must also include *evidence* in support of your claim. The evidence is your analysis of the data and your interpretation of what the analysis means. Finally, you must include a *justification* of the evidence in your argument. You will therefore need to use a scientific concept or principle to explain why the evidence that you decided to use is relevant and important. You will create your initial argument on a whiteboard. Your whiteboard must include all the information shown in Figure L29.2.

FIGURE L29.2 _____

Argument presentation on a whiteboard

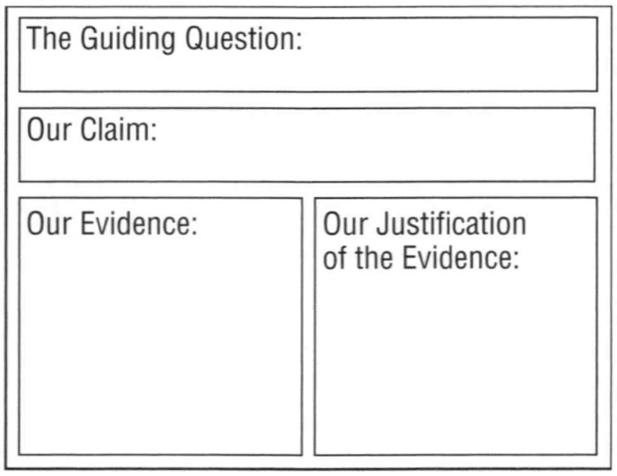

The Guiding Question:	
Our Claim:	
Our Evidence:	Our Justification of the Evidence:

Argumentation Session

The argumentation session allows all of the groups to share their arguments. One member of each group stays at the lab station to share that group's argument, while the other members of the group go to the other lab stations one at a time to listen to and critique the arguments developed by their classmates. The goal of the argumentation session is not to convince others that your argument is the best one; rather, the goal is to identify errors or instances of faulty reasoning in the initial arguments so these mistakes can be fixed. You will therefore need to evaluate the content of the claim, the quality of the evidence used to support the claim, and the strength of the justification of the evidence included in each argument that you see. To critique an argument, you might need more information than what is included on the whiteboard. You might, therefore, need to ask the presenter one or more follow-up questions, such as:

- How did your group collect the data? Why did you use that method?

- What did your group do to make sure the data you collected are reliable? What did you do to decrease measurement error?

- What did your group do to analyze the data, and why did you decide to do it that way? Did you check your calculations?

- Is that the only way to interpret the results of your group's analysis? How do you know that your interpretation of the analysis is appropriate?

- Why did your group decide to present your evidence in that manner?

- What other claims did your group discuss before deciding on that one? Why did you abandon those alternative ideas?

- How confident are you that your group's claim is valid? What could you do to increase your confidence?

Once the argumentation session is complete, you will have a chance to meet with your group and revise your original argument. Your group might need to gather more data or design a way to test one or more alternative claims as part of this process. Remember, your goal at this stage of the investigation is to develop the most valid or acceptable answer to the research question!

Report

Once you have completed your research, you will need to prepare an *investigation report* that consists of three sections that provide answers to the following questions:

1. What question were you trying to answer and why?

2. What did you do during your investigation and why did you conduct your investigation in this way?

3. What is your argument?

Your report should answer these questions in two pages or less. The report must be typed and any diagrams, figures, or tables should be embedded into the document. Be sure to write in a persuasive style; you are trying to convince others that your claim is acceptable or valid!

LAB 29

Lab Reference Sheet

The Differential Method for Determining a Rate Law

The following procedure can be used to determine the reaction order and the rate law for a reaction of the form $A + B \rightarrow C$.

1. The concentration of reactant A is held constant while the concentration of reactant B is varied, and the reaction time is measured in each condition. Then the concentration of reactant B is held constant while the concentration of reactant A is varied, and the reaction time is measured in each condition. The rate for each condition is then calculated by taking the inverse of the reaction time. Take the following data as an example:

Condition	[A] (moles)	[B] (moles)	Reaction time (sec)	Rate (1/time)
1	0.02	0.01	500	0.002
2	0.01	0.01	1,000	0.001
3	0.01	0.02	250	0.004

2. Determine the reaction order with respect to A (n) by comparing conditions 1 and 2. To do this, use the following equation:

$$Rate_{condition1}/Rate_{condition2} = ([A]_{condition1}/[A]_{condition2})^n$$

In this example,

$$0.002/0.001 = (0.02/0.01)^n$$

$$2 = 2^n$$

$$n = 1$$

3. Determine the reaction order with respect to B (m) by comparing conditions 2 and 3. To do this, use the following equation:

$$Rate_{condition3}/Rate_{condition2} = ([B]_{condition3}/[B]_{condition2})^m$$

In this example,

$$0.004/0.001 = (0.02/0.01)^m$$

$$4 = 2^m$$

$$m = 2$$

4. Write out the full rate law.

In this example,

$$Rate = k[A][B]^2$$

National Science Teachers Association

Checkout Questions

Lab 29. Rate Laws: What Is the Rate Law for the Reaction Between Hydrochloric Acid and Sodium Thiosulfate?

Use the following information to answer questions 1–4.

Iodide ions (I^-) react with persulfate ions ($S_2O_8^{2-}$) as follows:

$$2I^-(aq) + S_2O_8^{2-}(aq) \rightarrow I_2(aq) + 2SO_4^{2-}(aq)$$

The following rate data were collected by measuring the time required for the appearance of a dark blue color due to the interaction of iodine and starch.

Condition	[I^-] (moles)	[$S_2O_8^{2-}$] (moles)	Reaction time (sec)
1	0.04	0.04	270
2	0.08	0.04	138
3	0.04	0.08	142

1. What is the reaction rate for each condition?

2. What is the order of reaction for the iodide ions?

3. What is the order of reaction for the persulfate ions?

4. What is the rate law for this reaction?

5. "The reactant molecules collide during the reaction" is an observation.

    a. I agree with this statement.

    b. I disagree with this statement.

Explain your answer, using an example from your investigation about rate laws.

6. Theories can turn into laws.

   a. I agree with this statement.
   b. I disagree with this statement.

   Explain your answer, using an example from your investigation about rate laws.

7. Scientists often need to look for and use proportional relationships during an investigation. Explain why this is important, using an example from your investigation about rate laws.

8. Scientists often need to track how matter and energy move into, out of, and within a system. Explain why this is important, using an example from your investigation about rate laws.

LAB 30

Teacher Notes

Lab 30. Equilibrium Constant and Temperature: How Does a Change in Temperature Affect the Value of the Equilibrium Constant for an Exothermic Reaction?

Purpose

The purpose of this lab is to give students an opportunity to *apply* what they know about chemical equilibrium to determine the equilibrium constant for a reaction and to develop a rule that can be used to predict how the equilibrium constant will change in response to a change in temperature when a reaction is exothermic. This lab gives students an opportunity to use a colorimeter to measure the concentration of ions in a solution. Students will also learn about the role experiments play in science and why scientists need to be creative and have a good imagination.

The Content

Chemical equilibrium is the point in a reaction where the rate at which reactants transform into products is equal to the rate at which products revert back into reactants. When a reaction is in equilibrium, the concentration of the products and reactants is stable. At this point, there is no further net change in the concentrations of products and reactants unless the system is disturbed in some manner. The *equilibrium constant* provides a mathematical description of the equilibrium state for any reversible chemical reaction and gives the ratio of product-to-reactant concentration.

To illustrate this relationship between the concentration of products and the concentration of reactants at equilibrium, consider the following general equation for a reversible reaction:

$$aA + bB \leftrightarrows cC + dD$$

The equation for calculating the equilibrium constant, K_{eq}, for this general reaction is provided below. The square brackets refer to the molar concentrations of each chemical substance at equilibrium, and the exponents are the stoichiometric coefficients from the reaction equation.

$$K_{eq} = \frac{[C]^c[D]^d}{[A]^a[B]^b}$$

Notice that the products are found in the numerator and the reactants are found in the denominator in this ratio. This will always be the case. Also note that each concentration is raised to a power of its coefficient. The actual concentrations of the reactants and products that are present in the system at equilibrium will depend on the initial amounts of the reactants that were used at the beginning of the reaction and any extra reactants or products that were added to the system after the reaction started. The ratio of product-to-reactant concentration described by this equation and reflected in the value of the equilibrium constant, however, will always be the same as long as the system is in equilibrium and the temperature of the system does not change.

The value of the equilibrium constant allows chemists to determine the ratio of product-to-reactant concentration that will be present in the reaction mixture when it reaches equilibrium. When $K_{eq} > 1$, the concentration of products in the system will be greater than the concentration of the reactants. When $K_{eq} < 1$, the concentration of products will be less than the concentration of the reactants. Finally, when $K_{eq} = 1$, the concentration of products and reactants will be equal. A reaction with a large K_{eq} value, as a result, will have a greater product to reactant concentration ratio at equilibrium than a reaction with a smaller value.

The equilibrium constant of a reaction, as noted earlier, is dependent on temperature. A positive shift in the K_{eq} will cause the product to reactant concentration ratio at equilibrium to increase, whereas a negative shift will result in a decrease in the ratio. How the K_{eq} of a reaction will change in response to a change in temperature, however, is not uniform. The change in the K_{eq} of a reaction depends on the direction and magnitude of the temperature change. It will also depend on whether the reaction is exothermic or endothermic. In general, when a reaction is exothermic, an increase in temperature will decrease the value of the equilibrium constant. When a reaction is endothermic, in contrast, an increase in temperature will increase the value of the equilibrium constant.

Timeline

The instructional time needed to complete this lab investigation is 180–250 minutes. Appendix 2 (p. 501) provides options for implementing this lab investigation over several class periods. Option A (250 minutes) should be used if students are unfamiliar with scientific writing because this option provides extra instructional time for scaffolding the writing process. You can scaffold the writing process by modeling, providing examples, and providing hints as students write each section of the report. Option B (180 minutes) should be used if students are familiar with scientific writing and have the skills needed to write an investigation report on their own. In option B, students complete stage 6 (writing the investigation report) and stage 8 (revising the investigation report) as homework.

Materials and Preparation

The materials needed to implement this investigation are listed in Table 30.1 (p. 477). The consumables and equipment can be purchased from a science supply company such as Carolina, Flinn Scientific, or Ward's Science. You can purchase the calorimeter sensor and the interface that is needed to connect the sensor to a computer from lab supply companies such as Pasco and Vernier.

We recommend that you use a set routine for distributing and collecting the materials during the lab investigation. For example, the consumables and equipment for each group can be set up at each group's lab station before class begins, or one member from each group can collect them from a table or a cart as needed during class.

For best results, prepare all the solutions with analytical precision using an analytical balance and volumetric flasks. Prepare the solutions needed for this investigation as follows:

- *0.200 M iron(III) nitrate solution*: Add 20.20 g of ferric nitrate nonahydrate [$Fe(NO_3)_3 \cdot 9H_2O$] in 100 ml of 1 M nitric acid in a 250 ml volumetric flask. Mix thoroughly to dissolve, and dilute to the mark with 1 M nitric acid.

- *0.002 M iron(III) nitrate solution*: Use a volumetric pipette to transfer 5.00 ml of the 0.200 M iron(III) nitrate solution to a 500 ml volumetric flask that is half-filled with 1 M nitric acid. Dilute to the mark with 1 M nitric acid. Mix well.

- *1 M nitric acid:* Add 63 ml of concentrated nitric acid (15.8 M) to 750 ml of distilled water in a 1 L volumetric flask. Be sure to add the concentrated acid to the water (do not add the water to the concentrated acid). The heat of solution is exothermic so use borosilicate glassware. Mix the solution thoroughly. Once the solution has cooled to ambient temperature, dilute to 1 L with additional distilled water.

- *0.002 M potassium thiocyanate solution*: Dissolve 0.097 g of potassium thiocyanate in 250 ml of distilled water in a 500 ml volumetric flask. Mix thoroughly to dissolve, and then dilute to the mark with distilled water.

Safety Precautions

Remind students to follow all normal lab safety rules. Potassium thiocyanate is toxic by ingestion. The iron(III) nitrate solution contains 1 M nitric acid so it is corrosive; it will also stain clothes and skin. You will therefore need to explain the potential hazards of working with these chemicals and how to work with hazardous chemicals. In addition, tell students to take the following safety precautions:

- Wear indirectly vented chemical-splash goggles and chemical-resistant gloves and aprons when they are collecting their data.

- Use caution when working with hot plates, and keep them away from water and other liquids.

- Handle all glassware (including thermometers) with care.

- Wash their hands with soap and water when they are done collecting the data.

TABLE 30.1

Materials list

Item	Quantity
Consumables	
0.002 M iron(III) nitrate, $Fe(NO_3)_3$	60 ml per group
0.200 M iron(III) nitrate, $Fe(NO_3)_3$	20 ml per group
0.002 M potassium thiocyanate, KSCN	20 ml per group
Distilled water	100 ml per group
Ice (for ice baths)	As needed
Equipment and other materials	
Calorimeter sensor	1 per group
Sensor interface	1 per group
Cuvettes	4 per group
Pyrex test tubes	12 per group
Test tube rack	1 per group
Serological pipettes, 5 or 10 ml	3 per group
Pipette bulb	1 per group
Stirring rod	1 per group
Beakers, 50 ml	6 per group
Beakers, 250 ml	2 per group
Hot plate	1 per group
Thermometer (or temperature probe)	1 per group
Investigation Proposal A (optional but recommended)	3 per group
Whiteboard, 2' x 3' *	1 per group
Lab handout	1 per student
Peer-review guide and instructor scoring rubric	1 per student

* As an alternative, students can use computer and presentation software such as Microsoft PowerPoint or Apple Keynote to create their arguments.

LAB 30

Laboratory Waste Disposal

We recommend following Flinn laboratory waste disposal method 24b to dispose of the waste solutions. Information about laboratory waste disposal methods is included in the Flinn Catalog and Reference Manual; you can request a free copy at *www.flinnsci.com.*

Topics for the Explicit and Reflective Discussion

Concepts That Can Be Used to Justify the Evidence

To provide an adequate justification of their evidence, students must explain why they included the evidence in their arguments and make the assumptions underlying their analysis and interpretation of the data explicit. In this investigation, students can use the following concepts to help justify their evidence:

- Moles and molarity
- Endothermic and exothermic reactions
- Reaction rates
- Chemical equilibrium

We recommend that you discuss these fundamental concepts during the explicit and reflective discussion to help students make this connection.

How to Design Better Investigations

It is important for students to reflect on the strengths and weaknesses of the investigation they designed during the explicit and reflective discussion. Students should therefore be encouraged to discuss ways to eliminate potential flaws, measurement errors, or sources of bias in their investigations. To help students be more reflective about the design of their investigations, you can ask the following questions:

- What were some of the strengths of your investigation? What made it scientific?
- What were some of the weaknesses of your investigation? What made it less scientific?
- If you were to do this investigation again, what would you do to address the weaknesses in your investigation? What could you do to make it more scientific?

Crosscutting Concepts

This investigation is well aligned with two crosscutting concepts found in *A Framework for K–12 Science Education,* and you should review these concepts during the explicit and reflective discussion.

- *Systems and system models:* Scientists often need to use models to understand complex phenomenon. In this investigation, the students are directed to develop a model to help explain what is happening during a chemical reaction at the submicroscopic level.

- *Stability and change*: It is critical for scientists to understand what makes a system stable or unstable and what controls rates of change in a system (e.g., how changes in temperature affect the equilibrium constant of a chemical reaction).

The Nature of Science and the Nature of Scientific Inquiry

This investigation is well aligned with two important concepts related to the *nature of science* (NOS) and the *nature of scientific inquiry* (NOSI), and you should review these concepts during the explicit and reflective discussion.

- *The importance of imagination and creativity in science:* Students should learn that developing explanations for or models of natural phenomena and then figuring out how they can be put to the test of reality is as creative as writing poetry, composing music, or designing skyscrapers. Scientists must also use their imagination and creativity to figure out new ways to test ideas and collect or analyze data.

- *Nature and role of experiments:* Scientists use experiments to test the validity of a hypothesis (i.e., a tentative explanation) for an observed phenomenon. Experiments include a test and the formulation of predictions (expected results) if the test is conducted and the hypothesis is valid. The experiment is then carried out and the predictions are compared with the observed results of the experiment. If the predictions match the observed results, then the hypothesis is supported. If the observed results do not match the prediction, then the hypothesis is not supported. A signature feature of an experiment is the control of variables to help eliminate alternative explanations for observed results.

Hints for Implementing the Lab

- Demonstrate the reaction for the students as you introduce the investigation.

- We recommend that students fill out an investigation proposal for the second part of the investigation (the experiment about temperature). These proposals provide a way for you to offer students hints and suggestions without telling them how to do it. You can also check the proposals quickly during a class period.

- The students should fill out their investigation proposal after they have learned how to determine the equilibrium constant for the reaction at room temperature by following the directions on the handout. The students will be able to design a much better experiment once they learn how to determine an equilibrium constant in the lab.

LAB 30

- We recommend that you show students how to use a serological pipette and how to measure the absorbance of a solution using a colorimeter as part of the tool talk.

- For best results, set the wavelength on the colorimeter to 470 nm (blue).

- Students will need to test at least three different temperatures during their experiment.

- Remind students to include multiple test solutions at each temperature during their experiment and use the same concentrations for the test solutions.

- Students will need to average their results to help reduce measurement error.

- Show students how to make a hot-water bath and an ice bath as part of the tool talk. Make sure students do not heat any of the solutions directly on the hot plate. The temperature of the hot-water bath should not exceed 60°C.

Topic Connections

Table 30.2 provides an overview of the scientific practices, crosscutting concepts, disciplinary core ideas, and supporting ideas at the heart of this lab investigation. In addition, it lists NOS and NOSI concepts for the explicit and reflective discussion. Finally, it lists literacy and mathematics skills (*CCSS ELA* and *CCSS Mathematics*) that are addressed during the investigation.

TABLE 30.2

Lab 30 alignment with standards

Scientific practices	• Asking questions and defining problems • Planning and carrying out investigations • Analyzing and interpreting data • Using mathematics and computational thinking • Constructing explanations and designing solutions • Engaging in argument from evidence • Obtaining, evaluating, and communicating information
Crosscutting concepts	• Systems and system models • Stability and change
Core idea	• PS1.B: Chemical reactions
Supporting ideas	• Moles and molarity • Endothermic and exothermic reactions • Reaction rates • Chemical equilibrium
NOS and NOSI concepts	• Imagination and creativity in science • Nature and role of experiments
Literacy connections (CCSS ELA)	• *Reading:* Key ideas and details, craft and structure, integration of knowledge and ideas • *Writing:* Text types and purposes, production and distribution of writing, research to build and present knowledge, range of writing • *Speaking and listening:* Comprehension and collaboration, presentation of knowledge and ideas
Mathematics connections (CCSS Mathematics)	• Reason abstractly and quantitatively • Model with mathematics • Look for and make use of structure • Look for and express regularity in repeated reasoning

Lab Handout

Lab 30. Equilibrium Constant and Temperature: How Does a Change in Temperature Affect the Value of the Equilibrium Constant for an Exothermic Reaction?

Introduction

Chemical equilibrium is defined as the point in a reaction where the rate at which reactants transform into products is equal to the rate at which products revert back into reactants. When a reaction is in equilibrium, the concentration of the products and reactants is constant or stable. At this point, there is no further net change in the amounts of reactants or products unless the system is disturbed in some manner. The *equilibrium constant* provides a mathematical description of the equilibrium state for any reversible chemical reaction. To illustrate, consider the following general equation for a reversible reaction:

$$aA + bB \leftrightarrows cC + dD$$

The equation for calculating the equilibrium constant, K_{eq}, for this general reaction is provided below. The square brackets refer to the molar concentrations of each substance at equilibrium. The exponents are the stoichiometric coefficients of each substance found in the balanced chemical equation.

$$K_{eq} = \frac{[C]^c[D]^d}{[A]^a[B]^b}$$

The equilibrium constant describes the proportional relationship that exists between the concentration of the reactants and the concentration of the products for a specific chemical reaction when the reaction is in a state of equilibrium. The actual concentrations of the reactants and products that are present in the system at equilibrium will depend on the initial amounts of the reactants that were used at the beginning of the reaction and any extra reactants or products that were added to the system after the reaction started. The concentration ratio of products to reactants described by the equilibrium constant, however, will always be the same as long as the system is in equilibrium and the temperature of the system does not change.

The equilibrium constant is useful because it allows chemists to determine the product-to-reactant concentration ratio that will be present in the reaction mixture at equilibrium before the reaction begins. When $K_{eq} > 1$, the concentration of the products in the system will be greater than the concentration of the reactants. When $K_{eq} < 1$, the concentration of products will be less than the concentration of the reactants. Finally, when $K_{eq} = 1$, the concentration of products and reactants will be equal. A reaction with a large K_{eq} value, as a result, will have a greater product-to-reactant concentration ratio at equilibrium than a reaction with a smaller value.

The equilibrium constant of a reaction will change when the temperature changes. A positive shift in the K_{eq} will cause the product-to-reactant concentration ratio at equilibrium to increase, whereas a negative shift will result in a decrease in the concentration ratio. How the K_{eq} of a reaction will change in response to a change in temperature, however, is not uniform. The change in the K_{eq} of a reaction depends on the direction and magnitude of the temperature change. It will also depend on whether the reaction is exothermic or endothermic.

To control the amount of product or reactant present at the equilibrium point of a reaction in a closed system, chemists need to understand how the equilibrium constant will shift in response to changes in temperature for different types of reactions. You will therefore determine the equilibrium constant for an exothermic reaction and then explore how increases and decreases in temperature change the value of the equilibrium constant for this reaction. You will then develop a rule that you can use to predict how a change in temperature will affect the value of the equilibrium constant for other exothermic reactions.

Your Task

Determine the equilibrium constant for the reaction between iron(III) nitrate and potassium thiocyanate. Then determine how the equilibrium constant for this exothermic reaction is affected by a change in temperature. Once you understand how the equilibrium constant for this reaction changes in response to a change in temperature, you will need to develop a rule that you can use to predict how the equilibrium constant of other exothermic reactions will change in a response to a temperature change.

The guiding question of this investigation is, **How does a change in temperature affect the value of the equilibrium constant for an exothermic reaction?**

Materials

You may use any of the following materials during your investigation:

Consumables	Equipment
• 0.002 M iron(III) nitrate, $Fe(NO_3)_3$ • 0.200 M iron(III) nitrate, $Fe(NO_3)_3$ • 0.002 M potassium thiocyanate, KSCN • Distilled water • Ice	• Colorimeter sensor • Sensor interface • 4 cuvettes • 12 test tubes • Test tube rack • 3 serological pipettes (each 5 or 10 ml) • Pipette bulb • Stirring rod • 6 beakers (each 50 ml) • 2 beakers (each 250 ml, for water baths) • Hot plate • Thermometer

LAB 30

Safety Precautions

Follow all normal lab safety rules. Potassium thiocyanate is toxic by ingestion. Iron(III) nitrate solution contains 1 M nitric acid and is a corrosive liquid; it will also stain clothes and skin. Your teacher will explain relevant and important information about working with the chemicals associated with this investigation. In addition, take the following safety precautions:

- Wear indirectly vented chemical-splash goggles and chemical-resistant gloves and apron while in the laboratory.
- Use caution when working with hot plates because they can burn skin. Hot plates also need to be kept away from water and other liquids.
- Handle all glassware (including thermometers) with care.
- Wash hands with soap and water before leaving the laboratory.

Investigation Proposal Required?   ☐ Yes   ☐ No

Getting Started

The first step in this investigation is to determine the equilibrium constant for the reaction between iron(III) nitrate and potassium thiocyanate at room temperature. Iron(III) ions react with thiocyanate ions to form $FeSCN^{2+}$ complex ions according to the following reaction:

$$Fe^{3+}(aq) + SCN^-(aq) \leftrightarrows FeSCN^{2+}(aq)$$

$$\text{Yellow} \quad \text{Colorless} \quad \text{Orange-Red}$$

The equilibrium constant expression for this reaction is

$$K_{eq} = \frac{[FeSCN^{2+}]}{[Fe^{3+}][SCN^-]}$$

You can determine the value of K_{eq} by mixing solutions with known concentrations of Fe^{3+} and SCN^- and then measuring the concentration of the $FeSCN^{2+}$ ions in the mixture once the reaction is at equilibrium. The equilibrium concentration of the $FeSCN^{2+}$ ions in the solution can be determined by measuring the absorbance of the solution using a colorimeter. This is possible because the $FeSCN^{2+}$ ions produce a red color and the amount of light absorbed by the solution is directly proportional to the concentration of the $FeSCN^{2+}$ ions in it. You can, as a result, determine the $FeSCN^{2+}$ concentration of any solution by simply comparing the absorbance of that solution with the absorbance of a solution with a known $FeSCN^{2+}$ concentration (called a standard solution).

You will need to make a standard solution and five different test solutions to determine the equilibrium constant for the reaction between iron(III) nitrate and potassium thiocyanate at room temperature. Prepare the standard solution and five test solutions as described in Table L30.1.

TABLE L30.1

Components of the standard and test solutions

Sample	Reactants (ml)			
	0.200 M Fe(NO$_3$)$_3$	0.002 M Fe(NO$_3$)$_3$	0.002 M KSCN	Distilled water
Standard solution	9.00	0.00	1.00	0.00
Test solution 1	0.00	5.00	1.00	4.00
Test solution 2	0.00	5.00	2.00	3.00
Test solution 3	0.00	5.00	3.00	2.00
Test solution 4	0.00	5.00	4.00	1.00
Test solution 5	0.00	5.00	5.00	0.00

Once you have your solutions prepared, you can measure the absorbance of each one. Your teacher will show you how to use the calorimeter to measure the absorbance of the solutions. You will need to determine the concentration of FeSCN^{2+} in each test solution and then use this information to calculate an average equilibrium constant for the reaction. To accomplish this task, follow the procedure below:

1. Calculate the concentration of the Fe^{3+} and SCN$^-$ ions in the standard solution and the five test solutions using the dilution equation ($M_1V_1 = M_2V_2$).

2. Calculate the concentration of the FeSCN^{2+} ions in the standard solution and each test solution at equilibrium using the following equation:

$$[\text{FeSCN}^{2+}]_{\text{Equilibrium}} = \frac{A_{\text{TestSolution}}}{A_{\text{StandardSolution}}} \times [\text{FeSCN}^{2+}]_{\text{Standard}}$$

Assume the concentration of the FeSCN^{2+} in the standard solution at equilibrium is equal to the concentration of the SCN$^-$ ions in the standard solution at the start of the reaction. You can make this assumption because all of the SCN$^-$ ions should have been converted to FeSCN^{2+} ions due to the large amount of Fe^{3+} that was added to the standard solution.

3. Calculate the equilibrium concentration of Fe^{3+} ions in each test solution by subtracting the equilibrium concentration of FeSCN^{2+} from the initial concentration of Fe^{3+} ions using the equation

$$[\text{Fe}^{3+}]_{\text{TestSolutionEquilibrium}} = [\text{Fe}^{3+}]_{\text{TestSolutionInitial}} - [\text{FeSCN}^{2+}]_{\text{TestSolution}}$$

4. Calculate the equilibrium concentration of SCN^- ions in each test solution by subtracting the equilibrium concentration of $FeSCN^{2+}$ from the initial concentration of SCN^- ions using the equation

$$[SCN^-]_{TestSolutionEquilibrium} = [SCN^-]_{TestSolutionInitial} - [FeSCN^{2+}]_{TestSolution}$$

5. Calculate the value of the equilibrium constant for the five test solutions using the equation

$$K_{eq} = \frac{[FeSCN^{2+}]}{[Fe^{3+}][SCN^-]}$$

6. Calculate the average equilibrium constant for the reaction.

The second step in your investigation is to conduct an experiment to determine how a change in temperature affects the equilibrium constant for the reaction between iron(III) nitrate and potassium thiocyanate. To conduct this experiment, you must determine what type of data you need to collect, how you will collect the data, and how you will analyze the data.

To determine *what type of data you need to collect*, think about the following questions:

- What type of measurements or observations will you need to record during each experiment?
- When will you need to make these measurements or observations?

To determine *how you will collect the data*, think about the following questions:

- What will serve as your independent variable?
- What types of comparisons will you need to make?
- How will you change the temperature of the reaction?
- What will you do to reduce measurement error?
- How will you keep track of the data you collect and how will you organize it?

To determine *how you will analyze the data*, think about the following questions:

- What type of calculations will you need to make?
- What type of graph could you create to help make sense of your data?

The last step in this investigation is to develop a rule that you can use to predict how the equilibrium constant of other exothermic reactions will change in a response to a temperature change. This rule will serve as your answer to the guiding question of this investigation.

Connections to Crosscutting Concepts, the Nature of Science, and the Nature of Scientific Inquiry

As you work through your investigation, be sure to think about

- how scientists must define the system they are studying and then use models to understand it,
- why it is important to understand what makes a system stable or unstable and what controls the rates of change in a system,
- the importance of imagination and creativity in science, and
- the role of experiments in science.

Initial Argument

Once your group has finished collecting and analyzing your data, you will need to develop an initial argument. Your argument must include a *claim*, which is your answer to the guiding question. Your argument must also include *evidence* in support of your claim. The evidence is your analysis of the data and your interpretation of what the analysis means. Finally, you must include a *justification* of the evidence in your argument. You will therefore need to use a scientific concept or principle to explain why the evidence that you decided to use is relevant and important. You will create your initial argument on a whiteboard. Your whiteboard must include all the information shown in Figure L30.1.

Argumentation Session

The argumentation session allows all of the groups to share their arguments. One member of each group stays at the lab station to share that group's argument, while the other members of the group go to the other lab stations one at a time to listen to and critique the arguments developed by their classmates. The goal of the argumentation session is not to convince others that your argument is the best one; rather, the goal is to identify errors or instances of faulty reasoning in the initial arguments so these mistakes can be fixed. You will therefore need to evaluate the content of the claim, the quality of the evidence used to support the claim, and the strength of the justification of the evidence included in each argument that you see. To critique an argument, you might need more information than what is included on the whiteboard. You might, therefore, need to ask the presenter one or more follow-up questions, such as:

- How did your group collect the data? Why did you use that method?

FIGURE L30.1 _____

Argument presentation on a whiteboard

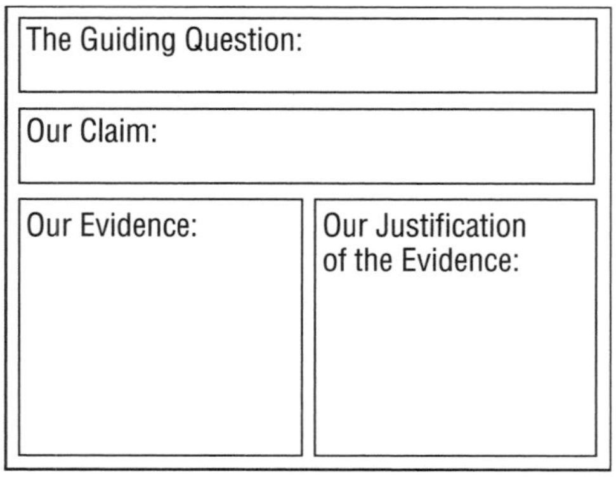

The Guiding Question:	
Our Claim:	
Our Evidence:	Our Justification of the Evidence:

LAB 30

- What did your group do to make sure the data you collected are reliable? What did you do to decrease measurement error?

- What did your group do to analyze the data, and why did you decide to do it that way? Did you check your calculations?

- Is that the only way to interpret the results of your group's analysis? How do you know that your interpretation of the analysis is appropriate?

- Why did your group decide to present your evidence in that manner?

- What other claims did your group discuss before deciding on that one? Why did you abandon those alternative ideas?

- How confident are you that your group's claim is valid? What could you do to increase your confidence?

Once the argumentation session is complete, you will have a chance to meet with your group and revise your original argument. Your group might need to gather more data or design a way to test one or more alternative claims as part of this process. Remember, your goal at this stage of the investigation is to develop the most valid or acceptable answer to the research question!

Report

Once you have completed your research, you will need to prepare an *investigation report* that consists of three sections that provide answers to the following questions:

1. What question were you trying to answer and why?

2. What did you do during your investigation and why did you conduct your investigation in this way?

3. What is your argument?

Your report should answer these questions in two pages or less. The report must be typed and any diagrams, figures, or tables should be embedded into the document. Be sure to write in a persuasive style; you are trying to convince others that your claim is acceptable or valid!

Checkout Questions

Lab 30. Equilibrium Constant and Temperature: How Does a Change in Temperature Affect the Value of the Equilibrium Constant for an Exothermic Reaction?

1. Determine the equilibrium constant, K_{eq}, for the following chemical reaction at equilibrium if the molar concentrations of the molecules are 0.20 M H_2, 0.10 M NO, 0.20 M H_2O, and 0.10M N_2:

$$2H_2(g) + 2NO(g) \rightarrow 2H_2O(g) + N_2(g)$$

Is there a greater concentration of total products or reactants in this equilibrium situation?

   a. Greater concentration of total products

   b. Greater concentration of total reactants

How do you know?

2. In the following reaction, the temperature is increased and the K_{eq} value decreases from 0.75 to 0.55:

$$N_2(g) + 3H_2 \leftrightarrow 2NH_3(g)$$

What kind of reaction is this?

a. Exothermic

b. Endothermic

How do you know?

3. In the following reaction, the enthalpy of reaction is $\Delta H = -92.5$ kJ and there is an increase in temperature:

$$PCl_3(g) + Cl_2(g) \leftrightarrow PCl_5(g)$$

How will the equilibrium constant shift?

a. It will increase.

b. It will decrease.

How do you know?

4. Scientists use experiments to prove that ideas are true.

a. I agree with this statement.

b. I disagree with this statement.

Explain your answer, using an example from your investigation about equilibrium constant and temperature.

5. Scientists need to be creative and have a good imagination to excel in science.

   a. I agree with this statement.
   b. I disagree with this statement.

   Explain your answer, using an example from your investigation about equilibrium constant and temperature.

6. An important goal in science is to understand what types of disturbances can make a system unstable and how a system will respond to a disturbance. Explain why this is important, using an example from your investigation about equilibrium constant and temperature.

7. Scientists often need to define a system under study and then use or develop a model to help them understand it. Explain why this is important, using an example from your investigation about equilibrium constant and temperature.

SECTION 4
Appendixes

APPENDIX 1

Standards Alignment Matrixes

Alignment of the Argument-Driven Inquiry Lab Investigations With the Scientific Practices, Crosscutting Concepts, and Core Ideas in A Framework for K–12 Science Education (NRC 2012)

Legend: ■ = filled; □ = open square; blank = no mark

Lab Investigation	Asking questions and defining problems	Developing and using models	Planning and carrying out investigations	Analyzing and interpreting data	Using mathematics and computational thinking	Constructing explanations and designing solutions	Engaging in argument from evidence	Obtaining, evaluating, and communicating information
1-Bond Character and Molecular Polarity	□	■	■	■	□	■	■	■
2-Molecular Shapes	□	■	■	■		■	■	■
3-Rate of Dissolution	□	■	■	■		■	■	■
4-Molarity	□	■	■	■		■	■	■
5-Temperature Changes Due to Evaporation	□	■	■	■		■	■	■
6-Pressure, Temperature, and Volume of Gases	□	■	■	■	□	■	■	■
7-Periodic Trends	□	■	■	□	■	■	■	■
8-Solutes and the Freezing Point of Water	□	■	■	□	■	□	■	■
9-Melting and Freezing Points	□	■	■	■		■	■	■
10-Identification of an Unknown Based on Physical Properties	□	■	■	■		■	■	■
11-Atomic Structure and Electromagnetic Radiation	□	■	■	■		■	■	■
12-Magnetism and Atomic Structure	□	■	■	■		■	■	■
13-Density and the Periodic Table	□	□	□	■	■	■	■	■
14-Molar Relationships			□	■	■	■	■	■
15-The Ideal Gas Law	■	■	■	■	■	■	■	■
16-Development of a Reaction Matrix	□		■	■		■	■	■
17-Limiting Reactants	□		■	■		■	■	■
18-Characteristics of Acids and Bases	□	■	■	■		■	■	■
19-Strong and Weak Acids	□	■	■	■		■	■	■
20-Enthalpy Change of Solution	□	■	■	■		■	■	■
21-Reaction Rates	□	■	■	■	■	■	■	■
22-Equilibrium	□	■		■		■	■	■
23-Classification of Changes in Matter	□		■	■		■	■	■
24-Identification of Reaction Products	□		■	■		■	■	■
25-Acid-Base Titration and Neutralization Reactions	□		■	■	■		■	■
26-Composition of Chemical Compounds	□		■	■	■	■	■	■
27-Stoichiometry and Chemical Reactions	□		■	■	■	■	■	■
28-Designing a Cold Pack	□	■	■	■	■	■	■	■
29-Rate Laws	□	■	■	■	■		■	■
30-Equilibrium Constant and Temperature	□	■	■	■	■	■	■	■

Aspect of the NRC Framework

Scientific practices

Lab Investigation

Aspect of the NRC Framework	Patterns	Cause and effect: Mechanism and explanation	Scale, proportion, and quantity	Systems and system models	Energy and matter: Flows, cycles, and conservation	Structure and function	Stability and change	PS1.A: Structure and properties of matter	PS1.B: Chemical reactions
Crosscutting concepts									
1-Bond Character and Molecular Polarity	■		■		■			■	
2-Molecular Shapes	■			■		■		■	
3-Rate of Dissolution		■		■				■	
4-Molarity			■		■			■	
5-Temperature Changes Due to Evaporation					■	■		■	
6-Pressure, Temperature, and Volume of Gases		■		■				■	
7-Periodic Trends	■		■					■	■
8-Solutes and the Freezing Point of Water	■				■			■	
9-Melting and Freezing Points				■	■			■	
10-Identification of an Unknown Based on Physical Properties			■		■			■	
11-Atomic Structure and Electromagnetic Radiation	■			■				■	
12-Magnetism and Atomic Structure				■		■		■	
13-Density and the Periodic Table	■		■					■	
14-Molar Relationships	■		■					■	
15-The Ideal Gas Law		■		■				■	
16-Development of a Reaction Matrix	■							■	■
17-Limiting Reactants			■	■					■
18-Characteristics of Acids and Bases	■					■		■	■
19-Strong and Weak Acids		■		■				■	■
20-Enthalpy Change of Solution				■	■			■	■
21-Reaction Rates		■		■					■
22-Equilibrium				■			■		■
23-Classification of Changes in Matter	■				■				■
24-Identification of Reaction Products	■				■				■
25-Acid-Base Titration and Neutralization Reactions			■		■			■	■
26-Composition of Chemical Compounds			■		■				■
27-Stoichiometry and Chemical Reactions		■		■		■		■	■
28-Designing a Cold Pack	■				■				■
29-Rate Laws			■		■				■
30-Equilibrium Constant and Temperature				■	■		■		■

strong alignment = ■; moderate alignment = □

Alignment of the Argument-Driven Inquiry Lab Investigations With the *Common Core State Standards*, in English Language Arts and Mathematics (NGAC and CCSSO 2010)

Standard	1	2	3	4	5	6	7	8	9	10	11	12	13	14	15	16	17	18	19	20	21	22	23	24	25	26	27	28	29	30
Reading																														
Key ideas and details	■	■	■	■	■	■	■	■	■	■	■	■	■	■	■	■	■	■	■	■	■	■	■	■	■	■	■	■	■	■
Craft and structure	■	■	■	■	■	■	■	■	■	■	■	■	■	■	■	■	■	■	■	■	■	■	■	■	■	■	■	■	■	■
Integration of knowledge and ideas	■	■	■	■	■	■	■	■	■	■	■	■	■	■	■	■	■	■	■	■	■	■	■	■	■	■	■	■	■	■
Writing																														
Text types and purposes	■	■	■	■	■	■	■	■	■	■	■	■	■	■	■	■	■	■	■	■	■	■	■	■	■	■	■	■	■	■
Production and distribution of writing	■	■	■	■	■	■	■	■	■	■	■	■	■	■	■	■	■	■	■	■	■	■	■	■	■	■	■	■	■	■
Research to build and present knowledge	■	■	■	■	■	■	■	■	■	■	■	■	■	■	■	■	■	■	■	■	■	■	■	■	■	■	■	■	■	■
Range of writing	■	■	■	■	■	■	■	■	■	■	■	■	■	■	■	■	■	■	■	■	■	■	■	■	■	■	■	■	■	■

Lab Investigation:
1-Bond Character and Molecular Polarity; 2-Molecular Shapes; 3-Rate of Dissolution; 4-Molarity; 5-Temperature Changes Due to Evaporation; 6-Pressure, Temperature, and Volume of Gases; 7-Periodic Trends; 8-Solutes and the Freezing Point of Water; 9-Melting and Freezing Points; 10-Identification of an Unknown Based on Physical Properties; 11-Atomic Structure and Electromagnetic Radiation; 12-Magnetism and Atomic Structure; 13-Density and the Periodic Table; 14-Molar Relationships; 15-The Ideal Gas Law; 16-Development of a Reaction Matrix; 17-Limiting Reactants; 18-Characteristics of Acids and Bases; 19-Strong and Weak Acids; 20-Enthalpy Change of Solution; 21-Reaction Rates; 22-Equilibrium; 23-Classification of Changes in Matter; 24-Identification of Reaction Products; 25-Acid-Base Titration and Neutralization Reactions; 26-Composition of Chemical Compounds; 27-Stoichiometry and Chemical Reactions; 28-Designing a Cold Pack; 29-Rate Laws; 30-Equilibrium Constant and Temperature

strong alignment = ■; moderate alignment = □

Lab Investigation

Lab Investigation	Speaking and listening: Comprehension and collaboration	Speaking and listening: Presentation of knowledge and ideas	Mathematics: Make sense of problems and persevere in solving them	Reason abstractly and quantitatively	Construct viable arguments and critique the reasoning of others	Model with mathematics	Use appropriate tools strategically	Attend to precision	Look for and make use of structure	Look for and express regularity in repeated reasoning
1-Bond Character and Molecular Polarity	■	■	■					■		
2-Molecular Shapes	■	■								
3-Rate of Dissolution	■	■		■		■		■		■
4-Molarity	■	■	■	■		■	■	■	■	■
5-Temperature Changes Due to Evaporation	■	■		■	■					■
6-Pressure, Temperature, and Volume of Gases	■	■		■	■	■			■	
7-Periodic Trends	■	■								■
8-Solutes and the Freezing Point of Water	■	■				■		■		■
9-Melting and Freezing Points	■	■		■	■					■
10-Identification of an Unknown Based on Physical Properties	■	■		■	■		■	■		■
11-Atomic Structure and Electromagnetic Radiation	■	■								■
12-Magnetism and Atomic Structure	■	■								■
13-Density and the Periodic Table	■	■								■
14-Molar Relationships	■	■		■	■		■			■
15-The Ideal Gas Law	■	■		■		■		■		
16-Development of a Reaction Matrix	■	■		■						■
17-Limiting Reactants	■	■		■		■				
18-Characteristics of Acids and Bases	■	■								
19-Strong and Weak Acids	■	■		■			■	■		■
20-Enthalpy Change of Solution	■	■		■				■		
21-Reaction Rates	■	■		■		■				
22-Equilibrium	■	■		■		■				
23-Classification of Changes in Matter	■	■								
24-Identification of Reaction Products	■	■		■						■
25-Acid-Base Titration and Neutralization Reactions	■	■	■	■		■	■	■	■	■
26-Composition of Chemical Compounds	■	■		■	■			■		■
27-Stoichiometry and Chemical Reactions	■	■		■	■			■		■
28-Designing a Cold Pack	■	■		■	■					■
29-Rate Laws	■	■		■	■	■				
30-Equilibrium Constant and Temperature	■	■		■		■			■	■

strong alignment = ■; moderate alignment = □

Alignment of the Argument-Driven Inquiry Lab Investigations With the Nature of Science (NOS) and the Nature of Scientific Inquiry (NOSI) Concepts*

Lab Investigation	Observations and inferences	Changes in scientific knowledge over time	Scientific laws and theories	Social and cultural influences	Difference between data and evidence	Methods used in scientific investigations	Imagination and creativity in science	Nature and role of experiments
1-Bond Character and Molecular Polarity		■			■			
2-Molecular Shapes		■				■		
3-Rate of Dissolution							■	■
4-Molarity				■	■			
5-Temperature Changes Due to Evaporation	■							■
6-Pressure, Temperature, and Volume of Gases			■				■	
7-Periodic Trends	■					■		
8-Solutes and the Freezing Point of Water							■	■
9-Melting and Freezing Points						■	■	
10-Identification of an Unknown Based on Physical Properties	■				■			
11-Atomic Structure and Electromagnetic Radiation			■			■		
12-Magnetism and Atomic Structure		■	■					
13-Density and the Periodic Table		■				■		
14-Molar Relationships		■			■			
15-The Ideal Gas Law			■					■
16-Development of a Reaction Matrix	■							■
17-Limiting Reactants			■				■	
18-Characteristics of Acids and Bases	■					■		
19-Strong and Weak Acids	■						■	■
20-Enthalpy Change of Solution	■		■					
21-Reaction Rates							■	■
22-Equilibrium							■	■
23-Classification of Changes in Matter	■				■			
24-Identification of Reaction Products	■		■					
25-Acid-Base Titration and Neutralization Reactions	■			■				
26-Composition of Chemical Compounds		■				■		
27-Stoichiometry and Chemical Reactions					■			■
28-Designing a Cold Pack				■			■	
29-Rate Laws	■		■					
30-Equilibrium Constant and Temperature							■	■

*The NOS/NOSI concepts listed in this matrix are based on the work of Abd-El-Khalick and Lederman 2000; Akerson, Abd-El-Khalick, and Lederman 2000; Lederman et al. 2002, 2014; and Schwartz, Lederman, and Crawford 2004.

References

Abd-El-Khalick, F., and N. G. Lederman. 2000. Improving science teachers' conceptions of nature of science: A critical review of the literature. *International Journal of Science Education* 22: 665–701.

Akerson, V., F. Abd-El-Khalick, and N. Lederman. 2000. Influence of a reflective explicit activity-based approach on elementary teachers' conception of nature of science. *Journal of Research in Science Teaching* 37 (4): 295–317.

Lederman, N. G., F. Abd-El-Khalick, R. L. Bell, and R. S. Schwartz. 2002. Views of nature of science questionnaire: Toward a valid and meaningful assessment of learners' conceptions of nature of science. *Journal of Research in Science Teaching* 39 (6): 497–521.

Lederman, J., N. Lederman, S. Bartos, S. Bartels, A. Meyer, and R. Schwartz. 2014. Meaningful assessment of learners' understanding about scientific inquiry: The Views About Scientific Inquiry (VASI) questionnaire. *Journal of Research in Science Teaching* 51 (1): 65–83.

National Governors Association Center for Best Practices and Council of Chief State School Officers (NGAC and CCSSO). 2010. *Common core state standards.* Washington, DC: NGAC and CCSSO.

National Research Council (NRC). 2012. *A framework for K–12 science education: Practices, crosscutting concepts, and core ideas.* Washington, DC: National Academies Press.

Schwartz, R. S., N. Lederman, and B. Crawford. 2004. Developing views of nature of science in an authentic context: An explicit approach to bridging the gap between nature of science and scientific inquiry. *Science Education* 88: 610–645.

APPENDIX 2
Options for Implementing ADI Lab Investigations

Option A

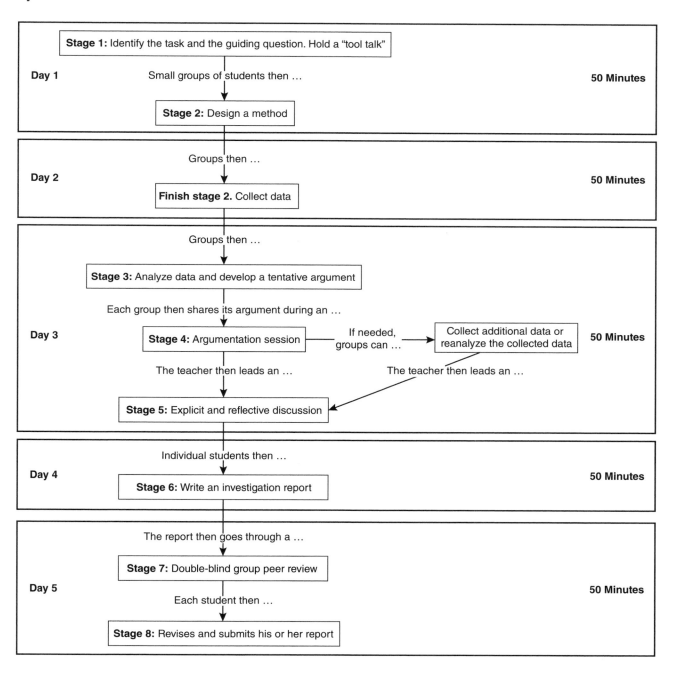

Day 1 — 50 Minutes

Stage 1: Identify the task and the guiding question. Hold a "tool talk"

Small groups of students then …

Stage 2: Design a method

Day 2 — 50 Minutes

Groups then …

Finish stage 2. Collect data

Day 3 — 50 Minutes

Groups then …

Stage 3: Analyze data and develop a tentative argument

Each group then shares its argument during an …

Stage 4: Argumentation session

If needed, groups can … → Collect additional data or reanalyze the collected data

The teacher then leads an … | The teacher then leads an …

Stage 5: Explicit and reflective discussion

Day 4 — 50 Minutes

Individual students then …

Stage 6: Write an investigation report

Day 5 — 50 Minutes

The report then goes through a …

Stage 7: Double-blind group peer review

Each student then …

Stage 8: Revises and submits his or her report

Option B

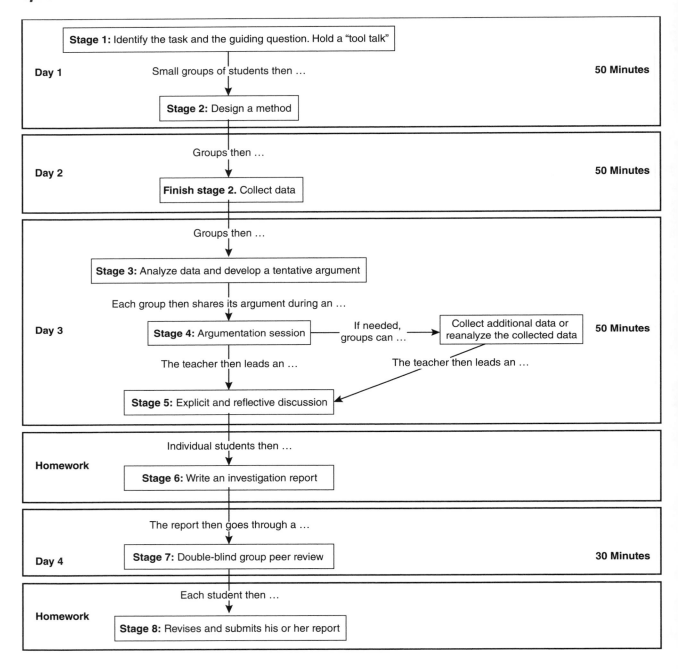

Option C

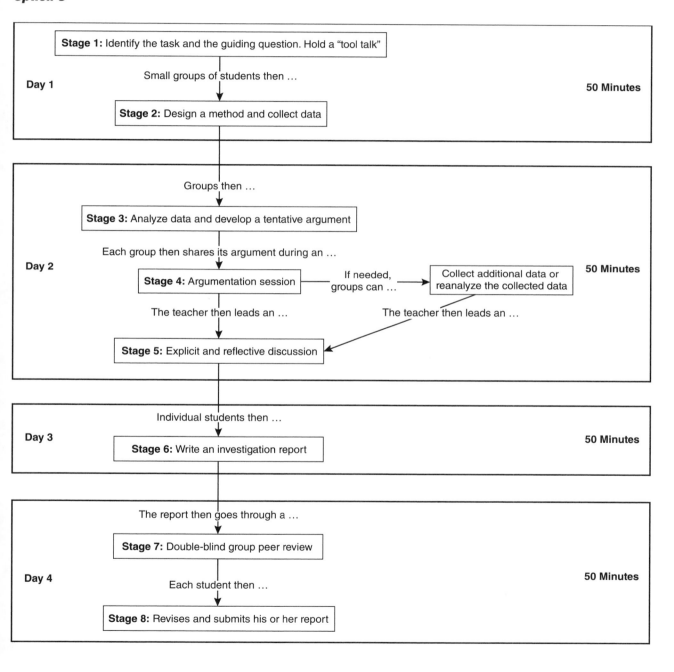

Option D

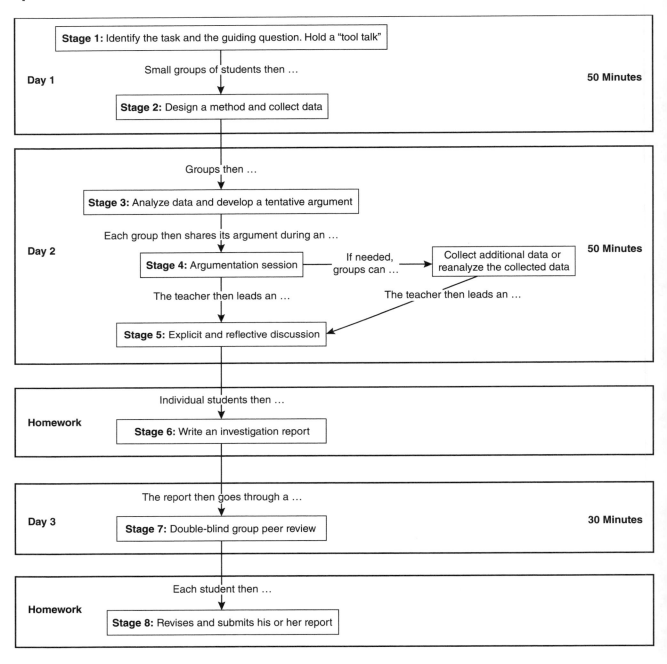

Option E

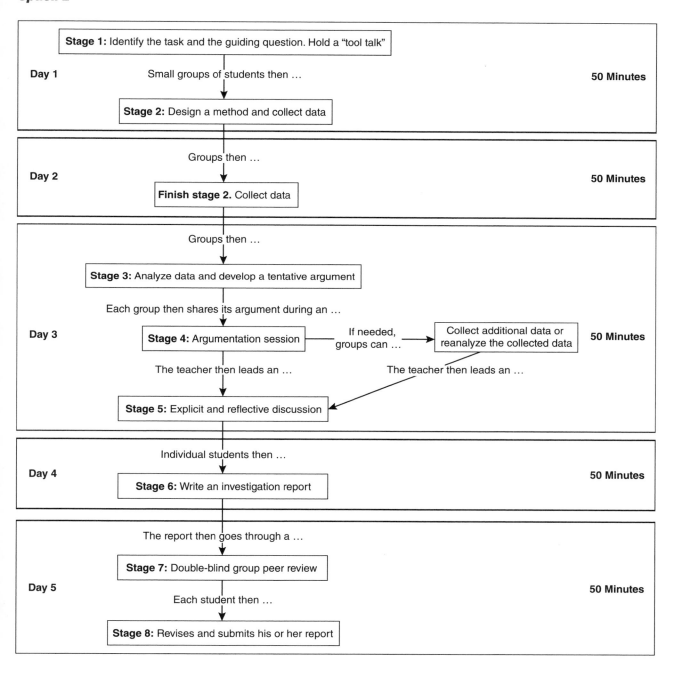

Option F

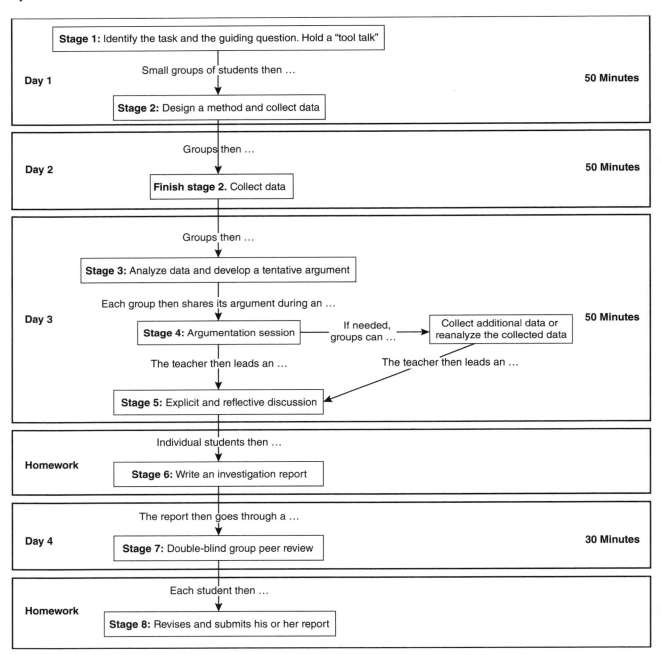

Option G

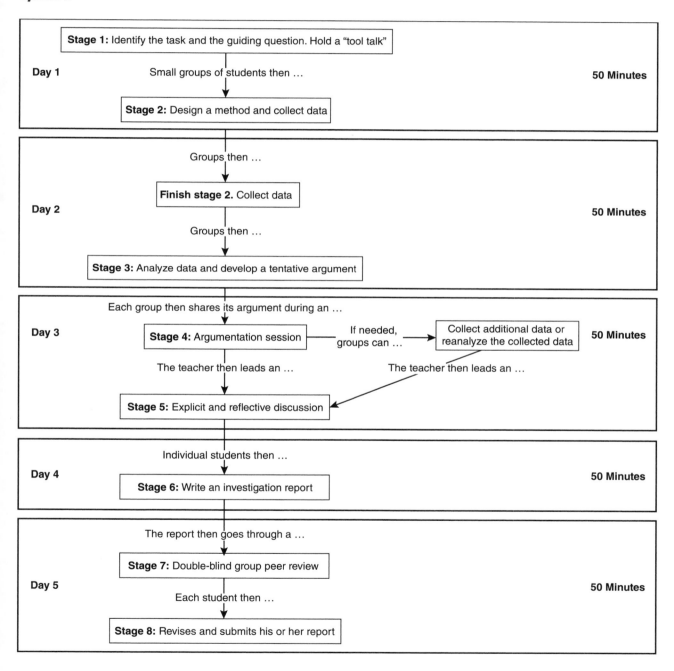

Option H

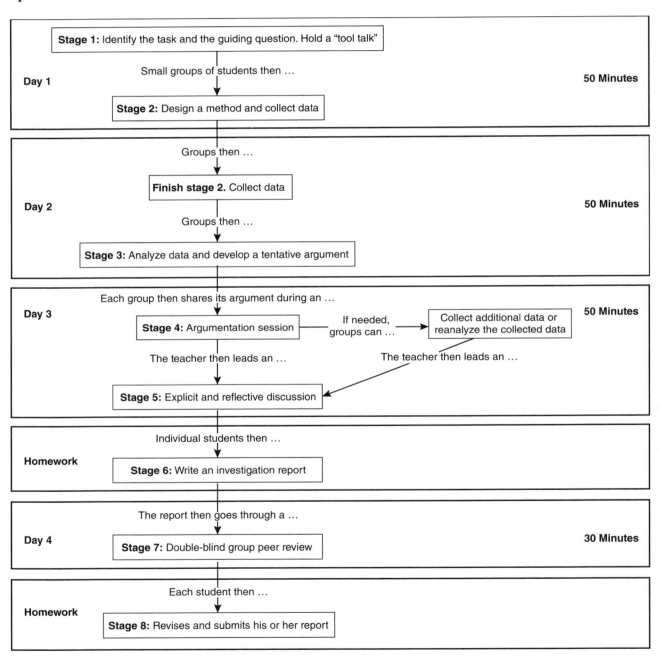

APPENDIX 3

Investigation Proposal Options

This appendix presents three investigation proposals that may be used in most labs. The development of these proposals was supported by the Institute of Education Sciences, U.S. Department of Education, through grant R305A100909 to Florida State University.

The format of investigation proposals A and B is modeled after a hypothetical deductive-reasoning guide described in *Exploring the Living World* (Lawson 1995) and modified from an investigation guide described in an article by Maguire, Myerowitz, and Sampson (2010).

References

Lawson, A. E. 1995. *Exploring the living world: A laboratory manual for biology.* McGraw-Hill College.

Maguire, L., L. Myerowitz, and V. Sampson. 2010. Diffusion and osmosis in cells: A guided inquiry activity. *The Science Teacher* 77 (8): 55–60.

Investigation Proposal A

The Guiding Question ...

Hypothesis 1 ← → Hypothesis 2

IF ...

IF ...

The Test

AND ...
Procedure

What data will you collect?

How will you analyze the data?

What safety precautions will you follow?

Predicted Result if hypothesis 1 is valid

Predicted Result if hypothesis 2 is valid

THEN ...

THEN ...

The Actual
Results

AND ...

I approve of this investigation. _____ _____

Instructor's Signature   Date

The development of this investigation proposal was supported by the Institute of Education Sciences, U.S. Department of Education, through Grant R305A100909 to the Florida State University. The format of the proposal is modeled after a hypothetical deductive-reasoning guide described in *Exploring the Living World* (Lawson 1995) and modified from an investigation guide described in Macquire, Myerowitz, and Sampson (2010).

Investigation Proposal B

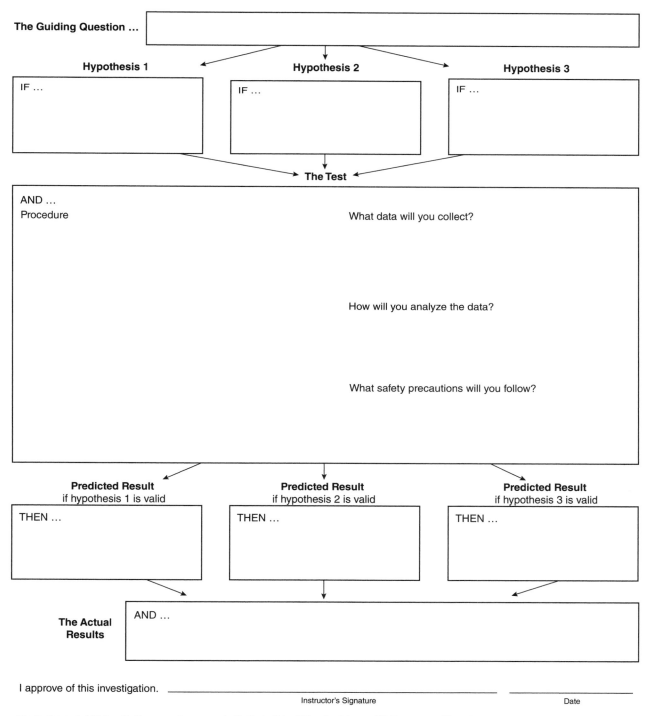

The Guiding Question ...

Hypothesis 1

IF ...

Hypothesis 2

IF ...

Hypothesis 3

IF ...

The Test

AND ...
Procedure

What data will you collect?

How will you analyze the data?

What safety precautions will you follow?

Predicted Result
if hypothesis 1 is valid

THEN ...

Predicted Result
if hypothesis 2 is valid

THEN ...

Predicted Result
if hypothesis 3 is valid

THEN ...

The Actual Results

AND ...

I approve of this investigation. _____   _____

Instructor's Signature   Date

The development of this investigation proposal was supported by the Institute of Education Sciences, U.S. Department of Education, through Grant R305A100909 to the Florida State University. The format of the proposal is modeled after a hypothetical deductive-reasoning guide described in *Exploring the Living World* (Lawson 1995) and modified from an investigation guide described in Macquire, Myerowitz, and Sampson (2010).

Investigation Proposal C

The Guiding Question ...	

What data will you collect?	

How will you collect your data?	Your Procedure What safety precautions will you follow?

How will you analyze your data?	

Your actual data	

I approve of this investigation. _____ _____

Instructor's Signature   Date

The development of this investigation proposal was supported by the Institute of Education Sciences, U.S. Department of Education, through Grant R305A100909 to the Florida State University.

APPENDIX 4
Peer-Review Guide and Instructor Scoring Rubric

Report By: _____
ID Number

Author: Did the reviewers do a good job? 1 2 3 4 5
Rate the overall quality of the peer review

Reviewed By: _____
ID Number

_____
ID Number

_____
ID Number

_____
ID Number

Section 1: Introduction and Guiding Question	Reviewer Rating			Instructor Score
1. Did the author provide enough background information?	☐ No	☐ Partially	☐ Yes	0 1 2
2. Is the background information accurate?	☐ No	☐ Partially	☐ Yes	0 1 2
3. Did the author describe the goal of the study?	☐ No	☐ Partially	☐ Yes	0 1 2
4. Did the author make the guiding question explicit and explain how the guiding question is related to the background information?	☐ No	☐ Partially	☐ Yes	0 1 2

Reviewers: If your group made any "No" or "Partially" marks in this section, please explain how the author could improve this part of his or her report.

Author: What revisions did you make in your report? Is there anything you decided to keep the same even though the reviewers suggested otherwise? Be sure to explain why.

Section 2: Method	Reviewer Rating			Instructor Score
1. Did the author describe the procedure he/she used to gather data and then explain why he/she used this procedure?	☐ No	☐ Partially	☐ Yes	0 1 2
2. Did the author explain what data were collected (or used) during the investigation and why they were collected (or used)?	☐ No	☐ Partially	☐ Yes	0 1 2
3. Did the author describe how he/she analyzed the data and explain why the analysis helped him/her answer the guiding question?	☐ No	☐ Partially	☐ Yes	0 1 2

Section 2: Method (*continued*)	Reviewer Rating			Instructor Score
4. Did the author use the correct term to describe his/her investigation (e.g., experiment, observations, interpretation of a data set)?	☐ No	☐ Partially	☐ Yes	0 1 2
Reviewers: If your group made any "No" or "Partially" marks in this section, please explain how the author could improve this part of his or her report.	**Author:** What revisions did you make in your report? Is there anything you decided to keep the same even though the reviewers suggested otherwise? Be sure to explain why.			

Section 3: The Argument	Reviewer Rating			Instructor Score
1. Did the author provide a claim that answers the guiding question?	☐ No	☐ Partially	☐ Yes	0 1 2
2. Did the author include high-quality evidence in his/her argument?	☐ No	☐ Partially	☐ Yes	0 1 2
Were the data collected in an appropriate manner?	☐ No	☐ Partially	☐ Yes	0 1 2
Is the analysis of the data appropriate and free from errors? Is the author's interpretation of the analysis valid?	☐ No	☐ Partially	☐ Yes	0 1 2
3. Did the author present the evidence in an appropriate manner by	☐ No	☐ Partially	☐ Yes	0 1 2
• using a correctly formatted and labeled graph (or table);	☐ No	☐ Partially	☐ Yes	0 1 2
• including correct metric units (e.g., m/s, g, ml); and • referencing the graph or table in the body of the text?	☐ No	☐ Partially	☐ Yes	0 1 2
4. Is the claim consistent with the evidence?	☐ No	☐ Partially	☐ Yes	0 1 2
5. Did the author include a justification of the evidence that explains why the evidence is important (why it matters) and	☐ No	☐ Partially	☐ Yes	0 1 2
defends the inclusion of the evidence with a specific science concept or by discussing his/her underlying assumptions?	☐ No	☐ Partially	☐ Yes	0 1 2
6. Is the justification of the evidence acceptable?	☐ No	☐ Partially	☐ Yes	0 1 2
7. Did the author discuss how well his/her claim agrees with the claims made by other groups and explain any disagreements?	☐ No	☐ Partially	☐ Yes	0 1 2
8. Did the author use scientific terms correctly (e.g., hypothesis vs. prediction, data vs. evidence) and reference the evidence in an appropriate manner (e.g., supports or suggests vs. proves) correctly?	☐ No	☐ Partially	☐ Yes	0 1 2

Section 3: The Argument (*continued*)

Reviewers: If your group made any "No" or "Partially" marks in this section, please explain how the author could improve this part of his or her report.

Author: What revisions did you make in your report? Is there anything you decided to keep the same even though the reviewers suggested otherwise? Be sure to explain why.

Mechanics	Reviewer Rating			Instructor Score
1. Organization: Is each section easy to follow? Do paragraphs include multiple sentences? Do paragraphs begin with a topic sentence?	☐ No	☐ Partially	☐ Yes	0 1 2
2. Grammar: Are the sentences complete? Is there proper subject-verb agreement in each sentence? Are there run-on sentences?	☐ No	☐ Partially	☐ Yes	0 1 2
3. Conventions: Did the author use appropriate spelling, punctuation, paragraphing, and capitalization?	☐ No	☐ Partially	☐ Yes	0 1 2
4. Word Choice: Did the author use the appropriate word (e.g., there vs. their, to vs. too, than vs. then)?	☐ No	☐ Partially	☐ Yes	0 1 2

Instructor Comments:

Total: _____ /50

IMAGE CREDITS

CHAPTER 1

Figure 5: Authors

Figure 6: Authors

Figure 7: Authors

LAB 1

Figure 1.1: Authors

Figure 1.2: Authors

Figure L1.1: PhET Interactive Simulations, University of Colorado, http://*phet.colorado.edu; http://phet. colorado.edu/en/simulation/molecule-polarity*

LAB 2

Table 2.1: a: User:Benjah-bmm27, Wikimedia Commons, Public domain. *http://en.wikipedia. org/wiki/File:Linear-3D-balls.png*; b: User:Benjah-bmm27, Wikimedia Commons, Public domain. *http:// en.wikipedia.org/wiki/File:Trigonal-3D-balls.png*; c: User:Benjah-bmm27, Wikimedia Commons, Public domain. *http://en.wikipedia.org/wiki/File:AX2E1-3D-balls.png*; d: User:Benjah-bmm27, Wikimedia Commons, Public domain. *http://en.wikipedia.org/ wiki/File:AX4E0-3D-balls.png*; e: User:Benjah-bmm27, Wikimedia Commons, Public domain. *http:// en.wikipedia.org/wiki/File:AX3E1-3D-balls.png*; f: User:Benjah-bmm27, Wikimedia Commons, Public domain. *http://en.wikipedia.org/wiki/File:AX2E2-3D-balls.png*.

Figure L2.1: Authors

Figure L2.2: PhET Interactive Simulations, University of Colorado, *http://phet.colorado.edu*; *http://phet. colorado.edu/en/simulation/molecule-shapes*

LAB 3

Figure 3.1: Authors

LAB 4

Figure L4.1: PhET Interactive Simulations, University of Colorado, *http://phet.colorado.edu; http://phet. colorado.edu/en/simulation/molarity*

LAB 5

Figure 5.1: Authors

Figure L5.1: Authors

LAB 6

Figure L6.1: Authors

Figure L6.2: Authors

LAB 7

Figure 7.1: User:Mirek2, Wikimedia Commons, CC0 1.0. *http://commons.wikimedia.org/wiki/File:Periodic_ trends.svg*

Figure 7.2: Swetha Ramireddy, Wikimedia Commons, Public domain. *http://commons.wikimedia.org/wiki/ File:Chart_of_Melting_Points_of_Elements.jpg*

Figure L7.1: Chris King, Wikimedia Commons, CC BY-SA 3.0, GFDL 1.2. *http://commons.wikimedia.org/ wiki/File:Atomic_%26_ionic_radii.svg*

LAB 8

Figure L8.1: Authors

Figure L8.2: Authors

LAB 9

Figure 9.1: Authors

LAB 10

Figure L10.1: Left: User:LHcheM, Wikimedia Commons, CC BY-SA 3.0, GFDL 1.2. *http://commons.*

wikimedia.org/wiki/File:Sample_of_Methyl_Blue.
jpg; Right: User:LHcheM, Wikimedia Commons, CC
BY-SA 3.0, GFDL 1.2. http://commons.wikimedia.org/
wiki/File:Methyl_Blue_aqueous_solution.jpg

LAB 11

Figure 11.1: User:Haade, Wikimedia Commons, CC
BY-SA 3.0, GFDL 1.2. http://commons.wikimedia.org/
wiki/File:Single_electron_orbitals.jpg

Figure 11.2: a: Richard Parsons, Wikimedia
Commons, CC BY-SA 3.0. http://commons.
wikimedia.org/wiki/File:Atomic_orbital_energy_levels.
svg; b: User:MovGP0, Wikimedia Commons, CC
BY-SA 3.0. http://commons.wikimedia.org/wiki/
File:Feynman-electron-photon-emission.svg

Figure L11.1: CK-12 Foundation, Wikimedia
Commons, CC BY-SA 3.0. http://commons.
wikimedia.org/wiki/File:Px_py_pz_orbitals.png

Figure L11.2: Authors

Figure L11.3: Authors

LAB 12

Figure 12.1: User:Haade, Wikimedia Commons, CC
BY-SA 3.0, GFDL 1.2. http://commons.wikimedia.org/
wiki/File:Single_electron_orbitals.jpg

Figure 12.2: CK-12 Foundation, Wikimedia
Commons, Public domain. http://commons.
wikimedia.org/wiki/File:Electron_configuration_
chlorine.svg;

CK-12 Foundation, Wikimedia Commons, Public
domain. http://commons.wikimedia.org/wiki/
File:Electron_configuration_iron.svg

Figure L12.1.: User:Sven, Wikimedia Commons, CC
BY-SA 3.0, GFDL 1.2. http://commons.wikimedia.org/
wiki/File:S-p-Orbitals.svg

Figure L12.2: CK-12 Foundation, Wikimedia
Commons, Public domain. http://commons.
wikimedia.org/wiki/File:Electron_configuration_iron.
svg

Figure L12.3: Will Thomas Jr.

LAB 13

Figure 13.1: User:Mirek2, Wikimedia Commons, CC0
1.0. http://commons.wikimedia.org/wiki/File:Periodic_
trends.svg

Figure L13.1: Wikimedia Commons, Public
domain. http://commons.wikimedia.org/wiki/
File:Mendeleev%27s_1869_periodic_table.png

LAB 14

Figure 14.1: Authors

LAB 15

Figure 15.1: Authors

Figure 15.2: Authors

LAB 17

Figure L17.1: Authors

Figure L17.2: Authors

Checkout Questions: Authors

LAB 19

Figure L19.1: Authors

Figure L19.2: Authors

LAB 20

Figure 20.1: Authors

Figure 20.2: Authors

Figure L20.1: Authors

Figure L20.2: Authors

LAB 21

Figure 21.1: Authors

Figure 21.2: Authors

LAB 22

Figure 22.1: Authors

Figure 22.2: Authors

Figure L22.1: Authors

LAB 24

Figure L24.1: User:PRHaney, Wikimedia Commons, CC BY-SA 3.0. *http://commons.wikimedia.org/wiki/File:Lead_%28II%29_iodide_precipitating_out_of_solution.JPG*

LAB 25

Figure L25.1: User:Liquid_2003, Wikimedia Commons, CC BY 2.0. *http://commons.wikimedia.org/wiki/File:Titrage.svg*

LAB 27

Figure L27.1: Authors

LAB 28

Figure L28.1: Authors

LAB 29

Figure 29.1: Authors

Figure L29.1: Authors

INDEX

Page numbers printed in **boldface** *type refer to tables or figures.*

Index

Index

Rate of Dissolution, 60–61, **62,** 66
Reaction Rates, 334, **336,** 341
Solutes and the Freezing Point of Water, 129, **130,** 134
Stoichiometry and Chemical Reactions, 431, **433,** 437
Strong and Weak Acids, 301–302, **303,** 307
Temperature Changes Due to Evaporation, 87, **89,** 94
Neutralization reactions. *See* Acid-Base Titration and Neutralization Reactions lab
Next Generation Science Standards, xi

O
Organisation for Economic Co-operation and Development (OECD), xv

P
Peer group review of investigation report, 12–13
    role of teacher in, **16**
Peer-review guide, 12–13, **16,** 19, 21, 22, 23, **513–515**
Periodic trends, 204, **204**
Periodic Trends lab, 112–123
    checkout questions for, 122–123
    lab handout for, 118–121
        argumentation session, 120–121
        connections, 119–120
        getting started, 119
        initial argument, 120, **120**
        introduction, 118, **118**
        investigation proposal, 119
        materials, 119
        report, 121
        safety precautions, 119
        task and guiding question, 118–119
    teacher notes for, 112–117
        content, 112–114, **113**
        hints for implementing lab, 116
        laboratory waste disposal, 115
        materials and preparation, 114, **114**
        purpose, 112
        safety precautions, 115
        timeline, 114
        topic connections, 117, **117, 494–498**
        topics for explicit and reflective discussion, 115–116
Personal protective equipment, xxi
PhET Interactive Simulations, 30, 44, 71, 517
Preparation for labs, 21. *See also specific labs*
Pressure, Temperature, and Volume of Gases lab, 98–111
    checkout questions for, 110–111
    lab handout for, 104–109
        argumentation session, 108
        connections, 106
        getting started, 105–106
        initial argument, 106–108, **107**
        introduction, 104
        investigation proposal, 105
        materials, 105
        report, 108–109
        safety precautions, 105
        task and guiding question, 104
    teacher notes for, 98–103
        content, 98–99
        hints for implementing lab, 102
        laboratory waste disposal, 101
        materials and preparation, 99, **100**
        purpose, 98
        safety precautions, 100

        timeline, 99
        topic connections, 102–103, **103, 494–498**
        topics for explicit and reflective discussion, 101–102
Proficiency in science, xi–xii, xvii–xx, 3, 19
Programme for International Student Assessment (PISA), xvii–xviii, **xviii**
Purpose of labs, 20. *See also specific labs*

R
Rate Laws lab, 456–472
    checkout questions for, 471–473
    lab handout for, 464–469
        argumentation session, 468–469
        connections, 468
        getting started, 466–468
        initial argument, 468, **468**
        introduction, **464,** 464–465
        investigation proposal, 466
        materials, 466
        report, 469
        safety precautions, 466
        task and guiding question, 466
    Lab Reference Sheet for, 470
    teacher notes for, 456–462
        content, 456–458
        hints for implementing lab, 461–462, **462**
        laboratory waste disposal, 459
        materials and preparation, 458, **459**
        purpose, 456
        safety precautions, 458
        timeline, 458
        topic connections, 462, **463, 494–498**
        topics for explicit and reflective discussion, 459–461
Rate of Dissolution lab, 56–69
    checkout questions for, 68–69
    lab handout for, 63–67
        argumentation session, 66–67
        connections, 66
        getting started, 65
        initial argument, 66, **66**
        introduction, 63
        investigation proposal, 65
        materials, 64
        report, 67
        safety precautions, 64
        task and guiding question, 64
    teacher notes for, 56–61
        content, 56–57, **57**
        hints for implementing lab, 61
        laboratory waste disposal, 58
        materials and preparation, 58, **59**
        purpose, 56
        safety precautions, 58
        timeline, 57
        topic connections, 61, **62, 494–498**
        topics for explicit and reflective discussion, 58, 60–61
Reaction matrix, 293, **293.** *See also* Development of a Reaction Matrix lab
Reaction products. *See* Identification of Reaction Products lab
Reaction Rates lab, 330–345. *See also* Rate Laws lab
    checkout questions for, 343–345
    lab handout for, 337–342
        argumentation session, 341–342
        connections, 341

Index